COSMIC EVOLUTION

COSMIC EVOLUTION

on the occasion of the 60th birthdays of Jean Audouze and James W. Truran

Institut d'Astrophysique de Paris   13–17 November 2000

Editors

Elisabeth Vangioni-Flam
Roger Ferlet
Martin Lemoine

Institut d'Astrophysique de Paris

World Scientific
New Jersey • London • Singapore • Hong Kong

Published by

World Scientific Publishing Co. Pte. Ltd.

P O Box 128, Farrer Road, Singapore 912805

USA office: Suite 1B, 1060 Main Street, River Edge, NJ 07661

UK office: 57 Shelton Street, Covent Garden, London WC2H 9HE

British Library Cataloguing-in-Publication Data
A catalogue record for this book is available from the British Library.

COSMIC EVOLUTION

ISBN 981-02-4786-9

Printed in Singapore by Uto-Print

CONTENTS

Origin and Evolution of the Light Elements

Cosmic Evolution

James W. Truran & Jean Audouze

PREFACE

This volume contains most of the papers presented at the Symposium "Cosmic Evolution" held at Institut d'Astrophysique de Paris (November 13-17, 2000), which celebrated the 60th birthday of Jean Audouze and James W. Truran, two leaders on both sides of the Atlantic of nucleosynthesis, galactic evolution, and the temporal aspect of universal structuration. This Conference addressed a wide range of cosmic phenomena, thus reflecting the large variety of subjects in which Jean and Jim have made major contributions: big-bang nucleosynthesis, origin and evolution of the light elements, study of chemical abundances at various epochs, stellar and cosmic ray nucleosynthesis, dynamical and chemical evolution of galaxies, and more recently cosmic evolution as a whole. This exciting Symposium has strengthened the conviction that: **(i)** abundance measurements, carried out through spectroscopic observations in the ultraviolet and the optical domains, and **(ii)** nuclear astrophysics and associated disciplines such as modeling of galactic evolution, are both essential in the scientific quest towards the understanding of the evolution of matter in our Universe.

The support of the following institutions was crucial for the success of this Conference and is here warmly aknowledged: the Commissariat l'Energie Atomique (SAp/DAPNIA/DSM/CEA), the Centre National d'Etudes Spatiales (CNES), the Centre National de la Recherche Scientifique (INSU/CNRS) and the European Space Agency (ESA). We also thank the personnel and staff of the Institut d'Astrophysique de Paris for help and material support. We finally thank the members of the Scientific Organizing Committee of the Conference for their help: Andreas Burkert, Alastair G. W. Cameron, Michel Cassé, Catherine Césarsky, Jean-Pierre Chièze, Bernard Fort, Wolfgang Hillebrandt, Keith A. Olive, Jean-Loup Puget, Hubert Reeves, Robert Rosner, Joseph Silk, Sumner Starrfield, Sylvie Vauclair, Alfred Vidal-Madjar, Laurent Vigroux and Donald G. York.

On behalf of the Organizing Committee, we would like to warmly thank again all the participants for giving us the opportunity to offer to Jean Audouze and Jim Truran, our friends and colleagues, the beautiful gift of an (almost) coherent picture of the history of the matter in the Universe.

Thank you Jean and Jim for building this bridge between astrophysics, particle and nuclear physics, on which so many of us now stand.

Happy Birthday Dear Jean and Jim!

With friendship and admiration,

Andreas Burkert, Roger Ferlet, Martin Lemoine, Elisabeth Vangioni-Flam.

We would like to express our deep sorrow after the death of Reuven Ramaty (Goddard Space Flight Center, NASA), a major actor in the study of solar and galactic non-thermal processes of nucleosynthesis. One of the last contributions of our late friend can be found in the present volume.

FORTY YEARS OF NUCLEAR ASTROPHYSICS OF LIGHT ELEMENTS

H. REEVES

1 rue Jacob 75006 Paris, France

The physical processes responsible for the formation of the light elements (A<12) have been gradually identified in the last forty years. These processes have influenced the course of stellar, galactic and cosmic events. A narration is presented of the history of this subject.

1 The origin of lithium, beryllium, and boron

The two founding papers of nucleosynthesis were published shortly before 1960. The authors were Burbidge, Burbidge, Fowler and Hoyle (BBFH) [1] for the first one and Al Cameron [3,4] for the second. The pionneering works in this field had been done before the war by Hans Bethe and later by Edwin Salpeter, my thesis chairman, at Cornell University.

One major question left unanswered by these papers was the nature of the processes responsible for the generation of the light elements from deuterium to boron. BBFH named it the x-process thereby admitting how little was known on this topic.

Another question was the astrophysical formation sites of the nuclei heavier than the iron peak. Jean Audouze worked mostly on the first question while Jim Truran worked mostly on the second question.

I remember a meeting at the Institute for Space Studies in 1962 in New York where the problem of the origin of the light elements was discussed by William Fowler. Jim Truran and Alistair Cameron, his thesis chairman, were also present. The proposed mechanism was neutron capture reactions following the bombardment of the protosolar nebula by fast particles emitted by the sun. To test this hypothesis, laboratory data on the relevant nuclear spallation cross-sections were needed. In the following years I visited many laboratories in North America and Europe including the ex-URSS, trying to convince the experimenters of the importance of this work, but without much success.

In the mean time a most tell-taling information came from the measurements of the relative abundances of fast atoms in the galactic cosmic rays (GCR) by Bernard Peters and his group. The abundance ratios of lithium, beryllium and boron to carbon and oxygen were found to be enormously larger than the corresponding stellar ratios (factors of 10^5 or so). In 1965, I came in contact with the Orsay group of René Bernas and his collaborators (Robert Klapisch, Marcelle Epherre, Elie Gradsztajn, Françoise Yiou). They had already started a program of measurements of the important spallation cross-sections. They were actively purifying sources, building highly sensitive mass-spectrometers and carrying targets to large accelerators, in particular at CERN, in Geneva.

I joined the Orsay group and we started to study the implications of these measurements for the problem of the origin of the light elements. Jean Audouze was one of my first graduate students. Putting together the GCR data, the spallation cross-sections and the stellar data on the abundances of the light elements it soon became clear that the bombardment of interstellar atoms by fast particles was the main process leading to the generation of lithium, beryllium and boron. Good agreement was obtained for ^6Li, ^9Be and ^{10}B. ^7Li and ^{11}B were somewhat underproduced and the generation of D and ^3He was entirely negligible. Together with another student, Maurice Meneguzzi , we published these results in 1970 (Meneguzzi, Audouze and Reeves 1971, MAR) [12].

2 The origin of deuterium and helium-3.

Around the same period, Johannes Geiss and his collaborators in Berne measured the solar wind $^3He/^4He$ ratio in aluminum foils deposited on the Moon by the Apollo astronauts. Comparison with the planetary helium isotopic ratio yielded the D/H ratio in the early solar system. The values obtained were quite compatible with the hypothesis of a Big Bang origin (Big-Bang Nucleosynthesis, BBN) of the deuterium and helium isotopes [9 20]. From the BBN computations of Wagoner, Fowler and Hoyle [29] one could then obtain for the first time an estimate of the baryonic density of the universe . It was found to be far too small to close the universe . These results were published by Reeves, Audouze, Fowler and Schramm in 1973 [19].

3 Galactic evolution models

In collaboration with the late Beatrice Tinsley at Yale University, Jean Audouze initiated a program of study of the galactic time evolution of the light elements. This program required the quantitative understanding of their formation and of their destruction mechanisms, such as astration. Numerical codes incorporating star formation rate (SFR) initial mass function (IMF) and stellar evolution models for different masses and initial chemical composition were built-up in order to compute the relevant abundances as a function of time, from the Big Bang till today.

4 More observations and more models

Throughout the following decades new observational data became progressively available. Most importantly they made possible the evaluation of the primordial values of the light elements, and hence of more precise evaluation of the cosmic baryonic density. The primordial abundance of lithium became possible after the celebrated measurements of its abundance in old stars (Pop II, metallicity independent) by François and Monique Spite and their collaborators [22 23 8]. The extent of surface depletion of these stars has been studied by many authors. Primordial deuterium values were obtained by the study of cloud absorptions lines in the light curves of distant quasars [25]. Measurements of ^4He abundances in low-metallicity dwarf galaxies [10 24 15] has yielded an estimation of its primordial abundance. Comparison of these primordial values with values pertaining to later period, such as the birth of the solar system or today, give information on the nature and intensity of the stellar processes responsible for these modifications specifically for ^7Li [2 6 28]. More recently another breakthrough came with the observations of beryllium and boron in low metallicity stars [16 5]. Their declining abundances as a function of metallicity give valuable insights on the properties of high energy fluxes responsible for their formation.

All these results were incorporated in galactic models by the collaborators and graduate students of Jean Audouze, in particular Michel Cassé, Elisabeth Vangioni-Flam, Nicolas Prantzos, Roland Lehoucq, Etienne Parizot, Martin Lemoine [11 14 26 27 17]. Active collaborations were pursued with many other groups. In particular the Chicago group (David Schramm, Keith Olive, Brian Field) [21 13 7] and the NASA group (Reuven Ramaty, Richard Lingenfelter, Benzion Kozlovski) [18]. These collaborations are going actively. We will hear some of their recent results during this symposium.

References

1. E.M. Burbidge, G.R. Burbidge, W.A. Fowler, and F.Hoyle, *Rev. Mod. Phys.* **29**, 547 (1957)
2. R. Cayrel, ISSI Workshop on Primordial Nuclei and Their Galactic Evolution. Berne Switzerland 6-10 May 1997, Space Science Review Kluwer Academic Press, May 1997 N. Prantzos, M. Tosi and R.von Steiger Editors, 1998, p. 145
3. A.G.W. Cameron, *ApJ* **130**, 452 (1959)

4. A.G.W Cameron, *ApJ* **130**, 916 (1959)
5. K. Cunha, 1999 IAU Symposium 198 Natal, Brazil, "The light elements and their evolution" edts Da Silva,L., Spite, M. and De Medeiros, J.R., 2000, p. 415
6. C.P.Delyiannis, NH. Pinsonneault and C. Charbonnel, IAU Symposium 198 Natal, Brazil, "The light elements and their evolution" edts Da Silva,L., Spite, M. and De Medeiros, J.R., 2000, p. 61
7. B.D. Fields and K. Olive *ApJ* **516**, 797 (1999)
8. P. François, V. Hill, M. Spite and F.Spite, IAU Symposium 198 "The light elements and their evolution" edts Da Silva,L., Spite, M., and De Medeiros, J.R., 2000, p.504
9. G. Gloekler and J. Geiss, IAU Symposium 198 "The light elements and their evolution" edts Da Silva,L., Spite, M., and De Medeiros, J.R., 2000, p.224
10. D. Kunth and W. Sargent *ApJ* **273**, 81 (1983)
11. M. Lemoine, E. Vangioni-Flam and M. Cassé *ApJ* **499**, 735 (1998)
12. M. Meneguzzi, J. Audouze and H. Reeves *A&A* **15**, 337 (1971)
13. K.A. Olive, IAU Symposium 198 "The light elements and their evolution" edts Da Silva,L., Spite, M. and De Medeiros, J.R., 2000, 547
14. E. Parizot, M. Cassé and E. Vangioni-Flam *A&A* **328**, 107 (1997)
15. M. Peimbert, and A. Peimbert, IAU Symposium 198 Natal, Brazil, "The light elements and their evolution" edts Da Silva,L., Spite, M., and De Medeiros, J.R., p. 194
16. F. Primas, IAU Symposium 198, "The light elements and their evolution" edts Da Silva,L., Spite, M., and De Medeiros, J.R., 2000, p. 405
17. N. Prantzos, M. Cassé and E. Vangioni-Flam, *ApJ* **403**, 603 (1993)
18. R. Ramaty, R.E. Lingenfelter and B. Kozlovsky *ApJ* **456**, 525 (1996)
19. H. Reeves, J. Audouze, W.A. Fowler and D.N. Schramm *ApJ* **177**, 909 (1973)
20. T.M. Bania, R.T. Rood and D.S. Balser, IAU Symposium198 Natal , Brazil, "The light elements and their evolution" edts Da Silva,L., Spite, M. and De Medeiros, J.R., 2000, p. 214
21. D.N. Schramm, in " On the Origin and Evolution of the elements", Paris, June 1992, N. Prantzos, E. Vangioni-Flam, and M. Cassé, (Cambridge University Press), p.112
22. M. Spite and F. Spite *Nature* **297**, 483 (1982)
23. F. Spite and M. Spite *A&A* **115**, 337 (1982)
24. T.X. Thuan and Y.I. Izotov, IAU Symposium 198 "The light elements and their evolution" edts Da Silva,L., Spite, M. and De Medeiros J.R., 2000 p.176

25. D.Tytler, J.M. O'Meara, N.Suzuki, D.Lubi and S.Burles, IAU Symposium 198 "The light elements and their evolution" edts Da Silva,L., Spite, M. and De Medeiros, J.R., 2000, p.125
26. E. Vangioni-Flam, M. Cassé, B. Fields and K. Olive *ApJ* **468**, 199 (1996)
27. E. Vangioni-Flam, M. Cassé and J. Audouze *Phys. Rep.* **333-334**, 365 (2000)
28. S. Vauclair, IAU Symposium 198 "The light elements and their evolution" edts Da Silva,L., Spite, M. and De Medeiros, J.R., 2000, p. 470.
29. R.V. Wagoner, W.A. Fowler and F. Hoyle, *ApJ* **148**, 3 (1967)

Part 1. Cosmology and Big-bang Nucleosynthesis

HIGH–ENERGY GAMMA–RAYS AS AN INDIRECT SIGNATURE OF SUPERSYMMETRIC DARK MATTER

P. SALATI

Laboratoire Ancilévien de Physique Théorique LAPTH
Chemin de Bellevue, LAPP
B.P. 110, 74941 Annecy-le-Vieux Cedex, France

Weakly–interacting massive particles may contribute a substantial fraction to the mass of our Milky Way as they may also be a major component of clusters of galaxies. Their mutual annihilations would yield several indirect signatures that are potentially detectable on Earth as extra–components to the various cosmic radiations. I will mostly concentrate here on gamma–rays and will shortly review the various strategies at hand – Cerenkov detectors or satellites – to find them. The clumpiness of the dark matter distribution – that plays a key role in boosting up the various signals – will also be addressed.

1 A few introductory remarks.

The search for remote supernovae SNIa [1] constrains the matter $\Omega_{\rm M}$ and the vacuum Ω_Λ energy densities to satisfy

$$1.3\,\Omega_{\rm M} \; - \; \Omega_\Lambda \simeq -0.4 \pm 0.2 \;\; . \tag{1}$$

Recent measurements [2] of the Cosmic Microwave Background (CMB) inhomogeneities are compatible with a flat universe where $\Omega_{\rm M} + \Omega_\Lambda = 1$. Together with the SNIa result, this leads to a vacuum density of $\Omega_\Lambda \sim 0.7$ and to a matter density $\Omega_{\rm M} \simeq 0.28 \pm 0.13$. This result is in agreement with the measurements of the gas fraction in X–ray clusters which also point toward a value of $\Omega_{\rm M} \simeq 0.36 \pm 0.05$. If on the one hand matter accounts for more than one–third of the closure density, on the other hand, primordial nucleosynthesis constrains the baryon density to be $\Omega_{\rm B} \simeq 0.05 \pm 0.005$. We are therefore lead to the conclusion that a large fraction of the matter in the universe is non–baryonic. This is in agreement with the scenarios of galaxy formation according to which baryons, at the end of recombination, fall inside the potential wells of a collapsing distribution of cold and non–interacting matter. One of the favoured candidates to account for that non–baryonic cold dark matter (CDM) is a neutral weakly interacting particle. Such a species is predicted in particular by supersymmetry, a theory that is actively tested at accelerators. It is conceivable therefore that most of the dark matter in the halo of the Milky Way is made of these so–called neutralinos. Their mutual annihilations would yield several indirect signatures that are potentially detectable on Earth as additional components to the various cosmic radiations

$$\chi + \chi \to q\bar{q}, W^+W^-, \ldots \to \gamma, \bar{p}, \bar{D}, e^+ \,\&\, \nu's \;\; . \tag{2}$$

Detection of these annihilation products relies on several experimental techniques which allow to probe different regions in the $(m_\chi \ , \ \sigma)$ plane. Searches for antiprotons, antideuterons and positrons are performed by balloon and satellite–borne devices. Because the flux depends on the square of the neutralino density n_χ, the limit which may be set on the annihilation cross section scales approximately as ${m_\chi}^2$. That type of search is mostly sensitive to low neutralino masses. High–energy photons are detected both by air Cerenkov telescopes (ACT) and by satellite–borne instruments. The neutralino annihilation rate, and hence the gamma–ray signal, also scale as ${m_\chi}^{-2}$. Because of the background in which that signal is swamped, the corresponding limit on the annihilation cross section approximately scales as the mass m_χ. This channel compares therefore with the direct searches, a technique that is sensitive to the neutralino density $n_\chi = \rho_{\rm DM}/m_\chi$ and that sets the same kind of limit on σ. On the contrary, the neutrino channel offers the unique opportunity to probe for large values of m_χ. The corresponding limit on the annihilation cross section scales this time as ${m_\chi}^{-1}$. The heavier the neutralino, the more stringent the bound. This channel is therefore complementary to the other searches. For simplicity, the archetypal example of annihilation gamma–rays will be discussed in the next section as it is also more relevant to astrophysicists than high–energy antideuterons or neutrinos.

2 High–energy photons.

Neutralino annihilations lead both to a continuum of high–energy photons as well as to monochromatic gamma–ray lines through the box–mediated direct production

$$\chi + \chi \to \gamma + \gamma \ \ \& \ \ \gamma + Z^0 \ . \tag{3}$$

The corresponding flux at the Earth may be expressed as

$$\Phi_\gamma^{\rm susy} \;=\; \frac{1}{4\pi} \, \frac{\langle \sigma v \rangle \, N_\gamma}{m_\chi^2} \int_{\rm los} \rho_\chi^2 \, ds \ . \tag{4}$$

That relation may be split into an astrophysical piece and a particle physics part. The former term merely amounts to the integral along the line of sight of the dark matter density squared ${\rho_\chi}^2$. For a neutralino distribution with total mass M and typical size R, it is approximately given by $M^2 \, R^{-5}$. Its bench mark value is $\sim 0.1 \, {M_\odot}^2 \ {\rm pc}^{-5}$ in the case of the Milky Way, for $R = 100$ kpc. The particle physics piece depends on the annihilation cross section $\langle \sigma v \rangle$ averaged over the distribution function of the neutralino velocities. In relation (4), N_γ denotes the total number of continuum or monochromatic photons that are produced in a single annihilation. In the latter case, $N_\gamma = 2$ for the two–photon reaction while $N_\gamma = 1$ for the γ–Z^0 channel.

Because the γ–ray signal from annihilating dark matter is faint, the most promising instruments are the atmospheric Cerenkov telescopes. When a high–energy photon impinges on the top of the atmosphere, it generates an electromagnetic shower that spreads over a significant area as it reaches the ground. The initial γ–ray energy is degraded into many optical photons that are potentially detectable by an array of telescopes. Analysis of the shape of that shower allows in principle to reconstruct the direction and the energy of the primary high–energy photon. ACT's get advantage of their large effective detection area. They suffer however from an insuperable glaucoma as they can only monitor small portions of the sky at the same time. They are perfectly suited for faint and point–like sources such as the galactic center or extra–galactic systems like M87. Satellite–borne detectors are complementary. They are well suited for large solid angle surveys but their collecting areas are small.

In any case, the detection is made difficult by the presence of various backgrounds in which the signal is swamped. The dominant source arises from the cosmic ray (CR) high–energy electrons that impact on the upper atmosphere and also generate electromagnetic showers. An ACT cannot actually distinguish between a photon and a CR electron–induced shower. Hadron–induced showers may also be mistaken for high–energy gamma–rays even if stereoscopy is a powerful discriminating tool. For satellite–borne instruments, the only background arises from the γ–ray diffuse emission of interstellar gas that shines under the action of the local CR protons.

The two–photon line seems to be the most promising reaction to look for as its spectral features are unique. The resulting monochromatic photons have an energy $E_{\gamma} = m_{\chi}$. Relation (4) translates into [3]

$$\Phi_{\gamma}^{\rm susy} \; = \; \left\{ 3.8 \times 10^{-11} \;\; {\rm photon \, cm^{-2} \, s^{-1} \, sr^{-1}} \right\} \, \langle \sigma v \rangle_{29} \, m_{10}^{-2} \, J(\mathbf{u}) \;\; , \qquad (5)$$

where $\langle \sigma v \rangle_{29}$ is the annihilation cross section for the two–photon line process expressed in units of 10^{-29} cm^3 s^{-1} while $m_{10} = m_{\chi}/10$ GeV. The line of sight integral J is expressed as a function of its bench mark value. The latter is given by the product $r_{\odot} \rho_{\odot}^2$ where a galactocentric distance of $r_{\odot} = 8.5$ kpc has been assumed together with a halo density of $\rho_{\odot} = 0.3$ GeV cm^{-3}, typical of the solar neighborhood. Because the neutralinos are widely spread in the galaxy, their gamma–ray emission is extremely faint except in the direction of the galactic center where their density profile should be quite peaked [4,5,6]. In particular, if a NFW [4] halo profile is assumed, the line of sight integral becomes $\langle J \rangle \sim 10^3 \; (1^{\circ}/\theta)$ – when averaged over an angle θ from the center – instead of the typical galactic value of ~ 1. The next generation of ACT's will typically have an effective collecting area of 0.1 km^2, an angular resolution of 0.1°, a threshold of 50 GeV as well as an energy resolution of $\pm$ 15 %. If the observation of the putative neutralino clump sitting at the galactic center is

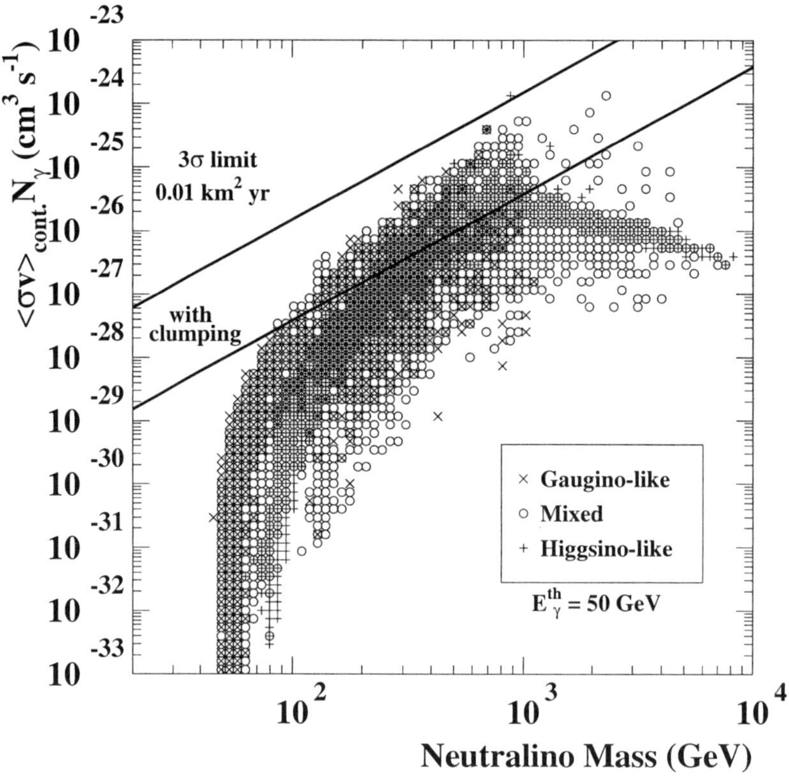

Figure 1. Annihilation rates in the continuum channels. The threshold has been set equal to $E_{\rm th} = 50$ GeV. Considering M87 as the source, the 3–σ detection limits for exposures of 0.01 km^2 yr are also presented. The region below the heavy solid lines will not be accessible, even with the next generation of Cerenkov telescopes. The lower solid line shows the region of accessibility if the annihilation rate is enhanced by a factor of 40 due to clumpiness.

performed during 0.1 yr – two or three months equivalent – a sensitivity of

$$2\,\langle\sigma v\rangle_{\gamma\gamma} \geq 5\times 10^{-29}\ \mathrm{cm^3\,s^{-1}}\ {m_{100}}^{0.85} \tag{6}$$

will be reached at the 3–σ level. By comparison, a 1 m^2 satellite–borne telescope with an energy resolution of 1 % would be sensitive down to

$$2\,\langle\sigma v\rangle_{\gamma\gamma} \geq 2\times 10^{-28}\ \mathrm{cm^3\,s^{-1}}\ {m_{100}}^{1.13}\ , \tag{7}$$

should it monitor for two entire years a region of the sky within 1° from the galactic center. The small number of collected events is nevertheless a problem. Whatever the method – ACT or satellite – only the upper fringe of the supersymmetric configurations in the $(\langle \sigma v \rangle_{\gamma\gamma}\ , m_{\chi})$ plane can be probed.

Annihilation photons make it also possible to investigate the presence of neutralinos in extra–galactic systems. The most promising site is the giant elliptical galaxy M87 at the center of the Virgo cluster, some 15 Mpc away. That galaxy is known to contain large amounts of dark matter. Its line of sight integral of $\sim 10\, {M_{\odot}}^2\, \mathrm{pc}^{-5}$ is two orders of magnitude larger than for the isothermal halo around our Milky Way. Furthermore, M87 extends on $\sim$ 30 arcmin and appears as a point–like source, well–suited for an ACT observation. Its continuum γ–ray signal is detectable by the next generation of instruments for the part of the supersymmetric configurations outlined [7] in the upper–left panel of Fig. 1. Because low–energy photons are predominantly produced in neutralino annihilations, the lower the detection threshold, the better the sensitivity.

3 Discussion and future prospects.

N–body simulations point toward the presence of a dense dark matter cusp at the galactic center – with radial profile [4]

$$\rho(r) \propto \frac{r^{-1}}{(1 + r/a)^2} \ , \tag{8}$$

where $\rho_{\odot} \sim 0.3$ GeV cm^{-3} and $a \sim 25$ kpc. The CDM density may even diverge more violently [6] as $r^{-3/2}$. In that context, the massive black hole [8] that sits at the galactic center should have dragged down – during its formation – some of the neutralinos floating around. A CDM spike should have therefore condensed [9] with a profile as steep as

$$\rho_{\,\mathrm{final}} \propto r^{-\gamma_{\mathrm{sp}}} \ , \tag{9}$$

where the index γ_{sp} varies from 1.4 for an initial isothermal distribution up to 2.5 in the most extreme situation. In the case of a precollapse NFW halo, the resulting neutralino spike would have now a core radius of ~ 0.01 pc together with a total mass amouting to a fraction $\lambda \sim 10^{-3}$ of the central massive black hole. An ACT would reach a 3–σ detection limit of

$$2\,\langle \sigma v \rangle_{\gamma\gamma} \geq 3 \times 10^{-34}\ \mathrm{cm}^3\, \mathrm{s}^{-1}\ {m_{100}}^{-0.15} \ , \tag{10}$$

for a detection cone encompassing a region within 0.1° from the galactic center. An impressive set of supersymmetric configurations would therefore become explorable.

Neutralinos may have also partially collapsed in domains of larger than average density. Inside those clumps, the rate of their annihilations – and hence the corresponding indirect signals – should be increased [10,11] by a factor of

$$\delta = \frac{1}{\rho_0} \left\{ \int \rho_{cl}^2 \, d^3x \right\} / \left\{ \int \rho_{cl} \, d^3x \right\}^{-1} . \tag{11}$$

Inside galactic haloes, the clumps should follow the CDM density fluctuations. Small structures went non–linear first – when the universe was denser than today. If a fraction f of the halo is made of small clumps whose mass does not exceed $M_{\mathrm{i}} \sim 10^8 \; M_{\odot}$, an enhancement clumpiness factor of $\mathcal{C} = 20 \, f$ is found for the Milky Way – reaching up a value of $\mathcal{C} = 40 \, f$ in the case of M87. If furthermore the neutralino density follows a $r^{-3/2}$ law inside each of these clumps – as recently shown by Calcáneo–Roldán & Moore [12] – the gamma–ray flux could be over two orders of magnitude brighter than in the smooth halo case. The various annihilation signals would be within reach of the near future instruments and solving the dark matter puzzle would no longer be a fantasy.

Acknowledgments : I would like to thank the organizers of the Cosmic Evolution Conference for their kind hospitality and warm welcome.

References

1. S. Perlmutter *et al.* (The Supernova Cosmology Project), Astrophys. J. **517**, 565 (1999).
2. P. de Bernardis *et al.* (The Boomerang Collaboration), New Astronomy Reviews **43**, 289 (1999).
3. L. Bergstrom, P. Ullio and J. Buckley, Astropart. Phys. **9**, 137 (1998).
4. J.F. Navarro, C.S. Frenk and S.D.M. White, Astrophys. J. **462**, 563 (1996).
5. A.V. Kravtsov, A.A. Klypin and A.M. Khokhlov, Astrophys. J. Suppl. **111**, 73 (1997).
6. Moore *et al.*, Astrophys. J. Letters **499**, L 5 (1998).
7. E.A. Baltz, C. Briot, P. Salati, R. Taillet and J. Silk, preprint lapth–763/99 and `astro-ph/9909112`, to be published in *Physical Review* **D**.
8. A.M. Ghez, B.L. Klein, M. Morris and E.E. Becklin, Astrophys. J. **509**, 678 (1998).
9. P. Gondolo and J. Silk, Phys. Rev. Lett. **83**, 1719 (1999).
10. L. Bergström, J. Edsjö and P. Ullio, Phys. Rev. D **58**, 083507 (1998).
11. L. Bergström, J. Edsjö, P. Gondolo and P. Ullio, Phys. Rev. D **59**, 043506 (1999).
12. C. Calcáneo–Roldán and B. Moore, `astro-ph/0010056` (2000).

EXTRACTION OF COSMOLOGICAL PARAMETERS FROM COSMIC BACKGROUND PRIMARY AND SECONDARY ANISOTROPIES

BIANCA AND FRANCESCO MELCHIORRI

Dept. of Physics, University La Sapienza, pz.le Aldo Moro 2, Rome, Italy
E-mail: francesco.melchiorri@roma1.infn.it

Cosmological models are characterized by few parameters, like the expansion rate (Hubble constant, H_0), the degree of isotropy (energy density in anisotropic stress Ω_{AN}), the curvature of the space-time (with a corresponding energy density Ω_k), the total mass density (Ω_m), the vacuum energy density (Ω_Λ), as well as the various matter and/or energy components (barionic, cold, hot, quintessence ....). Cosmic Background Anisotropy's data can help in limiting the allowed region in phase-space for these parameters, while a precise determination of them will require a wide set of different cosmological observations.

1 A Survey of Cosmological Parameters

1.1 Single Fluid, Homogeneous Cosmologies

During the first half of the last century the only relevant data for cosmology were the recession law for galaxies , some rough lower limit to the visible mass density, and an estimate of the mean abundance of light elements in the Universe. These data led G. Gamow to formulate his Big Bang Theory. At very basic level this Theory requires the validity of two principles:

[1] **The Principle of Homogeneity**: sometime referred to as The Cosmological Principle, it requires that the physical and chemical properties of the Universe at a given time are the same everywhere.

[2] **The Fluidity Principle**: the Universe is filled by a cosmic "ideal" fluid (matter and/or radiation), fully characterized by its density ρ and velocity field $\vec{\mathbf{v}}$. We neglect the dimensions and properties of isolated particles,as well as additional moments like viscosity etc.

Both the above principles have been questioned: the most widely diffused opinion is that [1] and [2] are valid beyond a certain scale, which however is not well determined, ranging from 10 Mpc up to 200 Mpc. In this context it is worthwhile to mention that CMB anisotropies pose a stringent upper limit to the maximum scale of inhomogeneities (of the order of 200 Mpc), through the so-called Sachs-Wolf second order effect (CMB photons suffer a frequency shift in their path through the time-variable gravitational potential associated to the evolving structure).

The first important parameter characterizing this oversimplified Big Bang model is the degree of possible anisotropy .

The isotropic Friedmann model has an expansion rate given by

$$H^2 = \left(\frac{\dot{R}}{R}\right)^2 = \frac{8\pi G\rho}{3} - \frac{\mathbf{k}}{R^2} \tag{1}$$

the quantity $\frac{\mathbf{k}}{H^2R^2}$ is some time referred to as the *curvature energy density* = $\Omega_{\mathbf{k}}$

In the IX Bianchi-type homogeneous but anisotropic Universes the above equantion may be written as

$$H^2 = \left(\frac{\dot{R}}{R}\right)^2 = \frac{8\pi G\rho}{3} + \mathbf{F}[\mathbf{R_i}] \tag{2}$$

where $\mathbf{F}[\mathbf{R_i}]$ is a quantity which accounts for the anisotropic expansion of the mean expansion rate. In analogy, we may introduce an *anisotropy energy density* Ω_{AN} such that

$$\Omega_{AN} + \Omega_k = \frac{\mathbf{F}[\mathbf{R_i}]}{H^2} + \frac{\mathbf{k}}{H^2 R^2_{mean}} \tag{3}$$

The best constraints are provided by the large scale isotropy of CMB, being Ω_{AN} linearly proportional to the amplitude of the Quadrupole Anisotropy:

The Quadrupole Anisotropy , as measured by COBE, is constrained by the limit

$$\left(\frac{\Delta T}{T}\right)_Q \leq 3 \times 10^{-6}$$

Since the amplitude of the Quadrupole anisotropy is linearly related to Ω_{AN}, we may conclude that Ω_{AN} was rather small at $Z \simeq 1000$: moreover, the quantity $\mathbf{F}$ is a rapidly decreasing function of the scale factor R ($\mathbf{F} \propto R^{-2} - R^{-6}$). It follows that $\Omega_{AN}$is even smaller today. In conclusion we have strong evidence that the first cosmological parameter we have considered, Ω_{AN} is close to zero , a fact which is in agreement with the predictions of Inflation where every primordial anisotropy is washed out by the exponential expansion.

If the Universe is homogeneous and isotropic at large, it has a Robertson-Walker metric

$$d\ell^2 = R^2(t)\left[\frac{dr^2}{1 - kr^2} + r^2 d\gamma^2\right]$$

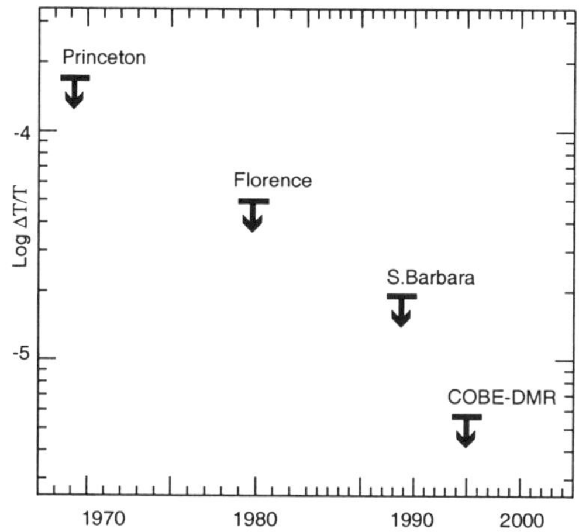

Figure 1. Upper limits on CMB Quadrupole Anisotropy: early claims of detection have been interpreted here as upper limits. Note that even the detection by COBE may be considered an upper limit if we take into consideration both statistical and systematic uncertainties

where $r = u, sinu, sinhu$ depending on the sign of k. Under these conditions Friedmann has found that the Einstein Field Equations reduce to two independent equations+ an Equation of State $\frac{p}{c^2} = w\rho$

The *Natural Parameters* of this Universe are (c=1):

a] The *Hubble Constant* today $H_0 = \dot{R}_0/R_0$ where R_0 is the value of the scale factor today

b] The *Curvature Energy Density* $\Omega_{\mathbf{k}} = k/R_0^2H_0^2$

c] The *Cosmological Constant* Λ and its associated energy density $\Omega_{\mathbf{\Lambda}} = \frac{\Lambda}{H_0^2R_0^2}$

d] the *state parameter* w, which is assumed to be =0 today (unless we introduce an additional fluid, like "quintessance").

A short rappresail of how to measure these parameters via CMB anisotropies follows:

1.2 CMB and the Hubble Constant

The basic idea is that of taking advantage of combined microwave and X-ray observations of clusters of galaxies.

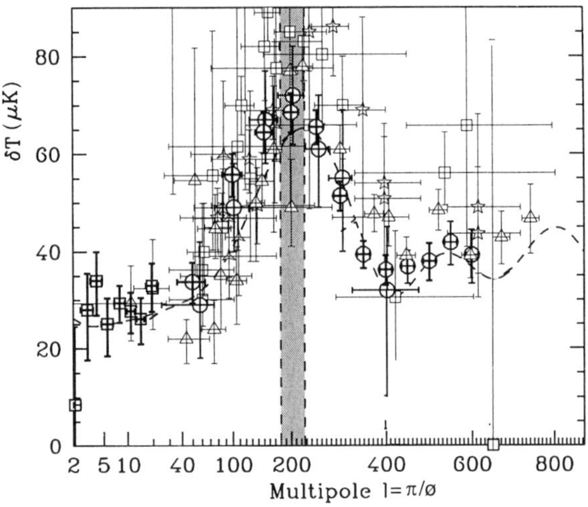

Figure 2. Hubble diagram (i.e. luminosity distance of the cluster d_L v/s the redshift) as derived by SZ and X-ray measurements: the lines refer to $H_0 = 50(upper), 75$ and $q_0 = 0, 1/2, 1$, adapted from [3].

The X-ray Brightness Φ_X of the hot gas in clusters is given by

$$\Phi_X = \frac{1}{4\pi(1+Z)^4} \int_{-\infty}^{+\infty} d\ell \int_{\nu_1(1+Z)}^{\nu_2(1+Z)} \frac{\partial^2 L_x}{\partial\nu\partial V} d\nu$$

$$\frac{\partial^2 L}{\partial\nu\partial V} = \alpha(T_{gas})\mathbf{n_e^2} g_{ff} \exp\left(\frac{-h\nu}{k_B T_{gas}}\right)$$

The Inverse Compton decrement in temperature of CMB , $\frac{\Delta T}{T_0}$ is given by

$$\frac{\Delta T}{T_0} = \zeta(x) \int_{-\infty}^{+\infty} \frac{k_B T_{gas}}{m_e c^2} \sigma_T \mathbf{n_e} d\ell$$

$$\zeta(x) = \frac{x^2 e^x}{(e^x - 1)^2} \left(x coth \frac{x}{2} - 4 \right)$$

In order to combine the two equations we need to know the spatial distribution of T_{gas} and n_e

The usual hypothesis are:

$T_{gas} = costant$ and

$n_e(r) = n_e(0)\left(1 + \frac{r^2}{d_A^2(z)\Theta_c^2}\right)^{-3\beta/2}$

These can be questioned: several clusters show significan deviations from spherical shape and from β model for electron density. Distant clusters observed with poor spatial resolution are **not useful** because X-ray observations do not allow to reconstruct precisely the shape of the cluster and correct for it. The result is a tendency to produce too low values for H.

1.3 CMB and Omega

It is well known that the position of the first Doppler peak in the angular power spectrum of CMB anisotropies is an approximate measure of the angular extent of the sound-horizon at the decoupling: its value is strongly dependent on the geometry of the Universe, i.e. on the value of Ω. On the other hand the amplitude of CMB anisotropies at the peak depends on the barionic content Ω_b. Therefore the CMB power spectrum offers an unique opportunity for determining accurately these two parameters. Note, however, that some degeneracy exists, which limits the accuracy to less than 10 % . The degeneracy could be also responsible of misinterpretations in data analysis. The first clean determination of the position of the Doppler peak is that of BOOMERANG NA [2]. The results obtained by BOOMERANG Antarctica [1] and MAXIMA seem to suggest a value of Ω slightly larger than one and Ω_b about two standard deviations above the value provided by the most recent primordial Nucleosynthesis data. All these results must be handled with care due to the presence of several possible spurious effects. If we combine CMB data with Supernovae results we find a detection of $\Omega_\Lambda \simeq 0.4 - 0.7$ at 7 sigma. This result, if confirmed will represent a great cosmological discovery.

The data relative to CMB anisotropy are summarized in Figure 3.

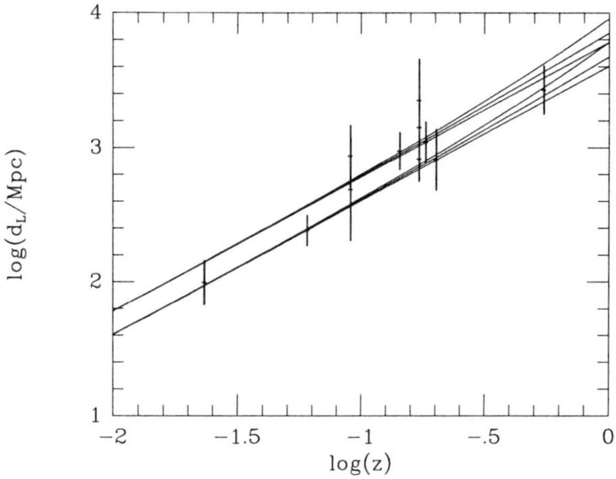

Figure 3. Collection of CMB anisotropy measurements: the central gray region correspond to $\Omega = 1$ (slightly dependent on the cosmological model) and the dashed curve is a best fit to the data. The data by BOOMERANG and MAXIMA are with thick lines, adapted from M. Tegmark, M. Zaldarriaga, astro-ph 0004393, 2000.

References

1. P. de Bernardis et al. , Nature, **404**, 955, 2000.
2. A. Melchiorri et al. , Ap.J.Lett. **L63**,536,2000.
3. M. Birkinshaw, in " 3K Cosmology", L.Maiani, F.Melchiorri, N. Vittorio Eds., AIP **476**, 298, 1998.

GALACTIC WINDS AT HIGH REDSHIFT AND SMALL SCALE CMB ANISOTROPY

SUBHABRATA MAJUMDAR

Indian Institute of Astrophysics, Bangalore - 34, India
E-mail: sum@iiap.ernet.in

BIMAN B. NATH

Raman Research Institute, Bnagalore - 80, India
E-mail: biman@rri.res.in

MASASHI CHIBA

National Astronomical Observatory, Mitaka, Tokyo,Japan
E-mail: chibams@gala.mtk.nao.ac.jp

The distortion of the Cosmic Microwave Background (CMB) due to galactic winds originating at high redshifts is considered. These winds have been hypothesized to be a possible source of metals in the intergalactic medium. We calculate the mean Compton distortion y and the angular power spectrum of the distortion. It is clear that the contribution to the angular power spectrum due to winds is important at small angular scales and we find that the clustering power spectrum dominates over Poisson power spectrum. Interestingly, the power spectrum from clustering falls in the l range where thermal SZ from clusters of galaxies peak, and thus may be measurable by 'Planck'. This may be used as a probe of evolution of bias at high redshifts.

1 Introduction

It is expected that dwarf starbursting galaxies will have winds driven by supernovae and winds from OB stars in them (Dekel & Silk 1986 [1]). This has been observed both locally and at high redshift. It is also possible that outflows from them provided the metals needed to explain the enrichment of the intergalactic medium (IGM) (Nath & Trentham [2]). These outflows could have distorted the microwave background, through the Sunyaev-Zel'dovich effect (SZE).

In this paper, we calculate the thermal and the kinetic SZE, taking into account both Poisson contribution and that due to clustering. We calculate the angular power spectrum of the distortions. Finally, we compare the distortion to that due to SZE from clusters of galaxies and comment on possible detection. More details can be found in Majumdar, Nath & Chiba (2000 [3]).

2 CMB distortion from winds

2.1 Modelling the ensemble of galactic outflows

The outflows are modelled following the approach of Tegmark, Silk & Evrard (1993, [4]). They have blastwaves which have thin shells sweeping up most of the ambient IGM gas (already ionised). The shells lose a small fraction $f_m = 0.1$ of their mass to the interior. Energy losses are mainly due to Bremsstrahlung and inverse Compton scattering off the CMB photons. Approximately 2% of the total luminosity of these galaxies goes into galactic winds.

We, numerically calculate the radius and velocity of the shell, and the energy and particle density inside the shell. We assume that the initial radius of the shell is the optical radius of the dwarf spheroidal. We calculate the wind parameters, for a given initial redshift z_{in}, as functions of the mass of the parent galaxy and redshift (for simplification, we consider outflows originating at a single epoch, at z_{in} between 3 & 15). The abundance of collapsed objects (normalised to COBE) is obtained by a modified version of the Press-Schechter mass function for low masses. The peculiar velocities of the objects are assumed to follow a Maxwellian distribution .

2.2 Distortion of the CMB background

The hot gas in the interior of the shells can distort the microwave background through the SZE. We calculate both the thermal SZE (which depends on the energy content of the gas) and the kinetic SZE (which is due to the peculiar motion of the gas). Once the distortion of the CMB due to the winds are calculated, we look at the power spectrum of the distortion. Along with the Poisson power spectrum, we also calculate the clustering power spectrum (following Komatsu and Kitayama ,1999 [5]). Note, that this depends on the bias of the objects.

3 Results and discussions

We confirm previous estimates of the mean y-distortion due to thermal SZ effect. We, however, find the kinetic SZE to be more important than the thermal SZE. The power spectrum due to kinetic SZE (Poisson) is found to be comparable or larger than the SZE from clusters of galaxies for $l > 10^5$. Clustering of galaxies at high redshift could increase the angular power spectum of distortions.It is seen (fig1) that the total power spectra (kinetic and thermal SZE, including both Poisson and clustering effects) for winds will

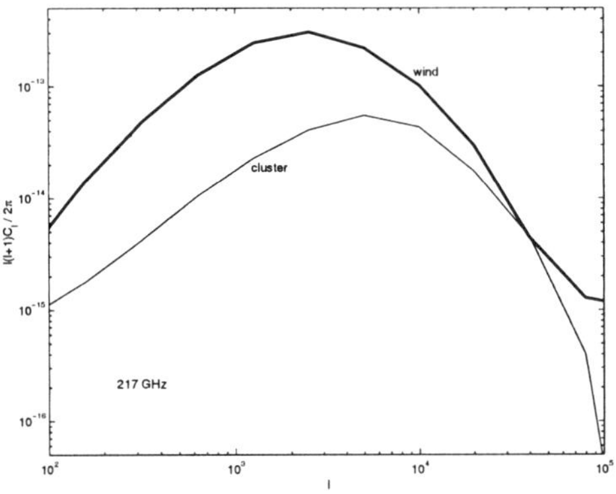

Figure 1. The composite power spectra (kSZ+tSZ, including contribution from both clustering and Poisson spectra) is shown for distortion from winds (thick line) and clusters (thin line), $z_{in} = 9$ and $v_r(z = 0) = 400 kms^{-1}$. The spectra are calculated for the proposed Planck observation frequency of 217 GHz, for ΛCDM cosmology.

dominate over the corresponding spectra for clusters, at 217 GHz, a proposed frequency for the PLANCK satellite mission. This brings about the interesting possibilty of probing the evolution of bias through the detection of this distortion.

We conclude that the SZ distortion from galactic winds at high redshift, if present, could constitute an important source of secondary CMB anisotropy on arc minute and sub-arcminute scales.

Acknowledgments

One of the authors (SM) would like to acknowledge IIA and DST, India for travel grant. He would also like to thank the organisers of the conference for financial support.

References

1. A. Dekel and J. Silk, ApJ **303**, 39 (1986)
2. B. Nath and N. Trentham, MNRAS **291**, 505 (1997)
3. S. Majumdar, B. Nath and M. Chiba astro-ph/ **0012016**, (2000)
4. M. Tegmark, J. Silk and A. Evrard ApJ **417**, 54 (1993)
5. E. Komatsu and T. Kitayama ApJ **526**, L1 (1999)

DETERMINATION OF THE COSMOLOGICAL PARAMETERS USING THE CMB RADIATION

J. ROLAND, J. FAIVRE, E. VANGIONI-FLAM
I.A.P., CNRS, 98 bis Bd Arago, 75014, Paris, France

M. CASSÉ
SAp/DAPNIA/DSM/CEA, Orme des Merisiers, 91 191 Gif/Yvette, France and I.A.P., CNRS, 98 bis Bd Arago, 75014, Paris, France

We present a method to determine the cosmological parameters using the CMB fluctuation observations. Taking the observational constraints from COBE, BOOMERANG and MAXIMA we get a spectral index of the power law of the primordial perturbations, $\alpha = 0.9 \pm 0.05$ (2σ) and the corresponding value of H_0 and η are respectively $H_o = 65 \pm 15\ km/s.Mpc$ (2σ) and $\eta = 7 \pm 1.5$ (2σ).

1 The $H_o - t_0$ and $H_o - \eta$ diagrams

We determine the cosmological parameters using observations of the CMB fluctation anisotropies. We use the CMBFAST program within the framework of standard cosmology, in this context, the temperature of the CMB is 2.728 K, there are only 3 massless neutrinos, there is no reionization, the primordial perturbations are scalar, the spectrum of the perturbation is a simple power law, the perturbations are adiabatic. The RECFAST correction is applied.

There are 5 free parameters, the Hubble constant H_o, the age of the Universe t_0, the matter density ρ_m, the parameter η related to the baryonic density $\Omega_b = 36.5\ \eta/H_0^2$ and the spectral index α of the power law spectrum of the primordial fluctuations. So, for given values of the parameters η and α, we calculated, for the different points of the $H_o - t_0$ plan, the matter density and the corresponding cosmological constant of the Friedmann solution which provide the minimum χ^2 estimated from the comparison of the CMB fluctuation calculations and the CMB fluctuation observations made by COBE, BOOMERANG and MAXIMA.

All regions of the $H_o - \eta$ diagram do not correspond to possible solutions, so we restrict the space of solutions to $\Omega_\Lambda < 0.7$ and $0.1 < \Omega_m < 0.6$. For a given value of the spectral index α of the power law spectrum of the primordial perturbations we can plot the $H_o - \eta$ diagram using the results obtained from the previous $H_o - t_0$ diagrams. Indeed, for a given value of H_0 we can deduce the minimum χ^2 corresponding to a possible solution and then we can plot the surface of $\chi^2_{\min}$ corresponding to the $H_o - \eta$ diagram for different values of the spectral index of the spectrum of the primordial perturbations. With present observations of COBE, BOOMERANG and MAXIMA the spectral index of the power law of the primordial perturbations is found to be $\alpha = 0.9 \pm 0.05$ (2σ) and the corresponding values of H_o and η are respectively $H_o = 65 \pm 15\ km/s.Mpc$ (2σ) and $\eta = 7 \pm 1.5$ (2σ). The derived value is higher than the baryon density calculated from the standard Big Bang nucleosynthesis [1]. However, other observations with the Cosmic background Imager (CBI, Chile) lead to a lower value of $\Omega_b\ h^2 = 0.009$ (compared the Boomerang/Maxima ones, i.e. 0.026). Consequently, it is important to stress that the CMB data are not precise enough to conclude, specially data concerning the second peak determining the value of Ω_b. Independant observations from future space missions (MAP, PLANCK-SURVEYOR) are needed to strengthen these new results.

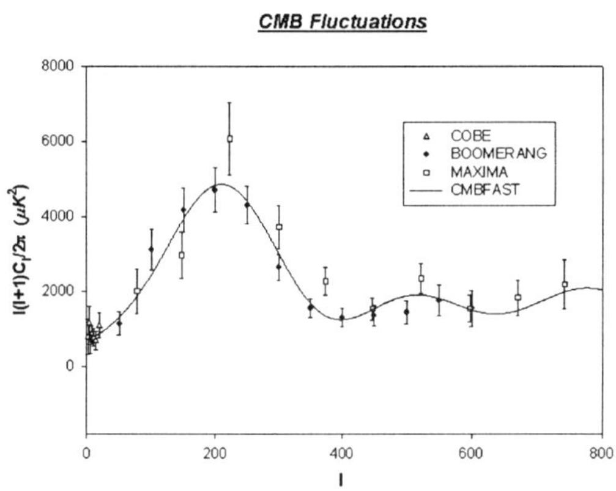

Figure 1. This fit of the CMB fluctuations is obtained with $H_0 = 67.5$ km/s.Mpc, the baryonic density is $\Omega_b = 0.056$, the age of the Universe is $t_0 = 13$ Gyr and the various densities are $\Omega_m = 0.37$ and $\Omega_\Lambda = 0.66$.

References

1. E. Vangioni-Flam, A. Coc and M. Cassé, *A& A* **360**, 15 (2000)

BIG BANG NUCLEOSYNTHESIS

KEITH A. OLIVE

TH Division, CERN, Geneva, Switzerland
also Theoretical Physics Institute, School of Physics and Astronomy
University of Minnesota, Minneapolis, MN 55455 USA
E-mail: olive@umn.edu

The current status of big bang nucleosynthesis is summarized. Particular attention is paid to recent observations of ^{4}He and ^{7}Li and their systematic uncertainties.

1 Introduction

The simplicity of the standard model of Big Bang Nucleosynthesis (BBN) and its success when confronted with observations place the theory as one of the cornerstones of modern cosmology[1]. The theory is a one parameter model, depending only on the baryon-to-photon ratio, η. Other factors, such as the uncertainties in reaction rates, and the neutron mean-life can be treated by standard statistical and Monte Carlo techniques[2,3]. The theory then allows one to make predictions (with specified uncertainties) of the abundances of the light elements, D, ^{3}He, ^{4}He, and ^{7}Li.

The dominant product of big bang nucleosynthesis is ^{4}He, resulting in an abundance of close to 25% by mass. Lesser amounts of the other light elements are produced: D and ^{3}He at the level of about 10^{-5} by number, and ^{7}Li at the level of 10^{-10} by number. The resulting abundances of the light elements are shown in Figure 1, over the range in $\eta_{10} = 10^{10}\eta$ between 1 and 10. The thickness of the curves show the theoretical uncertainties based on the input nuclear reaction rates[4]. This theoretical uncertainty is shown explicitly in the right panel as a function of η_{10}. The dark shaded boxes correspond to the observed abundances of ^{4}He and ^{7}Li and will be discussed below. The dashed boxes correspond to the ranges of the elements consistent with the systematic uncertainties in the observations. The broad band shows a liberal range for η_{10} consistent with the observations.

At present, there is a general concordance between the theoretical predictions and the observational data, particularly, for ^{4}He and ^{7}Li[5]. These two elements indicate that η lies in the range $1.7 < \eta < 4.7$, corresponding to a range in $\Omega_B h^2 = 0.006 - 0.017$. There is limited agreement for D/H as well, as will be discussed below. High D/H narrows the range to $1.6 < \eta < 3.3$ ($\Omega_B h^2 = 0.006 - 0.012$) and low D/H is compatible at the 2σ level in the range $4.7 < \eta < 6.3$ ($\Omega_B h^2 = 0.017 - 0.023$).

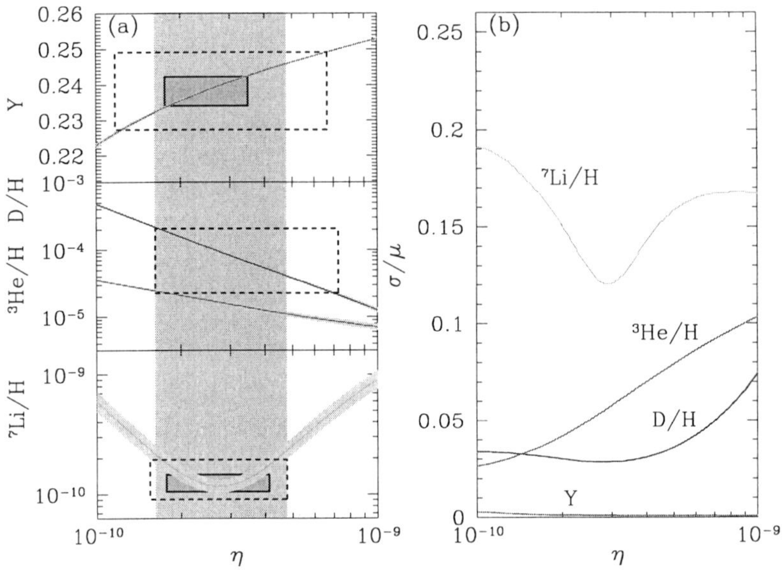

Figure 1. The light element abundances from big bang nucleosynthesis and their uncertainties as a function of η_{10}.

As discussed in more detail by A. Coc[6], there has recently been a new compilation of nuclear reaction rates which are relevant for big bang nucleosynthesis by the NACRE collaboration[7]. While the new rates lead to only very small changes with respect to the central predictions of BBN[8,4], the new analysis has greatly diminished the uncertainties in the BBN predictions[4]. This reduction is exemplified in Figure 2, showing the relative uncertainty for the d(d,p)t reaction as compared with the previous estimated uncertainty of Smith, Kawano, and Malany[3]. (Although not all rates show as dramatic a reduction and some rates are determined to be less certain, the overall net effect is a reduction in the theoretical errors for BBN.) The abundances and their uncertainties based on these new rates have been used in Fig. 1.

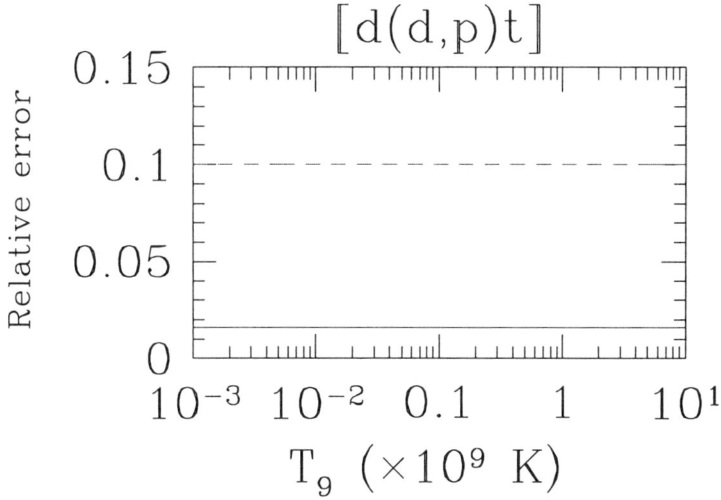

Figure 2. Fractional errors for the d(d,p)t reaction used in BBN. The solid curve shows the improvement based on the NACRE compilation taken from ref. [4] over the previous uncertainties[3] shown by the dashed curve.

2 Data

2.1 ^4He

The primordial ^{4}He abundance is best determined from observations of HeII $\rightarrow$ HeI recombination lines in extragalactic HII (ionized hydrogen) regions. There is a good collection of abundance information on the ^{4}He mass fraction, Y, O/H, and N/H in over 70 such regions[9,10,11]. Since ^{4}He is produced in stars along with heavier elements such as Oxygen, it is then expected that the primordial abundance of ^{4}He can be determined from the intercept of the correlation between Y and O/H, namely $Y_p = Y(\mathrm{O/H} \rightarrow 0)$. A detailed analysis[12] of the data found

$$Y_p = 0.238 \pm 0.002 \pm 0.005 \tag{1}$$

The first uncertainty is purely statistical and the second uncertainty is an estimate of the systematic uncertainty in the primordial abundance determination. The solid box for ^{4}He in Figure 1 represents the range (at $2\sigma_{\rm stat}$) from (1). The dashed box extends this by including the systematic uncertainty.

The helium abundance used to derive (1) was determined using assumed electron densities n in the HII regions obtained from SII data. Izotov, Thuan, & Lipovetsky[10] proposed a method based on several He emission lines to "self-consistently" determine the electron density. Their data using this method yields a higher primordial value

$$Y_p = 0.244 \pm 0.002 \pm 0.005 \tag{2}$$

One should also note that a recent determination[13] of the ^{4}He abundance in a single object (the SMC) also using the self consistent method gives a primordial abundance of 0.234 ± 0.003 (actually, they observe $Y = 0.240 \pm 0.002$ at [O/H] = -0.8).

As one can see, the resulting primordial ^{4}He abundance shows significant sensitivity to the *method* of abundance determination, leading one to conclude that the systematic uncertainty (which is already dominant) may be underestimated[14]. The status of our understanding of the systematic uncertainties in ^{4}He was discussed in detail by E. Skillman[15].

2.2 ^7Li

The abundance of ^{7}Li has been determined by observations of over 100 hot, population-II halo stars, and is found to have a very nearly uniform abundance[16]. For stars with a surface temperature $T > 5500$ K and a metallicity less than about 1/20th solar the abundances show little or no dispersion beyond that which is consistent with the errors of individual measurements. The Li data from ref. [17] indicate a primordial ^{7}Li abundance of[18]

$$\mathrm{Li/H} = (1.2 \pm 0.1) \times 10^{-10} \tag{3}$$

The small error is statistical and is due to the large number of stars in which ^{7}Li has been observed. The solid box for ^{7}Li in Figure 1 represents the $2\sigma_{\mathrm{stat}}$ range from (3).

There is, however, an important source of systematic error due to the possibility that Li has been depleted in these stars, though the lack of dispersion in the Li data limits the amount of depletion. In fact, a small observed slope in Li vs Fe and the tiny dispersion about that correlation indicates that depletion is negligible in these stars[17]. Models of galactic evolution predict such a small slope indicating a common origin of at least some of the observed Li with the other intermediate light elements of B and Be. This prediction stems back to the early work of Jean Audouze as reviewed by H. Reeves[19]. The slope of Li vs Fe has been used to extract the abundance in (3). The dashed box shows the estimated uncertainty with systematic errors included.

3 Concordance

Let us now turn to the question of concordance between the BBN predictions and the observations discussed above. This is best summarized in a comparison of likelihood functions as a function of η. By combining the theoretical predictions (and their uncertainties) with the observationally determined abundances discussed above, we can produce individual likelihood functions[5] which are shown in Figure 3. A range of primordial ^{7}Li values are chosen based on the the abundances in Eqs. (3) and the systematic uncertainties[18]. Also shown are both values of the primordial ^{4}He abundances discussed above. As one can see, at this level there is clearly concordance between ^{4}He, ^{7}Li and BBN.

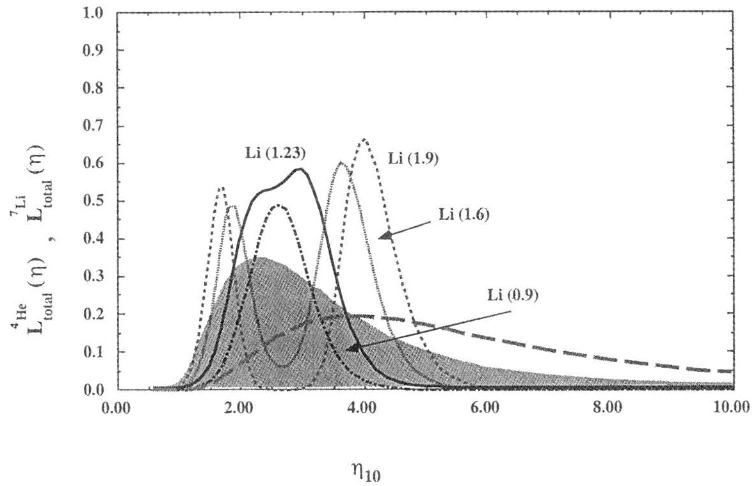

Figure 3. Likelihood distributions for four values of primordial ^{7}Li/H ($10^{10}\times$ ^{7}Li $= 1.9$ (*dashed*), 1.6 (*dotted*), 1.23 (*solid*), and 0.9 (*dash-dotted*)), and for ^{4}He (*shaded*) using $Y_p = 0.238 \pm 0.002 \pm 0.005$ (Eq. (1)). Also shown by the long dashed curve is the likelihood function based on the ^{4}He abundance from Eq. (2).

The combined likelihood, for fitting both elements simultaneously, is given by the product of two of the functions in Figure 3. The combined likelihood can be used for establishing confidence level intervals on η_{10} and $\Omega_B h^2$ as given in the introduction.

The remaining two light elements produced by BBN are D and ^{3}He. For the most part, the abundances of these elements are determined either in the local interstellar medium or in the in our own solar system. As such one needs a model of galactic chemical evolution to tie them to the BBN abundances. Deuterium is predicted to be a monotonically decreasing function of time. The degree to which D is destroyed is, however, a model dependent question and related to the production of ^{3}He. Stellar models predict that substantial amounts of ^{3}He are produced in stars between 1 and 3 $M_\odot$[20]. It should be emphasized that this prediction is in fact consistent with the observation of high ^{3}He/H in planetary nebulae[21]. However, the implementation of standard model ^{3}He yields in chemical evolution models leads to an overproduction of ^{3}He/H particularly at the solar epoch[22]. See ref. [23] for attempts to resolve this problem.

Deuterium however has been measured in quasar absorption systems. Several of these systems yield a low value of D/H $\simeq (3.0 \pm 0.3) \times 10^{-5}$(see ref. [24]). Such a low value may be problematic with respect to the high redshift luminosity function and cosmic chemical evolution[25] However, there remains an observation with D/H nearly an order of magnitude higher D/H $\simeq (2.0 \pm 0.5) \times 10^{-4}$ (see ref. [26]).

Because there are no known astrophysical sites for the production of deuterium, all observed D is assumed to be primordial[27]. As a result, any firm determination of a deuterium abundance establishes an upper bound on η which is robust. Thus, the recent measurements of D/H[24] at least provide a lower bound on D/H, D/H $> 2.1 \times 10^{-5} (2\sigma)$ and hence provide an upper bound to η, $\eta_{10} < 7.3$ and $\Omega_B h^2 < 0.027$.

Acknowledgments

I would like to thank both Jean Audouze and Jim Truran for their support and guidance over the many years that I have known them. This work was supported in part by DoE grant DE-FG02-94ER-40823 at the University of Minnesota.

References

1. T.P. Walker, G. Steigman, D.N. Schramm, K.A. Olive and K. Kang, *Ap.J.* **376** (1991) 51; S. Sarkar, *Rep. Prog. Phys.* **59** (1996) 1493; K.A. Olive, astro-ph/9901231; K.A. Olive, G. Steigman, and T.P. Walker, *Phys. Rep.* **333-334** (2000) 389.
2. L.M. Krauss and P. Romanelli, *Ap.J.* **358** (1990) 47; N. Hata, R.J. Scherrer, G. Steigman, D. Thomas, and T.P. Walker, *Ap.J.* **458** (1996) 637; G. Fiorentini, E. Lisi, S. Sarkar, and F.L. Villante, *Phys.Rev.* **D58** (1998)

063506; K. M. Nollett and S. Burles, *Phys. Rev.* **D61** (2000) 123505.
3. M. Smith, L. Kawano, and R.A. Malaney, *Ap.J. Supp.* **85** (1993) 219.
4. R.H. Cyburt, B.D. Fields, and K.A. Olive, astro-ph/0102179.
5. B.D. Fields, K. Kainulainen, D. Thomas, and K.A. Olive, *New Ast.* **1** (1996) 77.
6. A. Coc, E. Vangioni-Flam, and M. Casse, these proceedings, astro-ph/0101286.
7. C. Angulo, et al. *Nucl. Phys.* **A656** (1999) 3.
8. E. Vangioni-Flam, A. Coc, and M. Cassé, *A.A.* **360** (2000) 15.
9. B.E.J. Pagel, E.A. Simonson, R.J. Terlevich and M. Edmunds, *MNRAS* **255** (1992) 325; E. Skillman and R.C. Kennicutt, *Ap.J.* **411** (1993) 655; E. Skillman, R.J. Terlevich, R.C. Kennicutt, D.R. Garnett, and E. Terlevich, *Ap.J.* **431** (1994) 172.
10. Y.I. Izotov, T.X. Thuan, and V.A. Lipovetsky, *Ap.J.* **435** (1994) 647; *Ap.J.S.* **108** (1997) 1.
11. Y.I. Izotov, and T.X. Thuan, *Ap.J.* **500** (1998) 188.
12. K.A. Olive, E. Skillman, and G. Steigman, *Ap.J.* **483** (1997) 788; B.D. Fields and K.A. Olive, *Ap.J.* **506** (1998) 177.
13. M. Peimbert, A. Peimbert, and M.T. Ruiz, astro-ph/0003154.
14. K.A. Olive and E. Skillman, astro-ph/0007081.
15. E. Skillman, these proceedings.
16. F. Spite, and M. Spite, *A.A.* **115** (1982) 357.
17. S.G. Ryan, J.E. Norris, and T.C. Beers, *Ap.J.* **523** (1999) 654.
18. S.G. Ryan, T.C. Beers, K.A. Olive, B.D. Fields, and J.E. Norris, *Ap.J. Lett.* **530** (2000) L57.
19. H. Reeves, these proceedings; see also J. Audouze and B. Tinsley, *A.R.A.A.* **14** (1976) 43.
20. I. Iben, and J.W. Truran, *Ap.J.* **220** (1978) 980.
21. R.T. Rood, T.M. Bania, and T.L. Wilson, Nature **355** (1992) 618; D.S. Balser, T.M. Bania, R.T. Rood, T.L. Wilson, *Ap.J.* **483** (1997) 320.
22. K.A. Olive, R.T. Rood, D.N. Schramm, J.W. Truran, and E. Vangioni-Flam, *Ap.J.* **444** (1995) 680; D. Galli, F. Palla, F. Ferrini, and U. Penco, Ap.J. **443** (1995) 536;
D. Dearborn, G. Steigman, and M. Tosi, *Ap.J.* **465** (1996) 887.
23. S. Scully, M. Cassé, K.A. Olive, and E. Vangioni-Flam, *Ap.J.* **476** (1997) 521; K.A. Olive, D.N. Schramm, S. Scully, and J.W. Truran, *Ap.J.* **479** (1997) 752; D. Galli, L. Stanghellini, M. Tosi, and F. Palla *Ap.J.* **477** (1997) 218.
24. J.M. O'Meara, these proceedings; J.M. O'Meara et al., astro-ph/0011179.
25. M. Casse, K.A. Olive, E. Vangioni-Flam, and J. Audouze, *New Ast.* **3**

(1998) 259.
26. J.K. Webb, R.F. Carswell, K.M. Lanzetta, R. Ferlet, M. Lemoine, A. Vidal-Madjar, and D.V. Bowen, *Nature* **388** (1997) 250;
D. Tytler, S. Burles, L. Lu, X.-M. Fan, A. Wolfe, and B.D. Savage, *A.J.* **117** (1999) 63.
27. H. Reeves, J. Audouze, W. Fowler, and D.N. Schramm, *ApJ* **179** (1976) 909.

BIG-BANG NUCLEOSYNTHESIS WITH THE NACRE COMPILATION

ALAIN COC

CSNSM, Bât. 104, 91405 Orsay Campus, France

ELISABETH VANGIONI-FLAM

IAP, 98bis Bd Arago 75014 Paris, France

MICHEL CASSÉ

SAp, DAPNIA, DSM, CEA, 91191 Gif sur Yvette CEDEX France, and IAP

We update the Big Bang Nucleosynthesis (BBN) calculations on the basis of the recent NACRE compilation of reaction rates. In particular, we calculate the uncertainties related to the nuclear reaction rates on the abundances of ^7Li and compare our results with an other recent analysis.

Observations of light isotopes have flourished and in particular, high quality *Li* observations in the halo stars have been accumulated. Thus, it is timely to reassess the determination of the baryonic density of the Universe in the light of advances in nuclear physics and astronomical observations. (For details see K. Olive, these proceedings.)

Most of the important reactions for ^7Li production are available in the NACRE compilation of thermonuclear reaction rates[1] (others in our BBN network are adapted from earlier compilations[5,9] or more recent works[3,4]). One of the main innovative feature of NACRE with respect to former compilations[5] is that uncertainties are analyzed in detail and realistic lower and upper bounds for the rates are provided. Using these low and high rate limits, it is thus possible to calculate the effect of nuclear uncertainties on the light element yields. In previous works, we have studied the influence of individual reactions[10] or extreme yield limits[11]. Here we present results of Monte-Carlo calculations that allow to derive statistical limits on the yields. Other such calculations have been performed recently, based on a different compilation and analysis of nuclear data (Nollett & Burles[7], hereafter NB). The NB and NACRE compilations differ on several aspects. The NB compilation addresses primordial nucleosynthesis while the NACRE one is a general purpose compilation. Accordingly, the NB one contains a few more reaction of interest to BBN and a few more data in the energy range of interest. Also, *from the statistical point of view*, the rate uncertainties are better defined in NB. However, in NB, the astrophysical S–factors are fitted by splines which have no physical justification and can produce local artifacts by following to closely experimental data points. On the contrary, the NACRE compilation span wider energy ranges, and over these ranges, the S–factors are fitted to functions reflecting theoretical assumptions. This eliminates spurious local effects induced by a few erratic data points associated with experimental problems rather than genuine physical effect. Due to the difficulty of defining a universal statistical method for the wide set of reactions (each with its peculiarities) and range of temperature, the NACRE rate limits correspond to upper and lower bound rather than standard deviations. In a previous study[11], we considered all possible combinations of these high and low rates and derived extreme upper and lower limits for the ^7Li yields (Figure 1). Here, instead, we ran Monte-Carlo calculations assuming a log–uniform distribution for the rates between the low and high NACRE limits (keeping the mean rates equal to the recommended rates). For each η value, we calculated the mean value and standard deviation (σ) of the ^7Li yield distribution. The corresponding 2–σ limits are represented in Figure 1 together with the results obtained by Burles et al.[2] with the NB compilation. The results obtained from these two *independent* compilations agree very well reinforcing the confidence in these analyses.

The determination of Li in halo stars indicate that the Spite plateau is exceptionally thin (< 0.05 dex). We adopt the following range: $1.5\ 10^{-10} < {}^7Li/H < 2.2\ 10^{-10}$, taking into account a maximum destruction of 0.1 dex. Due to the valley shape of the Li curve, we obtain two possible η ranges which are consistent respectively with a high D and low ^4He observed values, and with a low D and high ^4He observed values. As presently, none of these solutions can be excluded, two possible ranges emerge: *i)* $1.35 < \eta_{10} < 2.0$ corresponding to $0.0049 < \Omega_B h^2 < 0.0073$ and *ii)* $3.4 < \eta_{10} < 4.9$ corresponding to $0.012 < \Omega_B h^2 < 0.018$. It is worth noting that these results (and also the extreme upper limit[11] of $\eta_{10} < 5.3$ or $\Omega_B h^2 < 0.02$) are hardly compatible with the recent MAXIMA and BOOMERANG baloons CMB observations[6] ($\Omega_B h^2 = 0.032 \pm 0.005$). However, other observations with the Cosmic Background Imager[8] (CBI, Chile) lead to a lower value of $\Omega_B h^2 = 0.009$ closer to the high D, low Ω_B range. Clarification will hopefully come from future space missions (i.e. MAP and PLANCK Surveyor).

References

1. C. Angulo, M. Arnould, M. Rayet et al., *Nucl. Phys.* **A656**, 3 (1999).
2. S. Burles, K.M. Nollett & M.S. Turner, submitted (astro-ph/0010171).
3. C.R. Brune, K.I. Hahn, R.W. Kavanagh & P.W. Wrean, *Phys. Rev. C* 60 (1999) 015801.

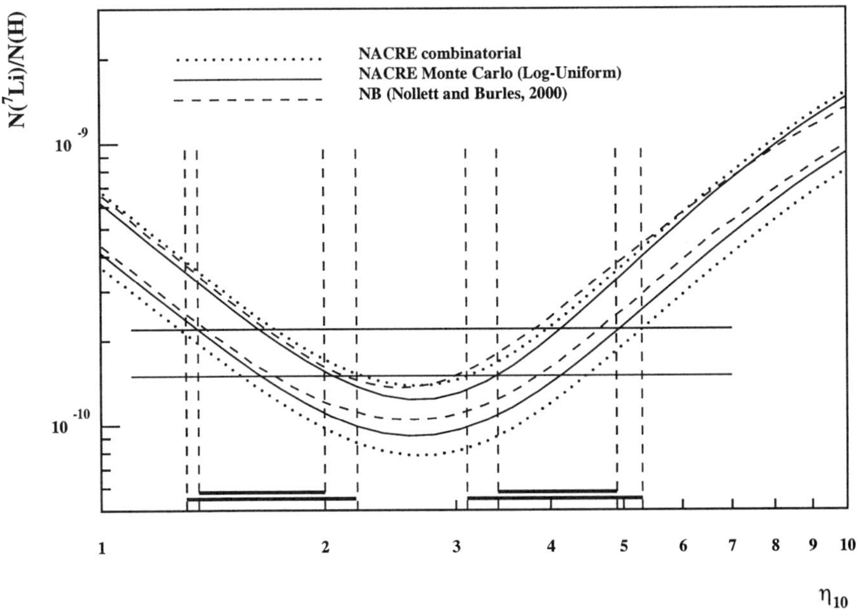

Figure 1. ^7Li limits obtained by combining NACRE limits[11], or Monte-Carlo calculations with the NACRE[1] or NB[7,2] compilations. Horizontal lines represent the ^7Li observational limits while vertical line represent the deduced η_{10} limits.

4. J.-W. Chen & M.J. Savage, *Phys. Rev. C* **60**, 065205 (1999).
5. G.R. Caughlan & W.A. Fowler, *Atomic Data and Nuclear Data Tables* **40**, 283 (1988).
6. A.H. Jaffe et al., PRL submitted (astro-ph/0007333).
7. K.M. Nollett & S. Burles, *Phys. Rev.* D **61**, 123505 (2000).
8. S. Padin et al., ApJ submitted (astro-ph/0012211).
9. M.S. Smith, L.H. Kawano & R.A. Malaney, *Astrophys. J. Sup.* **85**, 219 (1993).
10. E. Vangioni-Flam, A. Coc & M. Cassé, *Astron. Astrophys.* **360**, 15 (2000).
11. E. Vangioni-Flam, A. Coc & M. Cassé, in *Nuclei in the Cosmos 2000* to appear in *Nucl. Phys.* (2001).

ISSUES IN NON-STANDARD BIG BANG NUCLEOSYNTHESIS

KARSTEN JEDAMZIK
Max-Planck-Institut für Astrophysik,
85740 Garching bei München, Allemagne
E-mail: jedamzik@mpa-garching.mpg.de

I summarize results of two variants of a standard homogeneous Big Bang nucleosynthesis (BBN) scenario. First, I discuss trends in BBN scenarios which are characterized by the existence of small-scale baryonic inhomogeneity. Then I consider the decay of relic particles after BBN and show how such scenarios may lead to efficient production of ^{6}Li.

Big Bang nucleosynthesis is probably the earliest cosmic epoch where a comparison between observations and theoretical predictions may possibly yield detailed information about the state of the universe at early times. It is believed that the bulk of ^{4}He, essentially all D, and good fractions of ^{3}He and ^{7}Li as observed in the present universe originate from this freeze-out process from nuclear statistical equilibrium occurring at the approximate cosmic temperature of $\sim$ 100 keV. This is the reason why BBN has not only been investigated in a multitude of papers in its simplest, so called standard scenario, but also in a great number of variants of the standard scenario (hereafter, SBBN). For reviews on SBBN the reader is referred to Ref. [1] and on nonstandard BBN to Ref. [2,3]. In this talk I discuss results of studies of two such variants of SBBN. These studies are partially motivated by (a) recent observations of the anisotropies in the cosmic microwave background radiation [4] favoring slightly larger baryonic contribution to the critical density, Ω_b, than allowed by a standard BBN scenario and (b) new observational determinations of ^{6}Li and ^{9}Be in metal-poor halo stars, some of which have been announced first on this conference [5,6].

• **Inhomogeneous BBN**: The possibility that cosmic baryon number fluctuations may have existed on small scales in the early universe has received considerable attention between the late eighties and mid nineties [2]. Such fluctuations in baryon number would have impact on the production of light elements during BBN provided the baryonic mass of individual lumps exceeds $M_b \gtrsim 10^{-21} M_\odot$. It was speculated that production of inhomogeneities could result during a first-order QCD phase transition or even possibly during a scenario of electroweak baryogenesis. It was also speculated that inhomogeneous BBN (hereafter, IBBN) scenarios could be consistent with an Ω_b larger than that predicted by standard BBN ($\Omega_b h^2 \approx 0.02 \pm 0.002$ [7]), though detailed studies made clear that $\Omega_b = 1$ is incompatible even with IBBN. Furthermore, it was shown that IBBN may lead to significant production of the ^{6}Li, ^{9}Be, and ^{11}B isotopes. In light of recent observations of the cosmic microwave background radiation which favor slightly larger $\Omega_b h^2 \approx 0.03$ [4], and recent observational determinations of relatively high ^{6}Li and ^{9}Be abundances in metal-poor stars [5,6], it seems worth reinvestigating IBBN scenarios. We have performed detailed numerical simulations of BBN in the presence of an inhomogeneous baryon distribution for wide ranges of the parameters describing the inhomogeneities and for a few representive baryon-to-photon ratios [8]. Our choice of initial conditions is limited to scenarios where essentially all baryons are within overdense pockets and the remainder of the volume is initially void of baryons. We found that such scenarios may be consistent with observational limits on the primordial ^{4}He and D abundance for $\Omega_b h^2$ as large as ~ 0.05. However, they result in significantly overabundant production of ^{7}Li with respect to the ^{7}Li/H ratio as observed in stars belonging to the Spite plateau. Typical production of ^{6}Li, ^{9}Be, and ^{11}B in such scenarios are found to be still below the abundances of these isotopes observed in the most metal-poor stars to date. Unless ^{7}Li in stars on the Spite plateau has been significantly depleted, which seems unlikely, IBBN scenarios thus do not allow for a significant increase of Ω_b over that inferred from a SBBN scenario.

• **Relic Particles Decaying after BBN**: Many possible evolutionary scenarios of the very early universe ($T \gg 1\,\mathrm{MeV}$) predict production of relic particles, such as gravitinos or moduli, as well as topological defects, such as cosmic strings. Such relics may often be long-lived and decay after the BBN epoch. It has long been noted that their decay products may alter the light element abundance yields resulting from BBN. A comparison of predicted light-element abundances in such scenarios with observationally inferred primordial abundances may thus be used to constrain production mechanisms of such relics [3]. Arguments along these lines have been used, for example, to infer limits on the reheat temperature, T_R, of the universe after an inflationary epoch, since for $T_R \gtrsim 10^8 - 10^{10}\,\mathrm{GeV}$, gravitinos would be produced too abundantly. Their electromagnetically interacting decay products may initiate an electromagnetic cascade on the cosmic microwave background by $e^{\pm}$ pair creation and inverse Compton scattering. This cascade halts only when the energy of the cascade γ-rays has fallen below the $e^{\pm}$ pair creation threshold, $E_C(z) \approx 4.7 \times 10^7 z^{-1}\,\mathrm{MeV}$, where z is redshift. For decay at low enough redshift, $z \lesssim 2 \times 10^6$, such γ-rays may photodisintegrate ^{4}He, thereby producing ^{3}He and D in abundance which may be in conflict with observationally inferred upper limits on the primordial abundance of these isotopes. In recent work [9] I have shown that during this process efficient production of ^{6}Li also results. The ^{3}He (and ^{3}H) nuclei produced through the ^{4}He photodisintegration are initially energetic and may engage in non-thermal nuclear reactions, i.e. $^3\mathrm{H}\,(^3\mathrm{He}) + {}^4\mathrm{He} \rightarrow {}^6\mathrm{Li} + n\,(p)$, to form ^{6}Li. Note that ^{6}Li production in SBBN is very inefficient. The ^{6}Li produced in this process may be compared to observationally inferred upper limits on the pregalactic ^{6}Li abundance. If stellar depletion of the ^{6}Li isotope in the few stars where $^6\mathrm{Li}/^7\mathrm{Li}$ ratio determinations have been performed is negligible, such a comparison yields considerable improvement in the constraints on the abundance of relics decaying into electromagnetically interacting products after BBN. On the other hand, if a relatively high plateau of ^{6}Li abundances in metal-poor stars should be ever established, a possible explanation for such a plateau could be the pregalactic production of ^{6}Li by the decay of relics after the BBN epoch.

Acknowledgments

I am grateful to Martin Lemoine for his kind assistance of still sneaking me into the conference, though I had, as usually, missed the registration deadline by a long time.

References

1. Olive, K. A., Steigman, G., & Walker, T. P. 2000, Phys. Rept. 333, 389; Tytler, D., O'Meara, J. M., Suzuki, N., & Lubin, D. 2000, Phys. Rep. 333, 409.
2. Malaney, R. A. & Mathews, G. J. 1993, Phys. Rep. 229, 145.
3. Sarkar, S. 1996, Rept. Prog. Phys. 59, 1493.
4. de Bernardis, P. *et al.* 2000, Nature, 404, 955; Hanany, S. *et al.* 2000, Astrophys. J. Letters, 545, L5.
5. Primas, F., Asplund, M., Nissen, P. E., & Hill, V. 2000, Astr. & Astrophys. Lett. 364, L42.
6. Asplund, M. *et al.*, these proceedings.
7. Burles, S., Nollett, K. M., & Turner, M. S. 2000, astro-ph/0008495.
8. Jedamzik, K. & Rehm, J. B. 2001, submitted to Phys. Rev. D (astro-ph/0101292).
9. Jedamzik, K. 2000, Phys. Rev. Lett. 84, 3248.

LEPTON ASYMMETRIC UNIVERSE: NEW LIMITS FROM BBN AND THE CMB

T. KAJINO, M. ORITO
National Astronomical Observatory, Mitaka, Tokyo 181-8588, and Department of Astronomy, University of Tokyo, Bunkyo-ku, Tokyo 113-0033, Japan
E-mail: kajino@nao.ac.jp

G. J. MATHEWS
Department of Physics and Center for Astrophysics, University of Notre Dame, Notre Dame, IN 46556, U.S.A.

R. N. BOYD
Department of Physics and Department of Astronomy, Ohio State University, Columbus, OH 43210, U.S.A.

We study primordial nucleosynthesis in the presence of a net lepton asymmetry. We explore a previously unnoted region of the parameter space in which very large baryon densities $0.1 \leq \Omega_b \leq 1$ can be accommodated within the light-element constraints. This parameter space consists of large ν_μ and ν_τ degeneracies with a moderate ν_e degeneracy. Constraints on this parameter space from cosmic microwave background fluctuations are also discussed [2].

This short contribution is concerned with neutrinos during big-bang nucleosynthesis (BBN). In particular, we consider new insights into the possible role which degenerate neutrinos may have played in the early universe. The origin of the light elements has been a topic of interest to Jean Audouze for many years. Among other things, Jean has made important contributions toward constrainig neutrino decay[1] during BBN. Hence, a discussion of neutrinos and BBN is appropriate at a workshop honoring Jean's contributions.

There is no experimental reason to insist that the universal lepton number is zero. It is possible, for example, for the individual lepton numbers to be large compared to the baryon number of the universe, while the net total lepton number is small $L \sim B$. It has been proposed recently [3] that models based upon the Affleck-Dine scenario of baryogenesis might generate naturally lepton number asymmetry which is seven to ten orders of magnitude larger than the baryon number asymmetry. Neutrinos with large lepton asymmetry and masses ~ 0.07 eV might even explain the existence of cosmic rays with energies in excess of the Greisen-Zatsepin-Kuzmin cutoff [4]. It is, therefore, important for both particle physics and cosmology to carefully scrutinize the limits which cosmology places on the allowed range of both the lepton and baryon asymmetries.

Although lepton asymmetric BBN has been studied in many papers[5] (and references therein), there are several differences in the present work: For one , we have included finite temperature corrections to the mass of the electron and photon[6]. Another is that we have calculated the neutrino annihilation rate in the cosmic comoving frame, in which the Møller velocity instead of the relative velocity is to be used for the integration of the collision term in the Boltzmann equations[7,8].

Neutrinos and anti-neutrinos drop out of thermal equilibrium with the background thermal plasma when the weak reaction rate becomes slower than the universal expansion rate. If the neutrinos decouple early, they are not heated as the particle degrees of freedom change. Hence, the ratio of the neutrino to photon temperatures, T_ν/T_γ, is reduced. The biggest drop in temperature for all three neutrino flavors occurs for $\xi_\nu \sim 10$. This corresponds to a decoupling temperature above the cosmic QCD phase transition.

Non-zero lepton numbers affect nucleosynthesis in two ways. First, neutrino degeneracy increases the expansion rate. This increases the ^{4}He production. Secondly, the equilibrium n/p ratio is affected by the electron neutrino chemical potential, n/p $= \exp\{-(\Delta M/T_{n\leftrightarrow p}) - \xi_{\nu_e}\}$, where ΔM is the neutron-proton mass difference and $T_{n\leftrightarrow p}$ is the freeze-out temperature for the relevant weak reactions. This effect either increases or decreases ^{4}He production, depending upon the sign of ξ_{ν_e}.

A third effect emphasized in this paper is that T_ν/T_γ can be reduced if the neutrinos decouple early. This lower temperature reduces the energy density of neutrinos during BBN, and slows the expansion of the universe. This decreases ^{4}He production.

Figure 1 highlights the main result of this study, where we take $\xi_{\nu_\mu} = -\xi_{\nu_\tau}$. For low $\Omega_b h_{50}^2$ models, only the usual low values for ξ_{ν_e} and $\xi_{\nu_{\mu,\tau}}$ are allowed. Between $\Omega_b h_{50}^2 \approx 0.188$ and 0.3, however, more than one allowed region emerges. For $\Omega_b h_{50}^2 \gtrsim 0.4$ only the large degeneracy solution is allowed. Neutrino degeneracy can even allow baryonic densities up to $\Omega_b h_{50}^2 = 1$.

Several recent works[9,10,11] have shown that neutrino degeneracy can dramatically alter the power spectrum of the CMB. However, only small degeneracy parameters with the standard relic neutrino temperatures have been utilized. Here, we have calculated the CMB power spectrum to investigate effects of a diminished relic neutrino temperature.

The solid line on Figure 2 shows a $\Omega_\Lambda = 0.4$ model for which $n = 0.78$. This fit is marginally consistent with the data at a level of 5.2σ. The dotted line in Figure 2 shows the matter dominated $\Omega_\Lambda = 0$ best fit model with $n = 0.83$ which is consistent with the data at the level of 3σ. The main differences in the fits between the large degeneracy models and our adopted

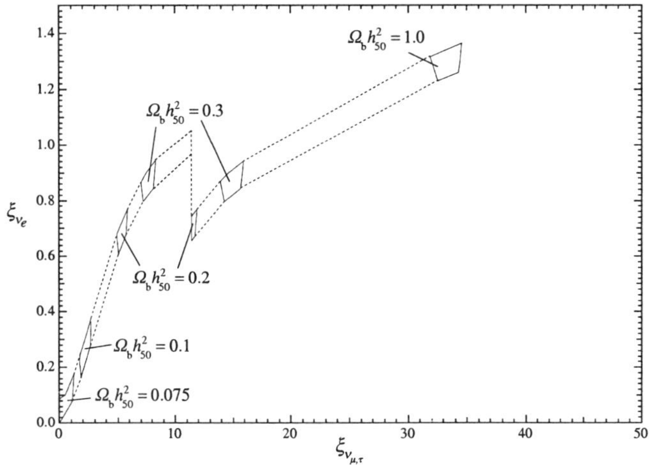

Figure 1: Allowed values of ξ_{ν_e} and $\xi_{\nu_{\mu,\tau}}$ for which the constraints from light element abundances are satisfied for values of $\Omega_b h_{50}^2 = 0.075$, 0.1, 0.2, 0.3 and 1.0 as indicated.

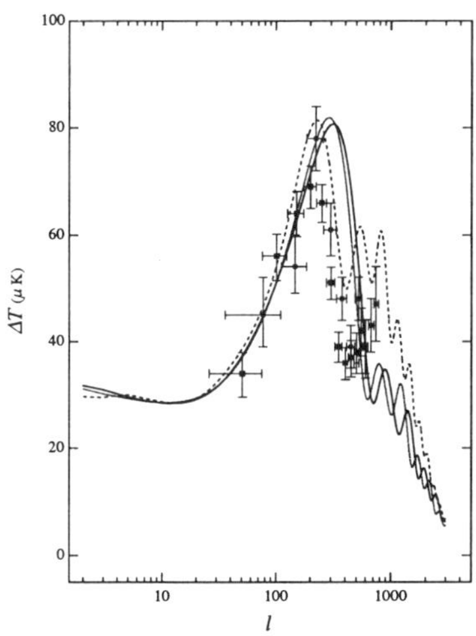

Figure 2: CMB power spectrum from MAXIMA-1 [13] (circles) and BOOMERANG [12] (squares) binned data compared with calculated $\Omega = 1$ models.

benchmark model are that the first peak is shifted to slightly higher l value and the second peak is suppressed. One can clearly see that the suppression of the second acoustic peak is consistent with our derived neutrino-degenerate models. In particular, the MAXIMA-1 results are in very good agreement with the predictions of our neutrino-degenerate cosmological models. It is clear that these new data sets substantially improve the goodness of fit for the neutrino-degenerate models [10]. Moreover, both data sets seem to require an increase in the baryonic contribution to the closure density as allowed in our neutrino-degenerate models.

1. P. Delbourgo-Salvador, P. Salati and J. Audouze, Phys. Lett., B276, 115 (1992).
astro-ph/0005446, submitted to ApJ (2000)
2. M. Orito, T. Kajino, G. J. Mathews, and R. N. Boyd, astro-ph/0005446, submitted to ApJ (2000)
3. A. Casas, W. Y. Cheng, & G. Gelmini, Nucl. Phys. B., **538**, 297, (1999)
4. G. Gelmini & A. Kusenko Phys. Rev. Lett., **82**, 5202, (1999)
5. H. Kang & G. Steigman Nucl. Phys. B., **372**, 494, (1992)
6. N. Fornengo, C. W. Kim, & J. Song, Phys. Rev. D., **56**, 5123, (1997)
7. P. Gondolo, & G. Gelmini, Nucl. Phys. B., **360**, 145, (1991)
8. K. Enqvist, K, K. Kainulainen, & V. Semikoz, Nucl. Phys. B., **374**, 392, (1992)
9. W. K. Kinney & A. Riotto, Phys. Rev. Lett., **83**, 3366, (1999)
10. J. Lesgourgues & S. Pastor, Phys. Rev. D., **60**, 103521,, (1999)astro-ph/0004412
11. S. Hannestad, Phys. Rev. Lett., submitted, (2000), astro-ph/0005018
12. P. Bernardls, et al. (Boomerang Collaboration) Nature., **404**, 955, (2000)
13. S. Hanany, et al. (MAXIMA-1 Collaboration), ApJL submitted, astro-ph/0005123

Part 2. Origin and Evolution of the Light Elements

D/H MEASUREMENTS

A. VIDAL-MADJAR
Institut d'Astrophysique de Paris, C.N.R.S./Paris VI, 98bis Boulevard Arago, F-75014, Paris, FRANCE
E-mail: alfred@iap.fr

Primordial evaluations of the deuterium abundance should provide one of the best tests of Big Bang nucleosynthesis models. Space as well as ground based observations seem however to result in different values. This asks for more observations in different astrophysical sites in order to link present day interstellar medium D/H evaluations to primordial ones. New investigations, made with FUSE (the *Far Ultraviolet Spectroscopic Explorer* launched in June 1999), are presented and in the case of the white dwarf G191-B2B line of sight a low D/H evaluation of $1.16 \pm 0.24 \times 10^{-5}$ (2σ) is confirmed. This seems to indicate that D/H variations are probably present in the nearby interstellar medium. The FUSE observations should help us reach in a near future a better global view of the evolution of that key element.

1 Introduction

During primordial Big Bang nucleosynthesis deuterium is produced in significant amounts and then destroyed in stellar interiors. It is thus a key element in cosmology and in galactic chemical evolution (see *e.g.* Audouze & Tinsley [1]; Boesgaard & Steigman [2]; Olive *et al.* [3]; Pagel *et al.* [4]; Vangioni-Flam & Cassé [5, 6]; Prantzos [7]; Scully *et al.* [8]; Cassé & Vangioni-Flam [9]).

Thanks to the *Copernicus* space observatory, the first direct measurement of the D/H ratio in the interstellar medium (ISM) representative of the present epoch (Rogerson & York [10]), was

$$(\mathrm{D/H})_{\mathrm{ISM}}^{Copernicus} \simeq 1.4 \pm 0.2 \times 10^{-5}.$$

More recently D/H evaluations were made in the direction of quasars (QSOs) in low metallicity media. They were completed toward three different QSOs' (Burles & Tytler [11, 12]; O'Meara & Tytler [13]) leading to a possible range of $2.4 - 4.8 \times 10^{-5}$ for the primordial D/H. These values correspond to a new estimations of the baryon density of the Universe, $\Omega_{\mathrm{b}}\mathrm{h}^2 = 0.019 \pm 0.0009$, in the frame of the standard BBN model (Burles *et al.* [14]; Nollett & Burles [15]). When compared to the recent $\Omega_{\mathrm{b}}\mathrm{h}^2$ evaluation made from the Cosmic Microwave Background (CMB) observations (see *e.g.* Jaffe *et al.* [30]) $\Omega_{\mathrm{b}}\mathrm{h}^2 = 0.032 \pm 0.005$, this seems to lead to a possible conflict. Note that another D/H measurement made toward a low redshift QSO leading to a D/H value possibly larger than 10^{-4} (Webb *et al.* [17]; Tytler *et al.* [18]) corresponds to an even stronger disagreement since it translates into $\Omega_b h^2 \leq 0.01$.

It is thus of prime importance to investigate the possibility of varying D/H ratios in different astrophysical sites (see *e.g.* Lemoine *et al.* [19]). If variations are indeed found, their cause should be investigated to deduce a more reliable primordial D/H evaluation.

2 Interstellar observations

Several methods were used to measure the interstellar abundance of deuterium. All will not be discussed here and for more details see *e.g.* Ferlet [20]. The more reliable approach is to observe in absorption, against the background continuum of stars, the atomic Lyman series of D and H in the far-UV.

Toward distant hot stars, with the *Copernicus* satellite, many important evaluations of D/H were obtained (see e.g. Rogerson and York [10]; York and Rogerson [23]; Vidal-Madjar *et al.* [24]; Laurent *et al.* [25]; Ferlet *et al.* [26]; York [27]; Allen *et al.* [28]) leading to the detection of variations recently enforced by IMAPS observations, one made toward δ Ori presenting a low value (Jenkins *et al.* [30]) confirming the previous analysis made by Laurent *et al.* [25] from Copernicus observations and the other one toward γ^2 Vel with a high value (Sonneborn *et al.* [31]). These observations seem to indicate that in the ISM, within few hundred parsecs, D/H may vary by more than a factor $\simeq 3$.

¿From published values, D/H ranges from

$$\sim 5 \times 10^{-6} < (\mathrm{D/H})_{ISM} < \sim 4 \times 10^{-5}.$$

This method also provided a precise measurement of D/H in the local ISM (LISM) in the direction of Capella using HST–GHRS (Linsky *et al.* [21]) :

$$(\mathrm{D/H})^{\mathrm{GHRS}}_{\mathrm{Capella}} = 1.60 \pm 0.09^{+0.05}_{-0.10} \times 10^{-5}$$

3 The nearby ISM

Observations of white dwarfs (WD) in the nearby ISM (NISM) for precise D/H evaluations were first proposed and achieved in the direction of G191–B2B by Lemoine *et al.* [32] using the HST–GHRS spectrograph at intermediate resolution. Follow up observations at higher resolution with the GHRS Echelle-A grating by Vidal–Madjar *et al.* [22] (the same instrument configuration used as in the case of the Capella study) lead to a precise D/H evaluation in the NISM along this line of sight within one HI region – the Local Interstellar Cloud (LIC) also observed toward Capella (these stars are separated by $\sim 8^o$ on the sky) – and within a more complex and ionized HII region presenting a double velocity structure. In these two main interstellar components the D/H ratio was found to be different particularly if one assumes that the D/H value

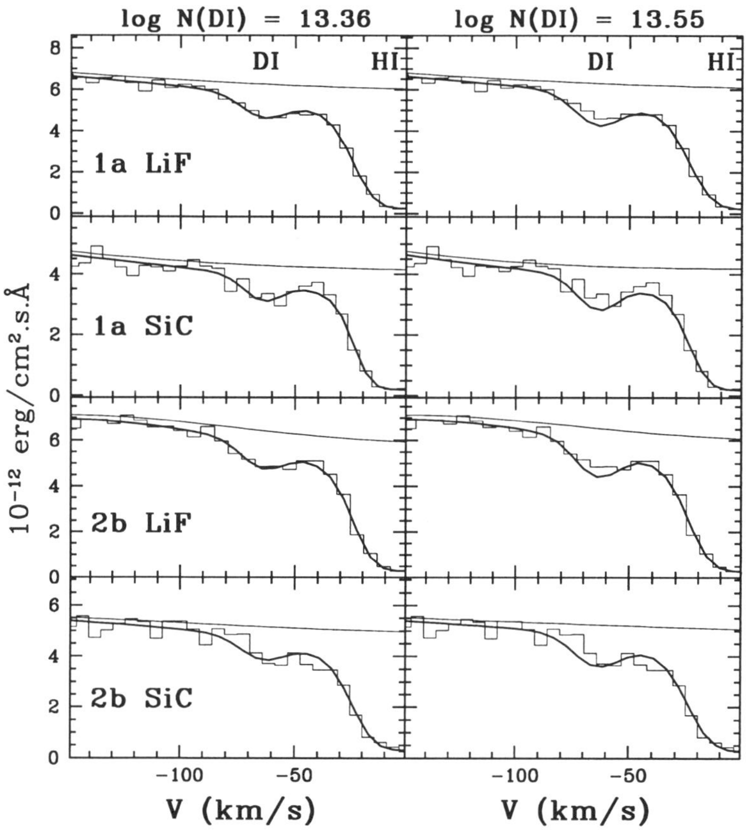

Figure 1. FUSE observations made in the direction of G191–B2B. The Lyman β line is shown in the four FUSE channels recorded simultaneously through the high resolution slit (two LiF and two SiC). On the left pannels the fits are shown with the best FUSE evaluation made by Vidal–Madjar *et al.* [35] and on the right ones with the D column densities evaluated by Sahu *et al.* [33] with the STIS instrumentation. These are more than 6σ incompatible with the FUSE observations.

within the LIC is the same as the one found in the direction of Capella. In that case D/H in the more ionized components has to be lower ($\sim 0.9 \times 10^{-5}$).

In any case a lower "average" D/H ratio was found (2σ error) :

$$(\mathrm{D/H})^{\mathrm{GHRS}}_{\mathrm{G191-B2B}} = 1.12 \pm 0.16 \times 10^{-5}$$

Although contested by Sahu *et al.* [33] who used new HST–STIS high resolution Echelle observations, Vidal–Madjar [34] showed that convergence in independant data analysis made with STIS as well as GHRS observations indicates that their original conclusion was extremely robust.

Since the disagreement between the two analysis was related to the D I column density estimation, FUSE data in the direction of G191–B2B was expected to clarify the situation because they give access to weaker deuterium Lyman lines less sensitive to systematics effects. Three independant data sets were obtained corresponding to three different FUSE entrance appertures (Vidal–Madjar *et al.* [35]). In Figure 1 are shown the fits obtained over the D Lyman β line in the different and independant FUSE channels with the total D column density deduced from the FUSE data (logN(D I)=13.36) or with the one evaluated by Sahu *et al.* [33] (logN(D I)=13.55) which stands at more than 6σ away. ¿From the FUSE data analysis alone representative of the D I column density and from the published H I values we can conclude (2σ error) :

$$(D/H)^{FUSE}_{G191-B2B} = 1.16 \pm 0.24 \times 10^{-5}$$

When comparing $(D/H)_{G191-B2B}$ with $(D/H)_{Capella}$, variations seem to be real in the NISM although an "average" value of 1.4×10^{-5} barely compatible with both evaluations is still possible.

The essential question is therefore : if D/H variations are however confirmed in the direction of more sightlines, what could be their cause ?

4 The FUSE observatory

The FUSE satellite starts to produce orders of magnitude more data on the distribution of D/H in the ISM. From the planned D/H survey, we should be able to evaluate the deuterium abundance in a wide variety of locations, possibly linked to the past star formation rate as well as to the supposed infall of less processed gas in our Galaxy, and thus better understand Galactic chemical evolution.

The FUSE sensitivity allows evaluations of the deuterium abundances in hundreds of lines of sights : i) in the direction of white dwarfs and cool stars in the NISM ; ii) toward hot sub-dwarfs in the more distant ISM and nearby Galactic halo ; iii) within the Galactic disk over several kilo-parsecs in the direction of O and early B stars ; iv) in the more distant Galactic halo, within high velocity cloud complexes as well as in intergalactic clouds in the direction of low redshift QSOs, AGNs and blue compact galaxies.

The first precise D/H evaluations were presented in early 2001 at the AAS meeting (Moos *et al.* [36]; Friedman *et al.* [37]; Hébrard *et al.* [38]; Kruk *et al.* [39]; Linsky *et al.* [40]; Sonneborn *et al.* [41]; Vidal–Madjar *et al.* [35]). The deuterium Lyman lines are clearly seen toward these few nearby white dwarfs

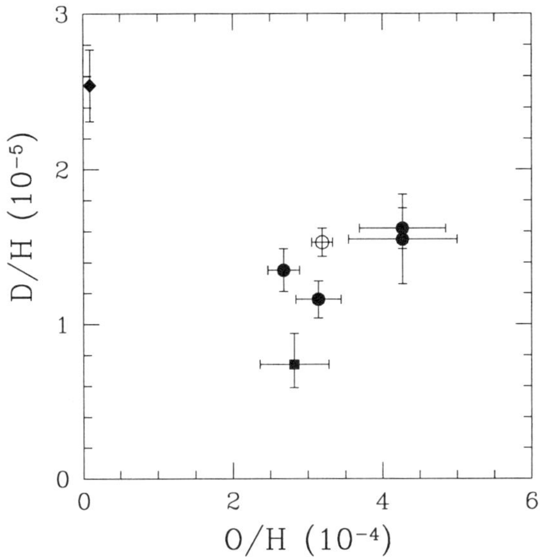

Figure 2. Different D/H evaluations as a function of O/H. The diamond is the observation made in the direction of a QSO by O'Meara and Tytler [13] corresponding to a low metallicity cloud; the square is the IMAPS observation in the direction of δOri by Jenkins *et al.* [30] in the ISM; the filled circles correspond to the FUSE observations and the open circle to the "average" D/H from Linsky and Wood [42] in relation to the ISM O/H evaluation made by Meyer *et al.* [29] from their survey.

and, as an example, the Lyman β line is shown in the case of G191–B2B as previously discussed (see Figure 1). In Figure 2 are shown few of these D/H evaluations made in the ISM with FUSE, HST, IMAPS along with one made recently in the direction of one QSO from ground based observations [13], as a function of the line of sight average metallicity as traced by O/H when available. It seems that the D/H variation does not anti–correlate with O/H. Thus a simple mechanism as astration, able to destroy D and produce O, seems not to be the one compatible with the observations. Other mechanisms should be investigated as the ones listed by *e.g.* Lemoine *et al.* [19].

5 Conclusion

In summary the status of the different – but discordant – D/H evaluations taken with no a priori bias to select one over another could be the following.

If the variations of the D/H ratio in the NISM are illusory, although observations show they exist at a 95% confidence level, one could quote an average value of $(D/H)_{NISM} \simeq 1.4 \times 10^{-5}$ barely compatible with all observations.

More in agreement with the present observations, D/H seems to vary in the ISM. One has thus to understand why.

Until then, *no measurement of the D/H ratio in the ISM or the IGM should be quoted as reliable.* This is true in particular for the "primordial" values found in the direction of QSOs since the physical state of the probed environment is more poorly known than the Galactic ISM.

Our hope is that the FUSE mission will solve these problems.

Acknowledgments

I am very grateful for the entire FUSE operation team for all the impressive work they are doing to make the FUSE observatory come true. I also thank the whole FUSE team for many positive interaction and comments. This work is based on data obtained by the NASA–CNES–CSA FUSE mission operated by the Johns Hopkins University under NASA contract NAS5–32985.

References

1. J. Audouze and B.M. Tinsley, *Ann. Rev. Astron. Astrophys.* **14**, 43 (1976).
2. A.M. Boesgaard and G. Steigman, *Ann. Rev. Astron. Astrophys.* **23**, 319 (1985).
3. K. Olive *et al.*, *Phys. Rev. Lett.* **B236**, 454 (1990).
4. B. Pagel *et al.*, MNRAS **255**, 325 (1992).
5. E. Vangioni-Flam and M. Cassé, *Ap.J.* **427**, 618 (1994).
6. E. Vangioni-Flam, K. A. Olive and N. Prantzos, *Ap.J.* **441**, 471 (1995).
7. N. Prantzos, AA **310**, 106 (1996).
8. S.T. Scully, *et al.*, *Ap.J.* **476**, 521 (1997).
9. M. Cassé and E. Vangioni-Flam in *Structure and Evolution of the Intergalactic Medium from QSO Absorption Line Systems*, eds. P. Petitjean and S. Charlot (IAP Conference, 331, 1998).
10. J. Rogerson and D. York, *Ap.J. Letters* **186**, L95 (1973).
11. S. Burles and D. Tytler, *Ap.J.* **499**, 699 (1998a).
12. S. Burles and D. Tytler, *Ap.J.* **507**, 732 (1998b).
13. J. O'Meara and D. Tytler, in these proceedings *Cosmic Evolution*, eds. M. Lemoine and R. Ferlet, 2001.
14. S. Burles *et al.*, *Phys. Rev. Lett.* **82**, 4176 (1999).

15. K.M. Nollett and S. Burles, *Phys. Rev. Lett.* **D61**, 123505 (2000).
16. A.H. Jaffe *et al.*, astro-ph/0007333, 2000.
17. J.K. Webb *et al.*, *Nature* **388**, 250 (1997).
18. D. Tytler *et al.*, *AJ* **117**, 63 (1999).
19. M. Lemoine *et al.*, *New Astronomy* **4**, 231 (1999).
20. R. Ferlet, in IAU#150 *Astrochemistry of Cosmic Phenomena*, eds. P.D. Singh, (Kluwer, 85, 1992).
21. J. Linsky *et al.*, *Ap.J.* **451**, 335 (1995).
22. A. Vidal–Madjar *et al.*, *Astron. Astrophys.* **338**, 694 (1998).
23. D. York and J. Rogerson, *Ap.J. Letters* **203**, 378 (1976).
24. A. Vidal–Madjar *et al.*, *Ap.J. Letters* **211**, 91 (1977).
25. C. Laurent, A. Vidal–Madjar and D.G. York, *Ap.J.* **229**, 923 (1979).
26. R. Ferlet *et al.*, *Ap.J. Letters* **242**, 576 (1980).
27. D.G. York, *Ap.J. Letters* **264**, 172 (1983).
28. M.M. Allen, E.B. Jenkins and T.P. Snow, *Ap.J. Suppl.* **83**, 261 (1992).
29. D.M. Meyer *et al.*, *Ap.J.* **493**, 222 (1998).
30. E.B. Jenkins *et al.*, *Ap.J.* **520**, 182 (1999).
31. G. Sonneborn, *et al.*, *Ap.J.* **545**, 277 (2000).
32. M. Lemoine *et al.*, *Astron. Astrophys.* **308**, 601 (1996).
33. M.S. Sahu *et al.*, *Ap.J. Letters* **523**, L159 (1999).
34. A. Vidal–Madjar, in *The Light Elements and Their Evolution*, eds. L. da Silva, M. Spite and J. R. de Medeiros (ASP Conference Series, 151, 2000).
35. A. Vidal–Madjar *et al.*, *Ap.J.* in preparation, (2001).
36. H.W. Moos *et al.*, *Ap.J.* in preparation, (2001).
37. S.D. Friedman *et al.*, *Ap.J.* in preparation, (2001).
38. G. Hébrard *et al.*, *Ap.J.* in preparation, (2001).
39. J.W. Kruk *et al.*, *Ap.J.* in preparation, (2001).
40. J.L. Linsky *et al.*, *Ap.J.* in preparation, (2001).
41. G. Sonneborn *et al.*, *Ap.J.* in preparation, (2001).
42. J.L. Linsky and B.E. Wood, in *The Light Elements and Their Evolution*, eds. L. da Silva, M. Spite and J. R. de Medeiros (ASP Conference Series, 141, 2000).

A NEW MEASUREMENT OF THE PRIMORDIAL D/H RATIO IN THE INTERGALACTIC MEDIUM: HS 0105+1619

J.M. O'MEARA, D. TYTLER, D. KIRKMAN, D. LUBIN, N. SUZUKI, A.M. WOLFE
University of California, San Diego,
MS0424, La Jolla, CA 92093-0424, USA
E-mail: jomeara@ucsd.edu

J. X. PROCHASKA
The Observatories of the Carnegie Institute of Washington,
313 Santa Barbara St., Pasadena, CA 91101, USA

We report the measurement of the primordial D/H abundance ratio towards QSO HS 0105+1619. The column density of the neutral hydrogen in the $z \simeq 2.536$ Lyman limit system is high, $\log N_{HI} = 19.422 \pm 0.009$ cm^{-2}, allowing for the deuterium to be seen in 5 Lyman series transitions. The measured value of the D/H ratio towards QSO HS 0105+1619 is found to be D/H$= 2.54 \pm 0.23 \times 10^{-5}$. Combined with prior measurements of D/H, the best D/H ratio is now D/H$=$ $3.0 \pm 0.4 \times 10^{-5}$, which is 10% lower than the previous value. The new values for the baryon-to-photon ratio, and baryonic matter density derived from D/H are $\eta = 5.6 \pm 0.5 \times 10^{-10}$ and $\Omega_b h^2 = 0.0205 \pm 0.0018$ respectively.

1 Introduction

The standard theory of big bang nucleosynthesis (SBBN) predicts the abundances of the light nuclei H, D, ^{3}He, ^{4}He, and ^{7}Li as a function of the cosmological baryon to photon ratio, $\eta = n_b/n_\gamma$ [1,2,3]. A measurement of the ratio of any two primordial abundances gives η, and hence the baryon density, while a second ratio tests the theory. In this paper, we summarize a fourth QSO, HS 0105+1619, which gives strong constraints on the primordial D/H ratio. A more complete discussion of this D/H system can be found in O'Meara *et al.*[4]

2 HS 0105+1619

2.1 The Hydrogen

We observe the entire Lyman series in hydrogen and determine a value for the column density of $\log N_{HI} = 19.422 \pm 0.009$ cm^{-2}. The hydrogen is readily fit using a single component of absorption at a redshift of $z \simeq 2.536$. The column density of the H I is tightly constrained by the damping wings of the Lyα line, whereas the velocity width, $b = 13.99 \pm 0.20$ km s^{-1}, is determined using the higher order Lyman series transitions.

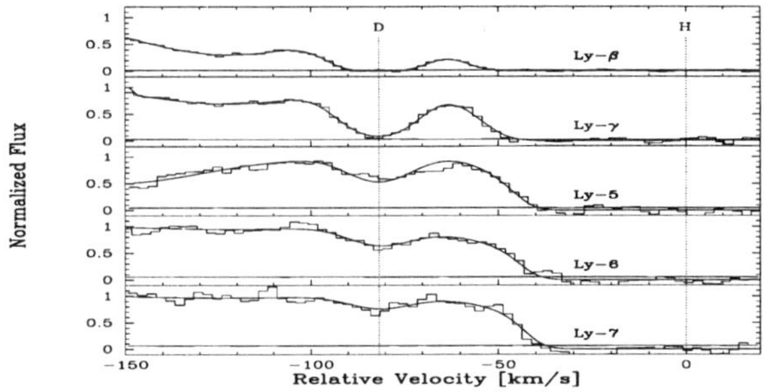

Figure 1: Deuterium absorption seen in the $z \simeq 2.536$ Lyman limit system towards HS 0105+1619.

2.2 *The Deuterium*

Since the column density of the hydrogen is so high, we expect the deuterium absorption to be strong and appear in many Lyman series transitions. This is indeed the case, as Figure 1 illustrates. We determine a best fit to the D I absorption of $\log N_{DI} = 14.826 \pm 0.039$ cm^{-2} and $b = 9.93 \pm 0.29$ km s^{-1}.

2.3 *Metal Lines*

A number of metal ions are observed at the same redshift of the D/H system. These transitions serve to confirm the single component model of the gas, and allow for a determination of the metallicity of the system, which is found to be approximately 1/100 solar.

3 New Values for D/H & Cosmological Parameters

For HS 0105+1619, we determine D/H $= 2.54 \pm 0.23 \times 10^{-5}$. Combined with previous measurements and limits to D/H[5,6,7] , which are summarized in Table 1, the new value for the primordial D/H ratio becomes D/H $= 3.0 \pm 0.4 \times 10^{-5}$, which is 10% lower than the previous value.

New values for cosmological parameters are as follows (using D/H and SBBN):

- $\eta = 5.6 \pm 0.5 \times 10^{-10}$
- $\Omega_b h^2 = 0.0205 \pm 0.0018$

Table 1 also illustrates why D/H has become more secure. We now have four places in the universe with varying astrophysical environments, such as

Table 1: PARAMETERS FOR THE D/H MEASUREMENTS

QSO	logD/H	z_{dh}	$\log N_{HI}$ (cm^{-2})	[Si/H]
PKS 1937–1009	-4.49 ± 0.04	3.572	17.86 ± 0.02	$-2.7, -1.9$
Q1009+2956	-4.40 ± 0.07	2.504	17.39 ± 0.06	$-2.4, -2.7$
Q0130–4021	< -4.17	2.799	16.66 ± 0.02	-2.6
HS 0105+1619	-4.60 ± 0.04	2.536	19.422 ± 0.009	-1.85

metallicity, ionization, and gas density, which are all consistent with a central value, and which show general agreement with the cosmological parameters as derived from other environments such as the CMB, and other light element abundances, many of which are discussed by other authors in these proceedings.

Acknowledgments

J.M.O. would like to thank the organizers for a wonderful conference. This work was funded in part by grant G-NASA/NAG5-3237 and by NSF grants AST-9420443 and AST-9900842.

References

1. Kolb, E.W., and Turner, M.S., "The Early Universe", (Addison Wesley 1990).
2. Nollett, K.M. & Burles, S. Phys.Rev.D in press astro-ph/0001440 (2000).
3. Olive, K.A., Steigman, G. & Walker, T.P. 2000, Phys. Rept., **333-334**, 389
4. O'Meara, J.M., Tytler, D., Kirkman, D., Suzuki, N., Prochaska, J.X., Lubin, D. & Wolfe, A.M. 2000, preprint, astro-ph/0011179
5. Burles, S. & Tytler, D. 1998a, ApJ, **499**, 699
6. Burles, S. & Tytler, D. 1998b, ApJ,**507**, 732
7. Kirkman, D., Tytler, D., Burles, S., Lubin, D., & O'Meara, J.M. ApJ **529**, 655–660, (2000) astro-ph/9907128

A D/H measurement in the damped Lyα system at $z_{\rm abs} = 3.025$ towards QSO 0347–3819

P. Molaro
Osservatorio Astronomico di Trieste, Via G.B. Tiepolo 11, I-34131, Trieste Italy

M. Dessauges-Zavadsky, S. D'Odorico
European Southern Observatory, Karl-Schwarzschild-Str. 2 D-85740 Garching Germany

We present the first D/H measurement in a damped Lyα (hereafter DLA) system at $z_{\rm abs} = 3.025$ towards QSO 0347–3819 obtained from the UVES-VLT spectra. The DLA absorber has a metallicity of [Zn/H] ≈ -1.25 and a relatively simple velocity structure, with two dominating components detected in several metal lines. The best fit of the Lyϵ, Ly8, Ly10 and Ly12 lines is obtained when the D I absorption is included in the two main components. The measured D column density yields $\frac{\rm D}{\rm H} = (2.24 \pm 0.67) \times 10^{-5}$ close to other low D/H values measured in Lyman limit systems.

1 Introduction

D is entirely produced by nuclear reactions in the Big Bang and is the most sensitive to the nuclear density at the nucleosynthesis epoch [2]. Direct D measurements of almost primordial material in high redshift objects are of high interest, since they do not depend critically on modeling of the galactic chemical evolution. However, the few systems studied so far provide highly discrepant D/H values differing by an order of magnitude.

The spectra of QSO 0347–3819 (V=17.3, $z_{\rm em} = 3.23$) were obtained during the Commissioning of the UVES on the VLT 8.2m Kueyen telescope at Paranal, in 1999. Two exposures of 4500 s each, covering the spectral range from 3650 to 4900 Å and from 6700 to 10000 Å were obtained with a resolution of 6–7 km s^{-1}.

The DLA system at $z_{\rm abs} = 3.025$ is a very good candidate for the D analysis, since it shows a relatively simple velocity structure dominated by two strong components, a low metallicity of [Zn/H] ≈ -1.25, indicating that the measured D/H will be representative of primordial D/H and the hydrogen Lyman series lines can be followed down to Ly12.

We modeled first the Lyman series with three H features corresponding to the main components of the metal lines and then we added D by assuming its contribution to the two dominant components only. The best fit performed over D I Ly8 and Ly10 reproduces nicely the blue wing of the Ly8, Ly10 and

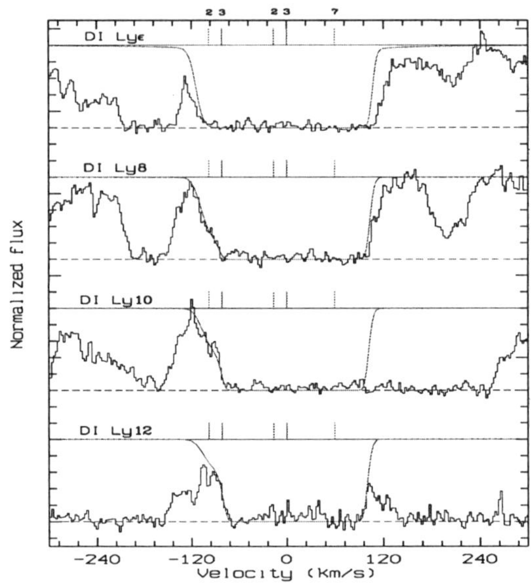

Figure 1: Lyman profiles plotted against velocity. The vertical scale goes from 0 to 1 for each transition. The fit (solid thin line) is performed by considering two components for D and three components for H.

Ly12 absorption profiles and is consistent with Lyϵ (see Fig. 1). It gives a D abundance of $\frac{\text{D}}{\text{H}} = (2.24 \pm 0.67) \times 10^{-5}$. The error follows from the fits and it includes the uncertainties in placing the continuum level and on the b-values.

The D and H column densities and b values resulting from the fits are consistent one relatively to other in the two fitted components. The derived D/H ratios are the same in the two components; and the $b(\text{D I})/b(\text{H I})$ ratios are equal to $\approx 1/\sqrt{2}$ showing that the D I absorption line has the expected width relatively to the H I width dominated by thermal motions. A detailed analysis of the data is presented by D'Odorico et al. [1].

This detection of D is the first made in a DLA system. It shows that the DLA systems are a very promising class of absorbers for systematic measurements of the D/H ratios at high redshifts.

References

1. D'Odorico S., Dessauges-Zavadsky M. & Molaro P., submitted *A&A* , (2000)
2. Schramm D.N. & Turner M.S., *Rev. Modern Phys.* **70**, 303 (1998)

DEUTERIUM NEAR AND FAR IN THE GALAXY

D. A. LUBOWICH[1]
Dept. of Physics and Astronomy, Hofstra U., Hempstead, NY 11550, USA
E-mail: dlubowic@aip.org

JAY M. PASACHOFF
Hopkins Observatory, Williams College, Williamstown, MA 01267, USA
E-mail: jmp@williams.edu

JASON A. OSTENSON
Williams College, Williamstown MA 01267 USA and
Keck Summer Fellow, Vassar College, Poughkeepsie NY 12604, USA
E-MAIL: *E-mail: Jaostenson@Vassar.Edu*

We report on our program to determine the deuterium abundance distribution in the Milky Way. We have reported D/H = 1.7 $\times 10^{-6}$ in a molecular cloud 10 pc from the Galactic Center, from which we infer recent continuous infall of pregalactic primordial gas. We have searched for the Balmer Dα lines with high S/N = 300 – 1000 observations of a halo star (HD 140283), slowly rotating B stars (ι Her and γ Peg), H II regions (Orion Nebula and M17), and the planetary nebula NGC 7027.

From observations of the J = 1-0 and 2-1 microwave spectral lines of DCN (Fig. 1a) we recently determined that D exists in the 50 km/s molecular cloud 10 pc from the Galactic Center (GC) with D/H = 1.7 $\times 10^{-6}$ (Lubowich, Pasachoff, Balonek, Millar, Tremonti, Roberts, and Galloway. Nature, 405, 102, 2000). We concluded that significant amounts of D are not produced via any Galactic process, the D observed in the GC is the result of infall from low-metallicity gas representing primordial or pre-Galactic matter, and the GC has not had a recent quasar or AGN phase within the past 5 Gyr. Our results are consistent with a cosmological D/H = 5 $\times 10^{-5}$, which implies that most of the baryons are in the form of dark matter and that most of this dark matter is non-baryonic. We will use this technique in 2001 to determine D/H in molecular clouds in the GC and edge of the Galaxy (2 pc and 28 kpc from the GC).

We are continuing our search for the Balmer Dα lines in stars and nebulae. We have not Dα in stars or nebulae. We have taken high S/N echelle spectra of the Orion Nebula at three positions (one position is shown in Figs. 1b and 1c), the slowly rotating sharp-lined B4 IV stars ι Her and γ Peg, the low-metallicity halo star HD 140283, and the planetary nebula NGC 7027. The spectra were taken with the NSO McMath 1.5-m and KPNO coudé feed 0.9-m telescopes. Our future work includes searching for D/H in stars and H II regions 14 - 18 kpc from the Galactic Center.

[1] DAL is supported by an NSF international travel grant administered by the AAS.

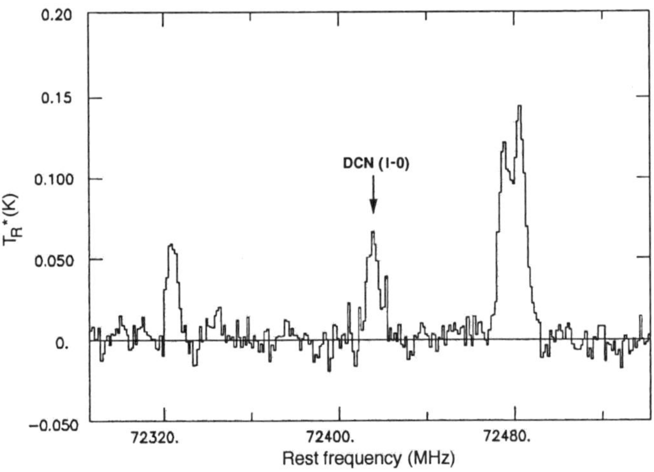

1a. DCN spectrum of the Galactic Center 50 km/s molecular cloud. (NRAO 12 m radio telescope)

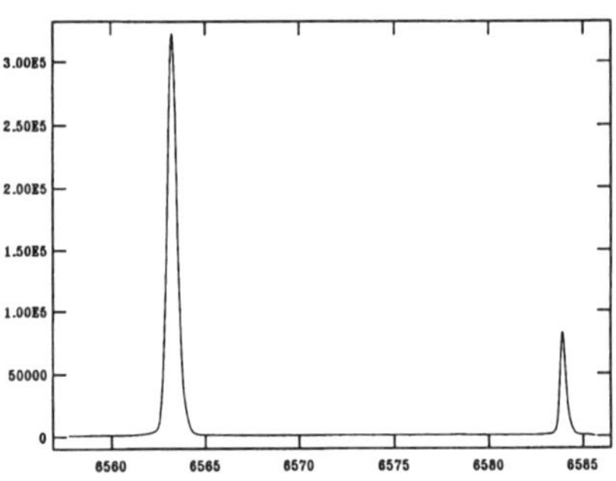

1b. Echelle spectrum of Orion Nebula 15 arcsec west of the brightest Trapezium star HD 37022. 1800 s exposure taken on 21 Dec. 1999 (NSO McMath-Pierce 1.5 m telescope, resolution = 84,000)

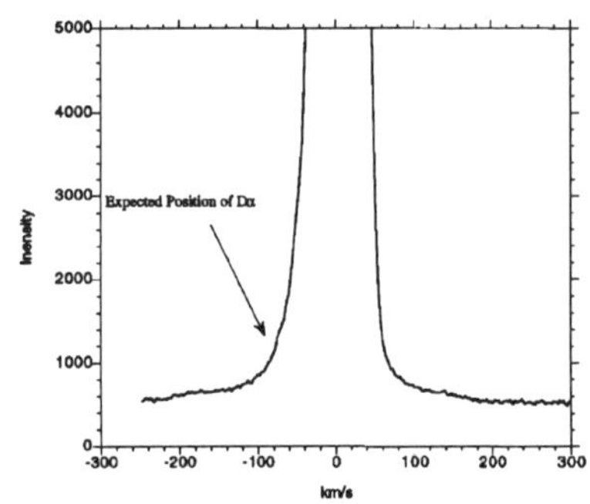

1c. Expanded echell spectrum of Orion Nebula 15 arcsec west of the brightest Trapezium star HD 37022. Combined nine 1800 s exposure taken on 21 Dec. 1999.

THE OBSERVATIONAL DETERMINATION OF THE PRIMORDIAL HELIUM ABUNDANCE: A Y2K STATUS REPORT

EVAN D. SKILLMAN

Astronomy Department, University of Minnesota, 116 Church St. SE, Minneapolis, MN 55455, USA

E-mail: skillman@astro.umn.edu

I review observational progress and assess the current state of the determination of the primordial helium abundance, Y_p. At present there are two determinations with non-overlapping errors. My impression is that the errors have been underestimated in both studies. I review recent work on errors assessment and give suggestions for decreasing systematic errors in future studies.

1 Background

In order to be a useful cosmological constraint, ^{4}He needs to be measured with a precision at the few percent level. Thus, the determination of the ^{4}He abundance with improved accuracy continues to be of prime importance to cosmology, and it is imperative to establish a comprehensive technique for determining ^{4}He abundances and reliable estimates of the true uncertainties. With the general availability of large format, linear CCD detectors, the accuracy of relative emission line ratios has improved to the point where several terms in the error budget which were assumed to be negligible may now be important. Indeed, the determination of the ^{4}He abundance has become limited by systematic uncertainties.

To date, the most useful ^{4}He abundance determinations are made by observing helium emission lines in HII regions of metal-poor dwarf galaxies. These measurements, with metallicities ranging down to 1/50th of the solar abundance, can be used to determine Y_p by means of a linear regression with respect to oxygen[11,12] or nitrogen[9]. The two most recent studies yield discrepant results: $Y_p = 0.2452 \pm 0.0015$ (Izotov et al.[2]) and $Y_p = 0.2345 \pm 0.0026$ (Peimbert et al.[10]).

2 Recent Progress

Izotov et al. (1994, 1997, 1998a,b, 1999): Izotov and collaborators have used large format CCD detectors to add new observations of 45 objects[5] (IT98), and these observations can be considered to replace most of the observations reported in the compilation of Pagel et al.[9] Izotov has emphasized the importance of underlying stellar absorption in the NW component of IZw18 [4], leading to an underestimate of the helium abundance in that object by Skillman & Kennicutt[13] and Pagel et al.[9] They have introduced and refined the "self-consistent" method of density determination, such that the density and optical depth estimates come directly from the helium line emission.

Peimbert, Peimbert, & Ruiz (2000): Peimbert and collaborators have devised their own self-consistent method and included both density and temperature as physical parameters to be derived from the helium emission line ratios. In all previous studies, the temperature had been derived from the [O III] lines, but photoionization modeling has shown that this will overestimate the appropriate temperature for the HeI lines. Additionally, they have shown a coupling between the density and temperature determinations resulting in higher densities (and thus lower helium abundances).

3 Current Concerns

With Izotov et al. (1994, 1997, 1998a,b, 1999): In some regards, the self-consistent method of Izotov et al. is not truly self-consistent in that exceptions are frequently made in the treatment of the data. For example, in many cases, the helium abundances derived from the density sensitive λ7065 line and the optical depth sensitive λ3889 line are not in agreement with the mean of the three main lines $\lambda\lambda$ 4471, 5876, 6678.

Perhaps more importantly, underlying absorption is not allowed for in their χ^2 minimizations. When underlying absorption is suspected to affect the λ4471 line, or NaI absorption is suspected to affect the λ5876 line, these lines are rejected from the He abundance calculation, after the fact. As can be seen in Figure 1, underlying absorption is important to some degree in almost all of the targets in the compilation of IT98 (see also[14]). Figure 1 shows a plot of the ratios of the He abundances derived from the three strongest lines. Clearly the He abundances derived from λ4471 are systematically lower than those derived from λ5876 which points directly to strong underlying absorption. The λ6678, λ5876 comparison looks much better in this regard.

Another concern with the self-consistent method employed by Izotov et al. is that it often returns unrealistic parameters (e.g., vanishingly small densities). In such cases, IT98 arbitrarily set the densities to a value of 10, but the exceptionally low densities probably indicate either a problem with the data or the method (or both) and should not be ignored. This also precludes a proper treatment of the errors in the density determination.

Although a secondary effect, the treatment of radiative transfer of the HeI

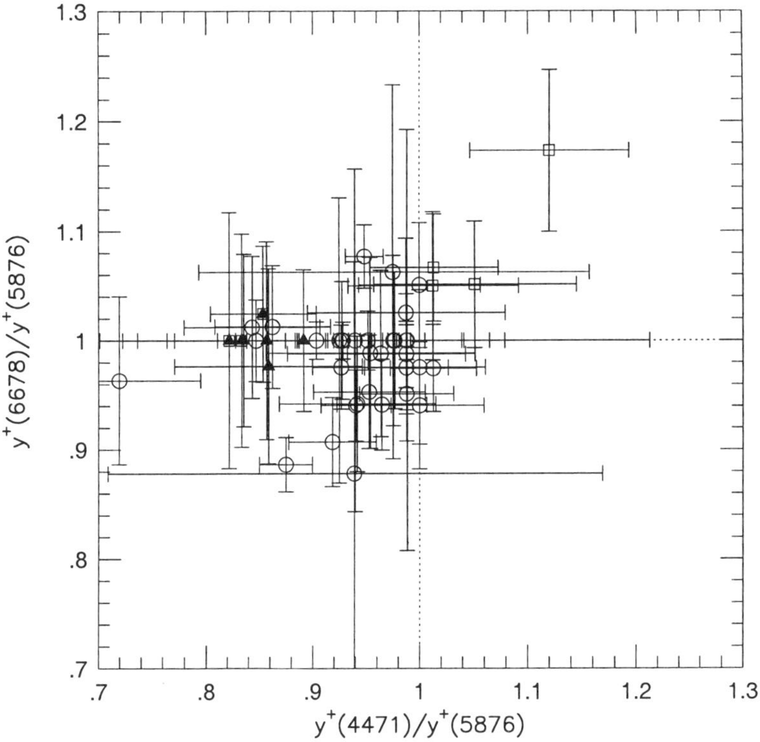

Figure 1. A comparison of the He abundances derived from the $\lambda4471$, $\lambda5876$, and $\lambda6678$ lines for the observations by IT98. The points are coded as: open squares for those objects where the $\lambda5876$ line has been weakened by NaI absorption, filled triangles where the $\lambda4471$ line has not been used by IT98 for He/H determination because of suspected underlying absorption, and open circles for the rest of the data. Note the systematic difference between the $\lambda4471$ and $\lambda5876$ measurements and the better agreement between the $\lambda6678$ and $\lambda5876$ results. The strong clustering of the points in the lower left quadrant is best understood if most of the observed targets have been affected by underlying stellar absorption.

lines is dealt with in a very simple manner by IT98 and relies on old models. New work in this regard is needed. Perhaps more importantly, de-blending of $\lambda3889$ HeI and HI lines and dealing with underlying absorption is problematic, and may lead to incorrect values of the optical depths in the helium lines.

With Peimbert, Peimbert, & Ruiz (2000): The inclusion of temperature fluctuations coupled to self-consistent solutions for density gives a strong

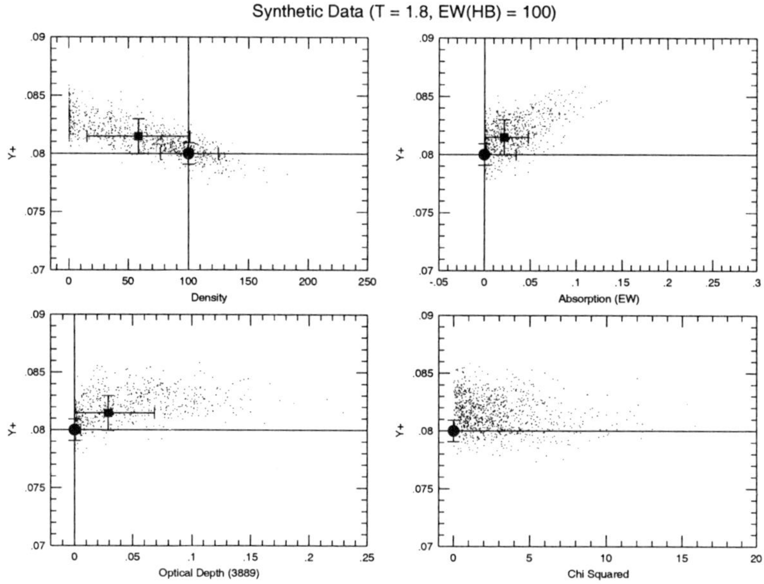

Figure 2. Results of modeling synthetic He I line observations. Starting with 6 He I line intensities, equivalent widths, and associated errors, we perform a Monte-Carlo simulation of the data by generating Gaussian distributions of the 6 He I lines, consistent with their errors. The four panels show the results of a density = 100, abs(HeI) = 0, and $\tau(3889) = 0$ model. The solid lines show the input values (e.g., He/H = 0.080) for the original calculated spectrum. The solid circles (with error bars) show the results of the χ^2 minimization solution (with calculated errors) for the original synthetic input spectrum. The small points show the results of Monte Carlo realizations of the original input spectrum. The solid squares (with error bars) show the means and dispersions of the output values for the χ^2 minimization solutions of the Monte Carlo realizations. Note the larger error bars for the Monte Carlo results and the upward bias in the helium abundance due to the coupling between the density and absorption estimates. (From Olive & Skillman 2001)

downward bias to the helium abundances. This may yield the correct answer, but measuring the proper value of the temperature fluctuations is not trivial. Peimbert et al. use a preferred value for the temperature fluctuations in their calculations, but if the estimated error on the temperature fluctuations is propagated through the abundance calculations, then the error in the final helium abundance would be larger than reported.

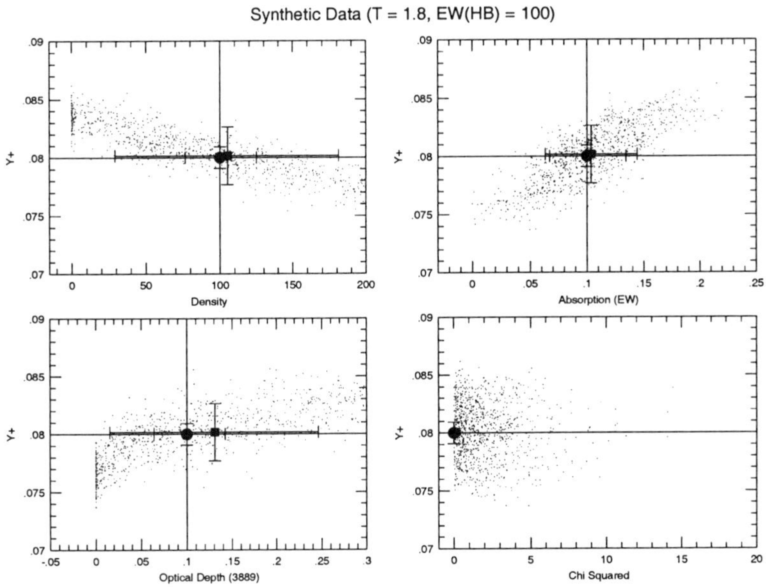

Figure 3. Similar plot to Figure 2 except that the underlying absorption is 0.1 Å and $\tau(3889) = 0.1$. This figure shows that if underlying absorption is present, self-consistent methods can identify and quantify it. Ignoring the presence of underlying absorption applies a bias to the resultant derived helium abundances. (From Olive & Skillman 2001)

4 Forward Look

As an observer, I emphasize that there is still a need for new low-metallicity targets. An abundance of these objects would allow us to concentrate on the ideal targets which have low metallicity, high equivalent width, and high excitation.

Keith Olive and I have been working on a robust method for dealing with underlying absorption self-consistently[8]. Results of this work are shown in Figures 2 and 3. Here it is shown that there are strong degeneracies in the sensitivity of the ^{4}He abundance to various physical parameters. Because of this, Monte Carlo simulations of the data are required to derive accurate estimates of both the ^{4}He abundance and the appropriate errors. Using Monte Carlo modeling of the errors in the derived helium abundances results in expansion of the errors from the χ^2 minimization by roughly a factor of two.

Further experimentation with self-consistent temperature solutions is required. Peimbert et al.[10] derive the value of the temperature fluctuations from temperatures derived from [O II] lines, [O III] lines, the ratio of the Balmer continuum flux to Balmer line emission, and the oxygen ionization fraction. It may be possible to derive the value of the temperature fluctuations directly from the helium lines themselves.

Benjamin et al.[1] presented an updated calculation of helium emissivities, and are currently working on updating the calculations of radiative transfer effects in the HeI lines.

Acknowledgments

I would like to thank the conference organizers for the invitation to very inspiring meeting. I am grateful to many colleagues and collaborators for informative and valuable discussions. Partial support from NASA LTSARP grant No. NAG5-9221 and the University of Minnesota is gratefully acknowledged.

References

1. Benjamin, R. A., Skillman, E. D., & Smits, D. P., ApJ **514**, 307 (1999)
2. Izotov, Y. I., Chaffee, F. H., Foltz, C. B., Green, R. F., Guseva, N., & Thuan, T. X., ApJ **527**, 757 (1999) (I99).
3. Izotov, Y. I., Lipovetsky, V. A., Chaffee, F. H., Foltz, C. B., Guseva, N., & Kniazev, A. Y., ApJ **476**, 698 (1997) (I97).
4. Izotov, Y. I., & Thuan, T. X., ApJ **497**, 227 (1998a).
5. Izotov, Y. I., & Thuan, T. X., ApJ **500**, 188 (1998b) (IT98).
6. Izotov, Y. I., Thuan, T. X., & Lipovetsky, V. A., ApJ **435**, 647 (1994).
7. Izotov, Y. I., Thuan, T. X., & Lipovetsky, V. A., ApJS **108**, 1 (1997).
8. Olive, K, & Skillman, E. D., *New Astr.*, sub., astro-ph/0007081.
9. Pagel, B. E. J., Simonson, E. A., Terlevich, R. J., & Edmunds, M. G., MNRAS **255**, 325 (1992) (PSTE).
10. Peimbert, M., Peimbert, A., & Ruiz, M. T., ApJ **541**, 688 (2000)
11. Peimbert, M., & Torres-Peimbert, S., ApJ **193**, 327 (1974)
12. Peimbert, M., & Torres-Peimbert, S., ApJ **203**, 581 (1976)
13. Skillman, E.D., & Kennicutt, R. C., Jr., ApJ **411**, 655 (1993).
14. Skillman, E.D., Terlevich, E., & Terlevich, R. Sp.Sci.Rev. **84**, 105 (1998).

THE ORIGIN AND EVOLUTION OF 3-HELIUM

ROBERT T. ROOD
Department of Astronomy, University of Virginia, P.O. Box 3818, University Station Charlottesville VA 22903, USA, E-mail: rtr@virginia.edu

THOMAS M. BANIA
Department of Astronomy, Boston University, 725 Commonweath Ave., Boston MA 02215, USA, E-mail: bania@ninkasi.bu.edu

DANA S. BALSER
National Radio Astronomy Observatory, P.O. Box 2, Green Bank WV 24944, USA Email: dbalser@nrao.edu

We have now detected the hyperfine line of $^3He^+$ in more than 30 H II regions. In a subset of "simple" H II regions which should have the most reliable abundances we find no trend in $^3He/H$ with position in the Galaxy or H II region metallicity. This suggests that stars neither create or destroy ^3He in large amounts. Given this we suggest that the mean abundance of our simple H II region sample, $^3He/H = 1.79 \pm 0.65 \times 10^{-5}$, is a reasonable approximation for the primordial value. Despite the fact that the H II regions suggest no stellar production, we do find high ^3He in some planetary nebulae. The scenario of Charbonnel in which most (but not all) low mass stars undergo rotationally driven mixing on the red giant branch above the luminousity function bump seems consistent with our observations and chemical evolution models.

1 Prolog

The realization (prompted by Gott et al. 1974) that one could do cosmology by determining the D/H ratio in a glass of *Chateau Lafite Rothschild* (if one were willing to give up a sample) was extraordinarily exciting to us. Could we become rich and famous cosmologists by extending the technique to, say, the $^3He/^4He$ in a helium filled balloon? No such luck—^3He is full of surprises. In the balloon we would have found $^3He/^4He \sim 10^{-6}$ or in continental rocks $\sim 10^{-8}$ (e.g., Anderson 1993). On the other hand the "cosmic abundance" (determined by the standard technique, i.e., looking up the last review article, Reeves 1974) was orders of magnitude higher. There was an exciting discovery at hand, unfortunately not cosmological and made decades earlier by others: the Earth formed with essentially no helium (or other chemically inert elements). Most terrestrial helium is the product of α decay of uranium, thorium, etc. Since there is no radioactive decay producing ^3He there should be no terrestrial ^3He. For that reason finding excess ^3He on the Earth is normally taken as an indication that something strange is going on: nuclear weapons, cold fusion (you can't win 'em all), or an infusion of extraterrestrial material. In the latter category, there is the exciting new result based in part on the $^3He/^4He$ ratio in gas trapped in fullerenes (aka buckyballs), that the massive Permian-Triassic extinction was caused by the impact of a carbonaceous asteroid (Becker et al. 2001).

While $^3He/^4He$ in a balloon was of no help to aspiring cosmologists, other solar system objects had yielded a cosmic abundance for Reeves' review. However, because ^3He is produced in stars as well as the Big Bang, one needs at least two abundances to extrapolate back to the primordial value. Thus we began our quest to determine the ^3He abundance beyond the Solar System, which is feasible only via observations of the 8.7 GHz (3.5 cm) hyperfine line of $^3He^+$, the analog of the 21 cm hyperfine line of ^1H. Galactic H II regions and planetary nebulae (PNe) made up our target list. Since the stellar contribution was potentially large we initially expected to find: 1) the ^3He in the current interstellar medium (ISM) would be (much?) larger than the protosolar system value, and 2) the $^3He/H$ abundance would be larger in places with enhanced stellar processing, i.e., in the inner Galaxy. We would not be at this meeting talking about ^3He if the surprises had not continued.

2 Status of the Observational Program

Here we address one recent criticism of our work, and describe the situation as we await full operation of the new Green Bank radio telescope (GBT).

First, *Do We Detect 3-Helium in the Outer Galaxy?* The quick answer is, most definitely yes. We raise this point only because Bell (2000) has recently questioned some of our results. Among other things he argues that we have not detected $^3He^+$ in any H II region outside the Solar circle and that our detections are created by bad baseline fits or interference. S209 was one of the sources he specifically questioned, yet from Fig. 1 it is clear that the line is not created by our baseline fit. Further, we now have significant integration time toward 28 outer Galaxy H II regions. These have yielded 15 solid detections, 3 probable detections, 2 possible detections, and 8 non-detections. *Not once* is there an absorption feature at either the H171η or $^3He^+$ frequency—if our 1–2 mK lines were being created by bad baselines there would be equal numbers of "absorption" and "emission" features. Interference is rare and seldom extends over more than one observing epoch (all sources have been observed for at least 3 epochs).

Updated results for simple H II Regions. One of the complicating factors in deriving the $^3He/H$ abundance ratio is that the intensity of the $^3He^+$ hyperfine

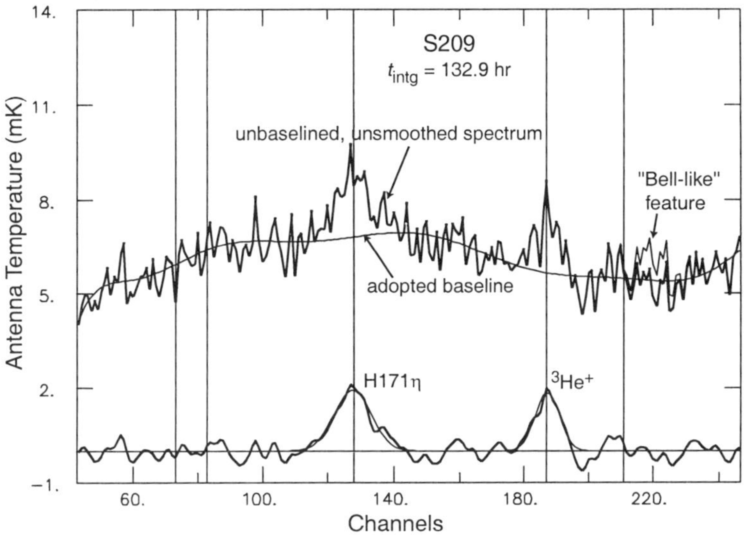

Figure 1. The upper spectrum is the raw, unbaselined, unsmoothed spectrum for the H II region S209. The adopted polynomial baseline is shown as the lighter line superimposed on the spectrum. Note that the baseline is entirely plausible and does not snake wildly around. The H171η recombination line and $^3He^+$ hyperfine line are readily visible in the raw spectrum. The lower spectrum shows result of our normal procedure, i.e., removing the baseline and smoothing with a gaussian weighting function to a velocity resolution of 8.1 km sec^{-1}. The lines are completely consistent with the raw spectrum and are obviously not a relic of our baseline removal. The data averaged in this figure was accumulated over a period of 15 years and the integration time is $t_{intg} = 132.9$ receiver-hours. Any adequately large subset of the data, e.g., the old maser-upconverter receiver only, new GBT HEMT receivers only, leads to the same conclusion. Interference at the level "observed" by Bell is indicated in the unsmoothed spectrum—such interference would be easily noticed in our raw spectrum.

line depends on the electron density, n_e, while hydrogen abundance is derived from the radio continuum which depends on n_e^2. Hence one must model the density structure of the H II region to derive an abundance. For many years our "modeling" consisted of assuming that each H II region was a homogeneous sphere. However, we ultimately developed fairly sophisticated models using higher resolution continuum mapping and radio recombination line strengths. This modeling yielded two interesting results: 1) whenever the correction factor (model abundance/homogeneous sphere abundance) was large (> few 10%) it was also uncertain; 2) there was a subset of H II regions which are well described by homogeneous spheres—we call these "simple." Low density H II regions tend to be simple.

Since their ^3He abundances should be more accurate, in recent years we

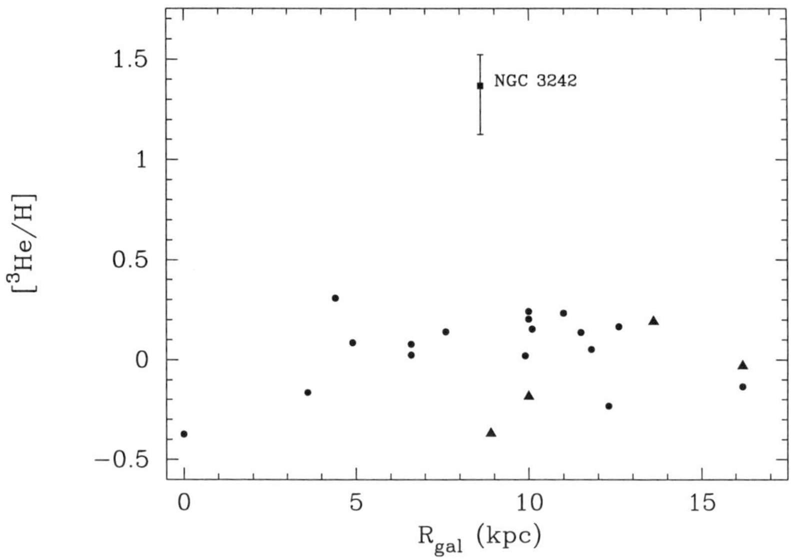

Figure 2. The 3-helium abundance for simple H II regions as a function of $R_{\rm gal}$ (kpc). Following conventional notation $[^3\mathrm{He/H}] = \log(x/\mathrm{H}) - \log(x/\mathrm{H})_\odot$ with $^3\mathrm{He/H})_\odot = 1.5\times10^{-5}$. Simple ionization corrections have been H II regions plotted as filled cirles; no ionization corrections have been made for sources plotted as triangles. The abundance for the planetary nebula NGC 3242 (BRR99) is shown to emphasize the magnitude of the "3-Helium Problem."

have concentrated our efforts toward increasing the sample of simple H II regions. Our current results are shown in Fig. 2. Contrary to naive expectations there is no trend in $[^3\mathrm{He/H}]$ with either position in the Galaxy (or nebular metallicity). Since the scatter in $^3\mathrm{He/H}$ abundance is comparable to that in the Li abundance within the Spite plateau, we refer to our result as the $^3\mathrm{He}$ plateau. Excluding the Galactic center source G1.1−0.1, the mean abundance $^3\mathrm{He/H} = 1.79 \pm 0.65 \times 10^{-5}$. This value is somewhat larger than the value reported in BRB00 because we include a simple helium ionization correction here not included in that paper.

3 Summary

We now have $^3\mathrm{He/H}$ abundance ratios for 21 "simple" H II regions throughout the Galaxy. The scatter in abundances is about a factor of two in each direction from the mean, but there is no trend in abundance with $R_{\rm gal}$ or [O/H]. The net stellar contribution to the $^3\mathrm{He}$ abundance appears to be close to zero.

On the other hand we have detected $^3\mathrm{He}$ in the PN NGC 3242 with two completely different radio telescope systems (RBW92, BBRW97, BRB99). The derived abundance is consistent with the large value predicted by standard stellar models. There is reasonably strong evidence for high ^{3}He/H in other PNe. Our PNe sample is biased toward PNe with low progenitor mass and no evidence for extra-mixing during the red giant (RGB) or asymptotic giant branch (AGB) phases. As suggested by Galli et al. (1997) our PNe observations are consistent with our H II region abundances only if our PNe are typical of a small fraction of the total PNe population.

Extra-mixing processes and hot and cool bottom envelope burning operating along the RGB and AGB can substantially reduce the ^{3}He which built up during main sequence hydrogen burning (Charbonnel et al. 1998, 2000; Sackmann & Boothroyd 1999, 2000). In almost all cases the ^{3}He is brought down to abundances comparable to that initially present in the stars, i.e., low mass stars are non-producers (as opposed to destroyers) of ^{3}He. Charbonnel links ^{3}He lowering to well studied anomalies in the ^{12}C/^{13}C isotope ratios in RG stars. She suggests that rotationally driven mixing operates in RGB stars more luminous than the luminosity function bump. The rather gentle burning lowers ^{3}He and ^{12}C/^{13}C. The mixing and abundance changes would not operate in all stars thus allowing the high ^{3}He abundance found in some PNe. The observed fraction ($\sim$ 10%) of upper RGB stars with high ^{12}C/^{13}C gives the fraction of stars enriching the ISM with ^{3}He. Chemical evolution models incorporating this factor into ^{3}He yields (Tosi 2000, Chiappini & Matteucci 2000) give ^{3}He abundances roughly consistent with our simple H II region abundances, the Solar System, and LISM abundances.

Since stars seem to be neither producing nor destroying ^{3}He in large amounts we can take the 3-Helium plateau value as an approximation to the primordial ^{3}He. Excluding the Galactic center source G1.1−0.1, we find $^3\mathrm{He/H} = 1.79 \pm 0.65 \times 10^{-5}$. The fact that we find ^{3}He in some PNe while at the same time there is no evidence for substantial input of ^{3}He into the ISM, fits in with the Charbonnel scenario. We suspect that net stellar processing of ^{3}He is positive, although less than we can now detect. If this the case the true primordial value will be less than the plateau value.

Acknowledgments

Jean Audouze's and Jim Truran's interest in 3-helium predates our own (e.g., Audouze & Tinsley 1973; Truran & Cameron 1971); we appreciate their encouragement over more than two decades. We thank Tom Wilson for his contributions to the ^{3}He project over the years. The research was supported by the U.S. National Science Foundation (AST 97-31484).

References

1. Anderson, D. L. 1993, Science, 261, 170
2. Audouze, J. and Tinsley, B. M. 1976, ARA&A, 14, 43
3. Balser, D. S., Bania, T. M., Brockway, C. R., Rood, R. T., & Wilson, T. L. 1994, ApJ, 430, 667 (BBBRW94)
4. Balser, D. S., Bania, T. M., Rood, R. T., & Wilson, T. L. 1997, ApJ, 483, 320 (BBRW97)
5. Balser, D. S., Bania, T. M., Rood, R. T., & Wilson, T. L. 1995, ApJS, 100, 371 (BBRW95)
6. Balser, D. S., Bania, T. M., Rood, R. T., & Wilson, T. L. 1998, ApJ, 510, 759 (BBRW98)
7. Balser, D. S., Bania, T. M., Rood, R. T., & Wilson, T. L. 1999, ApJ, 510, 759 (BBRW99)
8. Balser, D. S., Rood, R. T., & Bania, T. M. 1999, ApJ, 522, L73 (BRB99)
9. Bania, T. M., Balser, D. S., Rood, R. T., Wilson, T. L., & Wilson, T. A. 1997, ApJS, 113, 353 (BBRWW97)
10. Bania, T. M., Rood, R. T., & Balser, D. S., 2000, in The Light Elements and Their Evolution, Proc. I.A.U. Symposium 198, eds. L. da Silva, M. Spite, J. de Medeiros (Astron. Soc. Pacific: San Francisco), 214 (BRB00)
11. Becker, L., Poreda, R. J., Hunt, A. G., Bunch, T. E., & Rampino, M. 2001, Science, 291, 1530
12. Bell, M. B. 2000, ApJ, 531, 820
13. Chiappini, C. & Matteucci, F. 2000, in IAU 198, 540
14. Charbonnel, C., & do Nascimento Jr, J. D. 1998, A&A, 336, 915
15. Charbonnel, C., Delyannis, C. P., & Pinsonneault, M. 2000, in IAU 198, 87 (CDP00)
16. Galli, D., Stanghellini, L., Tosi, M., Palla, F. 1997, ApJ, 477, 218
17. Gott, J. R., III, Schramm, D. N., Tinsley, B. M., & Gunn, J. E. 1974, ApJ, 194, 543
18. Reeves, H. 1974, ARA&A, 12, 437
19. Rood, R. T., Bania, T. M., & Wilson, T. L. 1992, Nature, 355, 618 (RBW92)
20. Rood, R. T., Steigman, G., & Tinsley, B. M. 1976, ApJ, 207, L57 (RST76)
21. Sackmann, I.-J. & Boothroyd, A. I. 1999, ApJ, 510, 217
22. Sackmann, I.-J. & Boothroyd, A. I. 2000, in IAU 198, 98
23. Shaver, P. A., McGee, R. X., Newton, L. M., Danks, A. C., & Pottasch, S. R. 1983, MNRAS, 204, 53
24. Tosi, M. 2000, 2000, in IAU 198, 525
25. Truran, J.W., & Cameron, A.G.W. 1971, Ap&SS,14, 179

GALACTIC EVOLUTION OF D AND ^{3}HE

M. TOSI

Osservatorio Astronomico di Bologna, via Ranzani 1, I-41027 Bologna, Italy
E-mail: tosi@bo.astro.it

Recent results on the observed abundances of D and ^{3}He are discussed in the framework of their Galactic chemical evolution.

1 D and ^{3}He one year later

A wealth of accurate observational data on D and ^{3}He were presented one year ago at the IAU Symposium 198[1], where theoretical models on their stellar production/depletion were also discussed. The combination of the observational data and theoretical results available at that time allowed to put stringent constraints on Galactic evolution models of the two elements and suggested that (Tosi[2] and references therein): 1) the current models able to reproduce the largest set of observed Galactic properties all predict a D destruction lower than a factor of 3 during the whole Galaxy lifetime and, hence, a low primordial D, of about $(D/H)_p \simeq 3 \times 10^{-5}$; 2) all the observed ^{3}He abundances can be self-consistently explained[3] if $(D/H)_p \leq 3 \times 10^{-5}$, $(^3He/H)_p \leq 1 \times 10^{-5}$, and an extra mixing[4 5] occurs during the red giant phase of ~90% of low-mass stars, preventing them to significantly enrich the interstellar medium with ^{3}He.

At, and after, that Symposium, new data have been presented, and I shall comment them here in the framework of Galactic chemical evolution.

The left panel of Fig.1 shows the time behaviour from the Big Bang to the present epoch attributed to D from observational data and Galactic chemical evolution models. The latter were independently computed by different groups and share only the characteristics of being able to reproduce the largest set of observed features of our Galaxy (see [2] for references and details). The novelties reside in the D/H abundances inferred by O'Meara et al.[6] from a fourth QSO absorber and by Sonneborn et al.[7] in four directions of the local interstellar medium. The former brings the average D/H at high redshift down to 3×10^{-5}, and the latter confirms that there are variations by a factor of 3 in the local abundances of D.

Another interesting result is that no anticorrelation was found[7] between D and heavy element abundances. This is in principle contrary to standard expectations, but may simply result from small number statistics, combined with the spread affecting the abundances of all elements: if we just consider

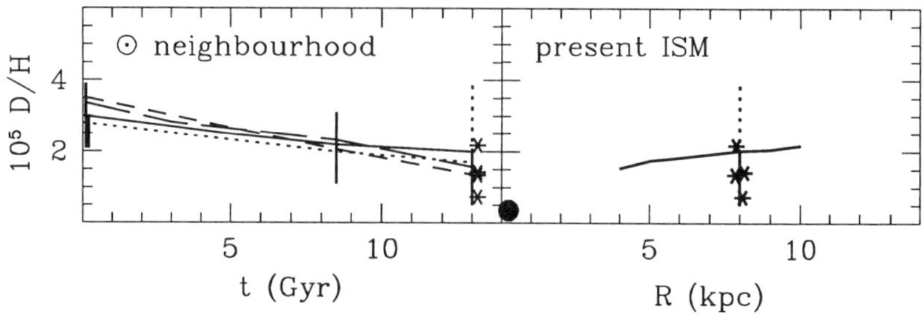

Figure 1. Observed D abundances and predictions from Galactic chemical evolution models. Left: evolution in the solar neighbourhood. The solid and dotted vertical bars refer to the observed abundances, and the curves to models by different groups (all quoted in Tosi 2000). The thick vertical bar corresponds to O'Meara et al. (2000) QSO absorber and the asterisks to Sonneborn et al. (2000) determinations in the local interstellar medium. Right: radial distribution at the present epoch. The symbols at 8 kpc are the same local values as in the left panel, the dot at the Galactic center refers to Lubovich et al. (2000) value and the solid line to Tosi's evolutionary models.

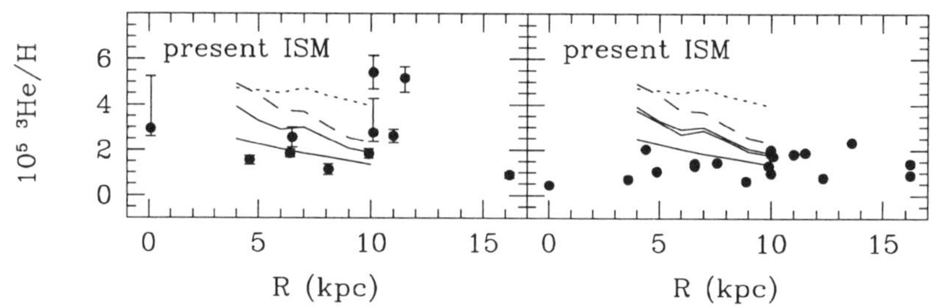

Figure 2. Observed ^{3}He abundances at the present epoch and predictions from Galactic chemical evolution models. Left: situation before the IAU Symposium 198 (see Tosi 2000 for references). Right: current situation, with the same model predictions, but the updated HII region abundances provided by Bania, Rood & Balser (2000). Only a model with extra mixing in all low-mass stars (lowest curve) is compatible with the new distribution, but this model makes it impossible to explain the high abundances measured in PNe.

the oxygen abundances of the four closest HII regions, we also find a variation of a factor of 2.5 and no indication of a radial abundance gradient, which is instead known to exist.

These data show that standard chemical evolution models reproduce large-scale, long-term phenomena, but don't account for small-scale fluctuations, unless introducing additional parameters. This does not imply that they don't follow correctly the D evolution; simply they are limited to the overall trends. A better check on the model D predictions will be possible when abundances will be available also in other Galactic locations and will show whether or not a radial D gradient exists. By now (see right panel of Fig.1 we only have the point[8] right at the Galactic center suggesting that the predicted gradient may indeed be correct; but the Galactic center is not representative of the disk evolution.

The left panel of Fig.2 shows how our models[3] [2] were able to fit the radial ^{3}He distribution published by Rood et al. [10]. The right panel shows instead how the same model predictions overestimate ^{3}He if the abundances are taken from [9] who selected from their data only the safer measures: if the new low abundances are not the result of a bias introduced by selection effects, there is no way to self-consistently reproduce the HII region, PNe and local abundances with the current knowledge on stellar nucleosynthesis.

Acknowledgements

After Beatrice Tinsley, Jean Audouze and Jim Truran were the first scientists who showed me how to deal with galactic chemical evolution: I thank the organizers of this wonderful meeting for the opportunity to publicly express them my sincere gratitude. This work has been funded by the Italian MURST, through Cofin98.

References

1. L. da Silva, M. Spite, & J.R. de Medeiros, *IAU Symp.198* (2000)
2. M. Tosi, *IAU Symp.198* p.525 (2000)
3. F. Palla, R. Bachiller, L. Stanghellini, M. Tosi, D. Galli, *A&A*, **355**, 69 (2000)
4. C. Charbonnel, *ApJ*, **453**, L41 (1995)
5. G.J. Wasserburg, A.I. Boothroyd, & I.-J. Sackman, *ApJ*, **427**, 618 (1995)
6. J.M. O'Meara, D. Tytler, D. Kirkman, N. Suzuki, J.X. Prochaska, D. Lubin, A.M. Wolfe, astro-ph/0011179 (2000)
7. G. Sonneborn, T.M. Tripp, R. Ferlet, E.B. Jenkins, U.J. Sofia, A. Vidal-Madjar, P.R. Wozniak, astro-ph/0007238 (2000)
8. D.A. Lubovich, J.M. Pasachoff, R.P. Galloway, T.A. Balonek, C. Tremonti, T. Millar, H. Roberts *IAU Symp. 198* p.167 (2000)
9. T.M. Bania, R.T. Rood, D.S. Balser, *IAU Symp. 198* p.214 (2000)
10. R.T. Rood, T.M. Bania, T.L. Wilson, D.S. Balser, in *The light element abundances*, P.Crane ed. (Springer, De), p.201 (1995)

THE ENIGMA OF ^{3}HE

D. GALLI, F. PALLA
Osservatorio Astrofisico di Arcetri,
Largo E. Fermi 5, 50125 Firenze, ITALY

P. G. PRADA MORONI, S. ZAVATARELLI
Università di Genova, Dipartimento di Fisica, and INFN
via Dodecaneso 33, 16146 Genova, ITALY

F. FERRINI
Università di Pisa, Dipartimento di Fisica,
Piazza Torricelli 2, 56100 Pisa, ITALY

We briefly discuss two alternative solutions proposed so far to alleviate the so called "^{3}He problem" and we examine the astrophysical implications of recent laboratory data on the ^{3}He(^{3}He,2p)^{4}He cross section.

The usefulness of ^{3}He as a cosmic baryometer is limited at present by a poor understanding of its evolution in our Galaxy. Rood, Steigman & Tinsley [1] were the first to notice that the ^{3}He yields of low-mass stars [2] ($X[^3\mathrm{He}] \sim 10^{-3}$) are much higher than the measured abundances in HII regions [3] ($X[^3\mathrm{He}] \sim 10^{-5}$). This discrepancy still awaits a satisfactory explanation [4] (see also Tosi 2001, this volume).

Two solutions have been proposed to solve, at least partially, the "^{3}He problem". According to the *extra-mixing hypothesis* [5], during the post-main sequence evolution of low-mass stars ($M < 2M_\odot$), fluid elements circulate from the bottom of the convective zone down to some inner layer hot enough to destroy ^{3}He, and vice-versa. The *nuclear physics solution* [6] invokes the existence of a low-energy resonance in the cross section of the ^{3}He(^{3}He,2p)^{4}He reaction, responsible for the destruction of $\sim$ 80% of the solar ^{3}He, and was proposed originally by Fowler [7] and Fetysov & Kopysov[8] to solve the solar neutrino problem.

A possible way to discriminate among these two possibilities is the direct measurement of the ^{3}He abundance in the ejecta of low-mass stars, i.e. in Planetary Nebulae (PNe). The search for ^{3}He in the ejecta of PNe via the 8.667 GHz spin-flip transition of ^{3}He$^+$, carried out by Rood and coworkers [9] since 1992, has produced so far *one solid detection* (NGC 3242), *two tentative detections* (IC 289, NGC 6720) and *three upper limits* (NGC 7662, NGC 6543, NGC 7009). These objects can be placed in the progenitor mass–

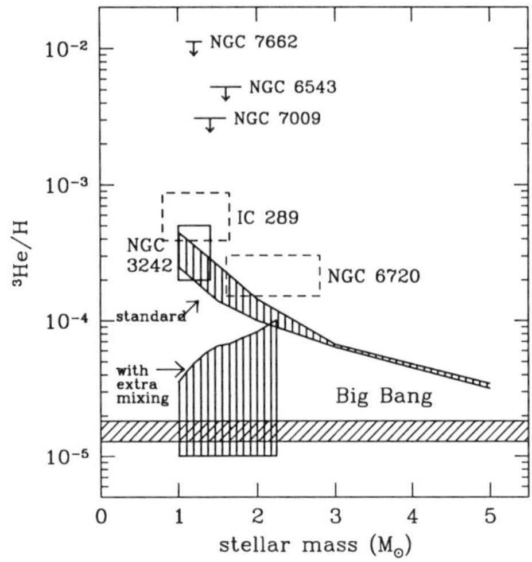

Figure 1. Abundance of ^{3}He in a sample of PNe vs. the main-sequence progenitor mass compared with the theoretical ^{3}He yields with and without extra-mixing.

^{3}He diagram [10] (Fig. 1), and compared with the predictions of stellar models with extra-mixing [11]. Consistency with the observed abundance of ^{3}He in the Galaxy (meteorites, local ISM, HII regions) is achieved only if the fraction of low-mass stars ($M < 2M_\odot$) undergoing extra-mixing is larger than $\sim 90\%$ [10]. One must then accept the unlikely conclusion that 1 to 3 PNe out of 6 objects of the PNe sample belong to the $\sim 10\%$ of low-mass stars which did *not* experience extra-mixing.

The *nuclear physics solution*, on the other hand, can now be re-examined in the light of the most recent laboratory data. The ^{3}He(^{3}He,2p)^{4}He cross section has been measured down to $E = 16.5$ keV by the LUNA experiment [12]. While no indication of a resonance has been found in this energy range, current data cannot rule out the existence of a narrow resonance at $E \simeq 12$ keV [12]. At present, the *extra-mixing hypothesis* seems to provide a better explanation of the ^{3}He problem than the *nuclear physics solution* (Fig. 2), but the former is far from offering a concordant interpretation of all the observational data and theoretical results. Additional determinations of ^{3}He abundances in PNe, and, even more important, direct measurements of the ^{3}He(^{3}He,2p)^{4}He cross section around $E \simeq 12$ keV would be of fundamental importance.

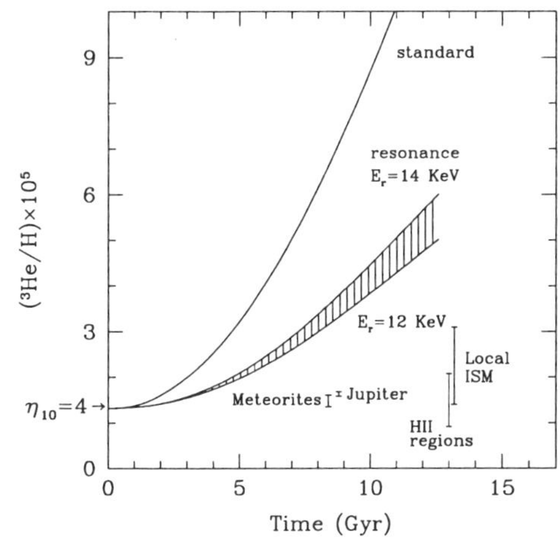

Figure 2. The evolution of ^{3}He obtained with stellar yields computed for a low-energy resonance in the ^{3}He(^{3}He,2p)^{4}He reaction with resonance energy and width compatible with the laboratory data (shaded area) compared with the result obtained with standard yields.

References

1. Rood, R.T., Steigman, G., Tinsley, B.M. ApJ, **207**, L57 (1976)
2. Iben, I., ApJ, **142**, 1447 (1965)
3. Balser, D., Bania, T.M., Rood, R.T., Wilson, T.L. ApJ, **510**, 759 (1999)
4. Galli, D., Palla, F., Ferrini, F., Penco, U. ApJ, **443**, 536 (1995)
5. Charbonnel, C. ApJ, **453**, L41 (1995)
6. Galli, D., Palla, F., Straniero, O., Ferrini, F. ApJ, **432**, L101 (1994)
7. Fowler, W.A. Nature, **238**, 24 (1972)
8. Fetysov, V.N., Kopysov, Y.S. Phys. Lett. B, **40**, 602 (1972)
9. Balser, D. S., Bania, T. M., Rood, R. T., Wilson, T. L. , ApJ, **483**, 320 (1997)
10. Galli, D., Stanghellini, L., Tosi, M., Palla, F. ApJ, **477**, 218 (1997)
11. Boothroyd, A., Sackmann, I.-J. ApJ, **510**, 232 (1999)
12. Junker, M. *et al.* Phys. Rev. C, **57**, 2700 (1998)

LI/H MEASUREMENTS IN STARS

R. CAYREL
Observatoire de Paris,DASGAL 61, av. de l'Observatoire F-75014 Paris, France

F. AND M. SPITE
Observatoire de Paris,DASGAL, section de Meudon 5, place Jules Janssen F-92195 Meudon Cedex, France

The abundance of Li in stars is still the object of intense activity. New results are available on several type of stars. Abundances have been measured in many extremely young or even pre-main sequence stars, and these data rule out important pre-main sequence depletion of ^{7}Li. The status of Li-rich giants and AGB or post AGB stars has been somewhat clarified, but the discovery of an extremely lithium-rich bright red giant in M3 located in the HR diagram among other lithium normal giants remains a challenge. Finally the case of metal-poor stars has been actively further investigated. A significant amount of ^{6}Li is now measured in 6 stars, and there is now general agreement on a very low level of scatter of Li abundance on the Spite plateau. The connection between the measured abundance of ^{7}Li on the Spite plateau and its cosmological abundance remains a subject of intensive research. An upper limit of the depletion of ^{7}Li of the order of 0.3 dex is suggested by the amount of ^{6}Li present in the post-turnoff very metal-poor halo stars.

1 Lithium in very young stars

The litterature on lithium abundance in stars of the last two years amounts to about 150 papers. Here we shall report only a few major results, and we apologize for the many interesting papers not quoted in this review.

One of the area which has made an important progress is the determination of Li/H in very young stars. Soderblom [1] (1999) has determined the Li abundance in stars and pre-main-sequence stars of the very young cluster NGC 2264. The abundance of Li stays at the level 3.2 dex (in the usual scale where log(nH)=12.0) from T_{eff}= 6000 to 3700 K, whereas log(Li) is down to 0.2 (depleted by a factor 1000) in the Pleiades already at T_{eff} = 4000 K, but still at 3.2 at T_{eff} = 6000 K. This clearly shows that the strong depletion after 100 Myr in the Pleiades does not occur during PMS phase, but after. Wichmann et al.[2](2000) have studied ROSAT X-ray sources in the Taurus-Auriga star-forming region. They find extremely strong Li resonance lines (EW > 0.6 Å) for most of their objects. This also shows that PMS stars do not deplete ^{7}Li. Unfortunately the line is much too saturated for learning something about the ^{6}Li component.

2 Lithium in evolved stars

As population I stars are 10-times more lithium-rich than subdwarfs, there must be some galactic enrichment of Li. A important step was made by Verne & Lambert [3](1989) , who found that many AGB stars of the small magellanic stars were quite Li-rich. This points towards the idea that evolved stars of intermediate metallicity may have been an important source of Li. The nature of the galactic lithium-rich giants or asymptotic branch giants has been investigated by Charbonnel & Balachandran [4] (2000). They show that there is a cluster of Li-rich giants at the so-called RGB bump. Several giants classified as Li-rich are indeed Li-normal, with just a moderate dilution factor, because they are more sub-giants than giants. Very few of these iron-normal stars seem to be able to contribute to the enrichment of the interstellar matter in lithium.

Balachandran et al. [5] (2000) have found two unusual Li-rich K giants (HD 233517 and HD 9746) with supermeteoritic Li abundances, the first one reaching $\log \epsilon(^7\text{Li}) = 4.2$. There is also some evidence of mass-loss for these star(IR excess), and they appear to be located at the RGB bump, as several other Li-rich stars studied in the Charbonnel & Balachandran paper. A very important result is that Li appears to be pure ^{7}Li, supporting a fresh formation of Li by the Cameron-Fowler [6](1971) mechanism.

Kraft et al. [7] have discovered an extremely Li-rich giant in the globular cluster M3. The star is neither an AGB nor a RGB bump star, but a cool bright (M_{bol}= -2.8 RGB star. The author hypothesize that the cool bottom burning (Sackmann & Boothroyd [8] 1999) might be responsible for this Li enhancement, likely of short duration.

Finally Reyniers & van Winckel [9] (2000) have found that the post-AGB SB2 binary star HD 172481 is very Li-rich. The object is especially interesting as it shows both supermeteoritic Li and important mass-loss, making it a good candidate for Li-enrichment of the ISM. Note its metallicity [Fe/H]= -0.55.

3 Lithium in halo stars

The subject is of course of particular interest because ^{7}Li is produced in the Big Bang Nucleosynthesis (BBN). Several issues have made important progresses in the last two years. First, thanks to the H& K survey (Beers et al. [10] 1992) of very metal-poor stars, an extension of the Spite plateau to much lower metallicity was made possible (fig. 1). The plateau now extends over three order of magnitudes in Fe/H. Second, the transformation of the equivalent width of the resonance line of Li at 670.78 nm into a Li

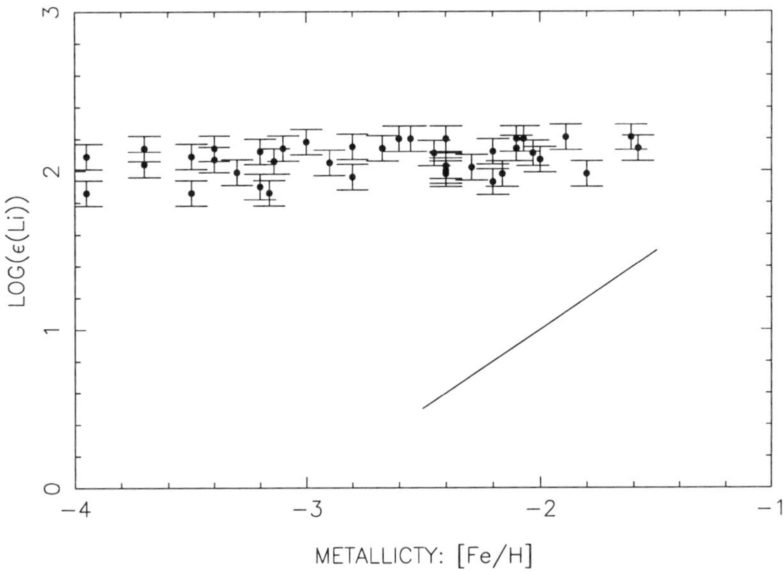

Figure 1. Lithium content of metal-poor stars as a function of metallicity. The abcissa [Fe/H] is the logarithmic difference between the abundance of iron with respect to hydrogen in the star and in the Sun. The ordinate is the logarithm of the ratio of the number of Li atoms to the number of hydrogen atoms, with the usual shift of 12.0, so the ordinate 2.0 means a ratio of 10^{-10}. Note that this diagramme is inappropriate to measure a scatter, as several sources are used, with different systematic errors. Elements not produced in the Big Bang have a slope of 1, as shown by the straight line drawn in the figure.

abundance has been put on a firmer basis. Kurucz [11] (1995) had claimed that 3-D NLTE effects would strongly overionize LiI in metal-poor stars, leading, to an underestimation of the Li abundance by 1 dex , in classical analyses, using 1-D models. Carlsson [12] 1994 had already computed NLTE effects in 1-D models, and found rather small NLTE corrections. Asplund [13] computed LTE Li abundance in 3-D models, and found a correction in the opposite direction as predicted by Kurucz. Finally, Cayrel & Steffen [14] (2000) and Asplund [15] (2000) computed independently NLTE Li abundances respectively on 2-D and 3-D models, and found correction to 1-D NLTE abundances, computed from model atmospheres having the same parameters metallicity, effective temperature and gravity, less than 0.1 dex. The conclusion is that the very large corrections suggested 5 years ago by Kurucz are not supported by actual detailed computations.

Table 1. Status of ^{6}Li/^{7}Li measurements

object	reference	[Fe/H]	^{6}Li/^{7}Li
HD 84937	Smith et al. [20](1993)	-2.3	0.05 ± 0.02
idem	Hobbs&Thorburn [21](1994)	—	0.07 ± 0.03
idem	Hobbs&Thorburn [22](1997)	—	0.08 ± 0.04
idem	Smith et al. (1998) [23]	—	0.06 ± 0.03
idem	Cayrel et al. (1999) [24]	—	0.052 ± 0.019
BD +26 3578 (≡ HDE 338 529)	Smith et al. [23](1998)	-2.4	0.05 ± 0.03
HD 140283	Ryan & Deliyannis(poster here)	-2.7	0.040 ± 0.015
G271-162	Nissen et al. (2000)[25]	-2.15	0.020 ± 0.010
HD 68 284	Nissen et al. [26](1999)	-0.59	0.041 ± 0.010
HD 130 551	idem	-0.62	0.056 ± 0.011

Another important issue is the question of the scatter of Li abundance on the Spite plateau. Most European astronomers have always considered that this scatter was small, practically within the observational errors. However, several other astronomers, mostly on the base of Thorburn [16] (1994) paper, have claimed that there was a significant intrinsic scatter of the order of 0.1 dex, and used this to support a rotationally induced Li depletion on the plateau. Recent papers, Spite et al. [17] (1996), Molaro [18] (1999), Ryan et al.[19] (1999) have convincingly shown that high S/N observations, used with uniform methods of reduction and a unique temperature calibration, result in a very small intrinsic dispersion of Li abundances on the plateau, just a few hundredths of a dex. We shall further discuss the implications of this in the next section.

Finally, one further progress is the determination of separate ^{6}Li and ^{7}Li abundances in more than one star. This is important in at least two respects (i) only ^{7}Li is produced in the standard BBN (ii) a significant amount of the very fragile ^{6}Li constrains the level of depletion of ^{7}Li. As ^{6}Li, if present, is supposed to be formed by spallation and slighlty more ^{7}Li is produced as well, one must subtract about 1.5 times the amount found in ^{6}Li from the ^{7}Li to have the primordial ^{7}Li. Table 1 gives the (hopefully) exhaustive list of ^{6}Li six positive abundance determinations, not upper limits. This is particularily for Jean Audouze, who used to say " Une hirondelle ne fait pas le printemps" (one swallow does not make a summer) when we were working together on HD 84 937.

4 The connection with cosmology: abundance of Li in the ISM at the birth of the star

We have already emphasized in the preceding section how important it was to measure ^{6}Li, in order to identify what should be deducted from total lithium for ^{6}Li, and ^{7}Li from spallation origin. Another relevant issue is the question of a possible depletion of ^{7}Li by nuclear burning below the convective zone, during the life-time of the metal-poor star. The scatter argument is no more valid for claiming an important depletion induced by rotation (see Ryan et al. [27] 2000). But the small scatter does not rule out other depletion mechanisms, as gravity waves (Montalbán & Schatzman [28] 2000), or microscopic diffusion (Thaedo & Vauclair, poster here). Microscopic diffusion alone does not work, as producing a signature at the hot end of the plateau (Proffit and Michaud [29] 1991). It is not an easy task to find a depletion mechanism transforming a plateau into a parallel plateau. The simplest mechanism is of course no depletion at all, but as long as identifed physical mechanisms of depletion are not shown to be inhibited by something else, the question cannot be considered as closed. We send back to the poster by Deliyannis & Ryan to convince you that ^{6}Li is a key factor in this affair. An upper limit of 0.3 dex for the depletion of ^{7}Li in very metal-poor stars is something which should rally a vast majority of us.

5 conclusion

A considerable work has been accomplished during the past two years on the abundance of Li in stars.

Among the main results let us recall:

- The lack of depletion of ^{7}Li during the PMS phase
- An improved understanding of the synthesis of ^{7}Li in evolved stars
- the derivation of Li abundance in 3-D + NLTE models
- the obtention of ^{6}Li abundances in more than one star
- general agreement that the scatter of Li abundance is small
- a narrowing down of the amount of acceptable depletion of ^{7}Li on the plateau, thanks to the observation of ^{6}Li

Let us hope that as much (or more!) be done in the next two years.

References

1. D.R. Soderblom, AJ 118, 1301 (1999)

2. R. Wichmann, G. Torres, C.H.F. Melo, S. Frink, S. Allain, J. Bouvier, J. Krautter, E. Covino, R. Neuhuser, A&A 359, 181 (2000)
3. V.V. Smith, D.L. Lambert, ApJ 345, L75 (1989)
4. C. Charbonnel & S. Balachandran, A&A 359, 563
5. S.C Balachandran, F.C. Fekel, G.W. Henry, H. Uitenbroek, ApJ 542, 978
6. A.G.W. cameron & W.A. Fowler, ApJ 164, 111
7. R.P. Kraft, R.C. Peterson, P. Guhathakurta, C. Sneden, J.P. Fulbright, G.E. Langer, ApJ 518, L53 (1999)
8. I.-J. Sackmann, A.I. Boothroyd, ApJ 510, 217 (1999)
9. M. Reyniers, H. Van Winckel, astro-ph/0010513 (2000)
10. T.C. Beers, G.W. Preston, S.A. Shectman, AJ 103, 1987 (1992)
11. R.L. Kurucz, ApJ 452, 102
12. M. Carlsson, R.J. Rutten, J.H.M.J. Bruls, N.G. Shchukina A&A 288, 860 (1994)
13. M. Asplund, ÅNordlund, R. Trampedach, R.F. Stein, A&A 346, L17
14. R. Cayrel, M. Steffen, in *The light Elements and their Evolution* IAU Symp. 198 L. Da Silva, M. Spite and J.R. de Medeiros eds. p. 437 (2000)
15. M Asplund in *The light Elements and their Evolution* IAU Symp. 198 L. da Silva, M. Spite and J.R. de Medeiros eds. p.448 PASP
16. J.A. Thorburn ApJ 421, 318 (1994)
17. M. Spite, P. François, P.E. Nissen, F. Spite, A&A 307, 172 (1996)
18. P. Molaro in *LiBeB, Cosmic rays, and related X-and gamma-rays* R. Ramaty, E. Vangioni-Flam, M. Cassé and Keith Olive eds. p. 6 ASP series 171(1999)
19. S. Ryan, J.E. Norris, T.C. Beers ApJ 523, 654 (1999)
20. V.V. Smith, D.L. Lambert, P.E. Nissen, ApJ 408, 262 (1993)
21. L.M. Hobbs, J.A. Thorburn, ApJ 428, L25
22. L.M. Hobbs, J.A. Thorburn, ApJ 491, 772
23. V.V. Smith, D.L. Lambert, P.E. Nissen, ApJ 506, 405 (1998)
24. R. Cayrel, M. Spite, F. Spite, E. Vangioni-Flam, M. Cassé, J. Audouze, A&A 343, 923 (1999)
25. P.E. Nissen, M. Asplund, V. Hill, S. D'Odorico A&A 357, L49 (2000)
26. P.E. Nissen, D.L. Lambert, F. Primas, V.V. Smith A&A 348, 211 (1999)
27. S.G. Ryan, T.C. Beers, K.A. Olive, B.D. Fields, J.E. Norris, ApJ 530, L57 (2000)
28. J. Montalbán, E. Schatzman, A&A 354, 943 (2000)
29. C.R. Proffitt & G. Michaud, ApJ 371, 584 (1991)

ASTROPHYSICAL RELEVANCE OF HOT BOTTOM BURNING IN AGB

F. D'ANTONA
Osservatorio Astronomico di Roma
I-00040 Monteporzio, Italy
E-mail: dantona@coma.mporzio.astro.it

I summarize the results of our recent modelling of Hot Bottom Burning in the thermally pulsing Asymptotic Giant Branch Stars and the consequences for the Galactic evolution of lithium and for the self-pollution model for Globular Cluster stars.

1 Introduction

The most luminous Oxygen rich Asymptotic Giant Branch (AGB) stars in the Magellanic Clouds are lithium rich (Smith and Lambert 1989), an evidence that they are subject to 'Hot Bottom Burning' (HBB) at the bottom of their convective envelope. Lithium is produced through the reaction chain $^3He +^4 He \rightarrow^7 Be \rightarrow^7 Li$, and the decay time of Be into Li by k-capture is long enough that berillium can be transported by convection to outer stellar layers, where Li does not burn (Cameron and Fowler 1971). Non-instantaneous turbulent mixing coupled to nuclear burning is necessary to describe this phase (Sackman and Boothroyd 1992, Mazzitelli et al. 1998). The models show that the intermediate mass stars (M $\simeq 3.5 - 6M_\odot$) are subject to HBB during their AGB evolution. Our models allow to follow *selfconsistently* the evolution of many the other nuclear species: Li production is important at $\sim 40 \times 10^6$K at the relatively low densities at the bottom of the AGB envelopes; full CN cycling is effective at $\sim 60 \times 10^6$K; at temperatures $\sim 85 \times 10^9$K or larger, the ON cycle is active and oxygen burns in the envelope.

2 Our recent studies on HBB – AGB stars

We have recently obtained a direct observational confirmation of the theory, by examining the spectra of early–AGB and AGB stars in the open cluster NGC 1866 of the LMC (Maceroni et al. 2001): of the two luminous AGB stars present in this cluster, one shows no Li at all (but a light Li line is present in our spectra of the early–AGB stars), the other one has a strong Li line. Models predict first destruction of the Li remnant of the previous evolutionary phases, and then production by the Cameron Fowler chain: this is the first time that this sequences of events is found in a LMC cluster.

Our stellar modelling and calibration of mass loss is described by Ventura et al. (2000 and 2001), A byproduct of these models is that it seems difficult that massive AGBs give an important contribution to the galactic evolution of Li (see Romano et al. 2001).

3 ON cycling in HBB and the self-pollution model for GCs

In many GCs, the stars show inhomogeneity in their surface composition, for what concerns the abundances of the light elements from Li to Al (see, for reviews, Smith 1987 and Kraft 1994). In particular, red giants in some GCs show a Na–O anticorrelation and a Mg–Al anticorrelation. These anomalies may be the sign of 'in situ' processes, by which some still unexplained deep mixing mechanism brings at the stellar surface the products of complete CNO cycling occurring deep in the stellar interiors However, Gratton et al. (2001) have recently found the same abundance anomalies also at the turnoff and in subgiants of the GC NGC 6752. A model able to explain the GC inhomogeneities in non evolved stars is based on the 'self–pollution' or 'self–enrichment' hypothesis: the surface composition of the GC stars which we see today evolving has been contaminated by accretion from the ejecta of more massive stars which have evolved during the past history of the cluster. Intermediate mass stars are the best candidates to produce accretion on *on the already formed stars*, an hypothesis was put forward first by D'Antona et al. (1983).

Full AGB models of metallicities $\leq 1 \times 10^{-3}$ have so efficient HBB that the temperatures at the bottom of the convective envelope become ($\geq 85 \times 10^6$K), the full CNO cycle operates, and the oxygen abundance in the envelope is reduced. We can predict that, at the same time, HBB will produce sodium by proton captures on the primordial neon, and aluminum by proton captures on magnesium, explaining the anticorrelations Na-O and Mg-Al. Our results (figure 1) are consistent with a large oxygen spread in moderately low metallicity clusters such as NGC 6752, and a smaller oxygen spread is predicted for clusters of larger metallicity.

In these models we also follow the Li evolution: it is first produced and then burned in the envelope. The net result is that the average abundance in the ejected envelope is smaller than the initial one, but there is not the complete Li destruction, which would be obtained if this type of advanced nucleosynthesis takes place in a stellar context different from HBB. We expect that polluted stars (which have lower oxygen) also have a lower Li. The spread should be smaller when the metallicity is larger (e.g. in the GC 47 Tuc). At this metallicity, the Li abundance in the ejected matter results to be very close to the Big Bang abundance. As we do not know the mass loss rate to better

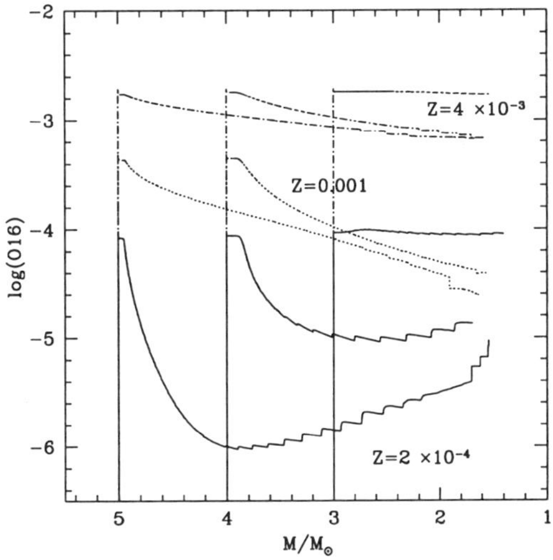

Figure 1: The Oxygen abundance in our models, as a function of the actual stellar mass along each sequence. The $3M_\odot$ does never burn oxygen. Notice the progressively larger oxygen depletion in the 5 and $4M_\odot$ models as the metallicity decreases.

than a factor $\simeq$ 2 or 3, it is well possible that the stars in which there is pollution have in fact a *larger* Li abundance than the Big Bang one.

References

1. Cameron A.G., Fowler W.A., ApJ **164**, 111 (1971)
2. D'Antona, F., Gratton, R., Chieffi, A., Mem.S.A.It. **54**, 17 (1983)
3. Gratton, R. et al., A&A in press (2001)
4. Kraft, R.P., PASP **106**, 553 (1994)
5. Maceroni, C., Testa, V., D'Antona, F., Plez, B. & Garcia Lario, P., in this book (2001)
6. Mazzitelli I., D'Antona F., Ventura P., A&A **348**, 846 (1999)
7. Romano, D., Matteucci, F., Ventura, P. & D'Antona, F., in this book (2001)
8. Sackmann I.J., Boothroyd A.I., ApJ **392**, L71 (1992)
9. Smith, G.H., PASP **99**, 67 (1987) 1992
10. Smith V.V., Lambert D.L., ApJ **345**, L75 (1989)
11. Ventura P., D'Antona F., Mazzitelli, I., A&A **363**, 605 (2000)
12. Ventura, P., D'Antona, F., Mazzitelli, I., in this book (2001)

LITHIUM ISOTOPE RATIOS IN METAL-POOR HALO STARS

MARTIN ASPLUND
Uppsala Astronomical Observatory, Sweden (martin@astro.uu.se)

DAVID L. LAMBERT
Dept. of Astronomy, University of Texas, Austin, USA (dll@as.utexas.edu)

POUL ERIK NISSEN
Inst. of Physics and Astronomy, Aarhus University, Denmark (pen@ifa.au.dk)

FRANCESCA PRIMAS
European Southern Observatory, Garching, Germany (fprimas@eso.org)

VERNE V. SMITH
Dept. of Physics, University of Texas, El Paso, USA (verne@physics.utep.edu)

Equipped with very high quality observations obtained from VLT/UVES and the new generation of 3D hydrodynamical model atmospheres, we have initiated a survey of ^{6}Li/^{7}Li ratios in metal-poor halo stars. Here we present the first preliminary results of this investigation, reporting four new likely detections at a similar level (^{6}Li/^{7}Li $= 2 - 5\%$) to that previously found for HD 84937, BD +26°3578 and G271-162.

1 Introduction

Much work has recently been devoted to the search for ^{6}Li in metal-poor halo stars due to its significance for astrophysics and cosmology. Positive detections of the isotope would limit the possible depletion of ^{7}Li and thus better constrain Big Bang nucleosynthesis. ^{6}Li also provide an additional test of theories for the production of the light elements (Li, Be and B) by cosmic ray spallation. Finally, due to its fragile nature ^{6}Li is a sensitive probe of stellar mixing events. To date, ^{6}Li has only been detected in the three halo stars HD 84937, BD +26°3578 and G271-162 (Smith et al. 1993[9], 1998[10]; Hobbs & Thorburn 1997[5]; Cayrel et al. 1999[3]; Nissen et al. 2000[7]) and in two disk stars with [Fe/H] $\simeq -0.6$ (Nissen et al. 1999[6]). A much larger sample for a range of metallicity is needed to properly address the above-mentioned topics, a task which is now becoming feasible with facilities like UVES on VLT. [a]

[a]The full version of the present talk with additional figures can be found at http://www.astro.uu.se/~martin/talks/Cosmic00

2 Observations and Analysis

The determinations of ^{6}Li are based on the increased width and asymmetry of the Li I 670.8 nm doublet introduced by the isotopic shift. Therefore both a high spectral resolution and very high S/N are needed. To obtain this we have made use of the excellent capabilities of UVES on VLT. During three nights in July 2000, 13 stars were observed with $\lambda/\Delta\lambda = 120,000$ and $S/N > 400$, which is sufficient for the purpose.

Besides excellent observations, a proper understanding of the intrinsic stellar line broadening due to convective Doppler shifts and rotation is required in order to obtain accurate ^{6}Li/^{7}Li determinations; ThAr calibration spectra provide knowledge of the also present instrumental broadening. Traditionally this is achieved from estimating the contribution from rotation and micro-/macroturbulence from other spectral lines using 1D hydrostatic model atmospheres. A novel feature of our investigation is the use of 3D hydrodynamical model atmospheres, which self-consistently predict the time-dependent temperature structure and convective velocity field giving rise to the broadening and asymmetry of spectral lines (e.g. Asplund et al. 1999[1], 2000[2]). Therefore the concepts of macro- and microturbulence are obsolete with 3D models. However, the stellar rotational velocity must still be determined, which is estimated from Fe I, Ca I and K I lines of similar strength and formation depth as the Li I line. In this preliminary analysis all 3D line calculations are based on the assumption of LTE.

With the rotational and instrumental broadening fixed, the ^{6}Li/^{7}Li ratios are determined from a χ^2-analysis where the ^{6}Li/^{7}Li ratios, total Li abundance and the observed wavelength zero-point are all free parameters, which is necessary in order to avoid spurious results. The excellent agreement between observed and predicted line profiles (cf. Nissen et al. 2000[7]) strongly support the realism of the 3D model atmospheres and increases the confidence in the final results. As a complement, we also carry out a traditional analysis with 1D MARCS model atmospheres.

3 Preliminary Results

Of the five here studied stars, a preliminary analysis has revealed a positive detection of ^{6}Li in four: CD$-30°$18140 (^{6}Li/^{7}Li $= 0.03 \pm 0.01$), G13 $-$ 9 (0.05 ± 0.02), HD 140283 (0.04 ± 0.01) and HD 160617 (0.04 ± 0.01 from a 1D analysis in the absence of a suitable 3D model atmosphere). Only in the case of HD 19445 is the result consistent with no ^{6}Li (0.00 ± 0.01), in agreement with the prediction of strong ^{6}Li-depletion at its location in the HR-diagram.

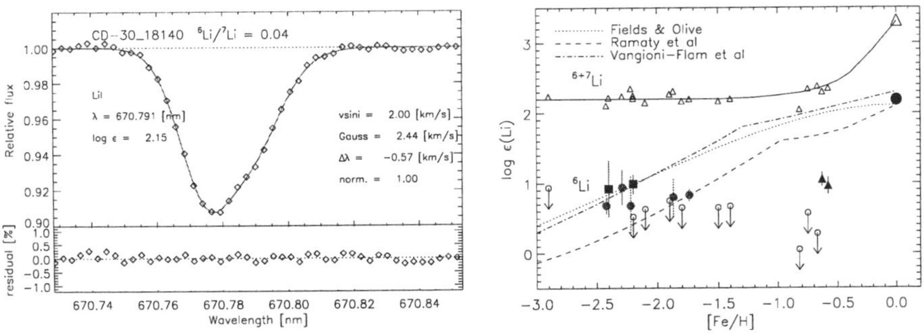

Figure 1. *Left panel*: The predicted (solid line) and observed (diamonds) Li I line in CD−30°18140. *Right panel*: The five ^{6}Li determinations reported here and in Nissen et al. (2000) from UVES observations (filled circles) together with previous results for halo (filled squares, Smith et al. 1998) and disk stars (filled triangels, Nissen et al. 1999)

Fig. 1 illustrates the excellent fits and the high quality of the observations.

As seen in Fig. 1, the sofar reported ^{6}Li determinations all have similar values, which, however, are in poor agreement with theoretical models for cosmic ray spallation (e.g. Fields & Olive 1999[4]; Ramaty et al. 2000[8]; Vangioni-Flam et al. 1999[11]). Indeed there are tantalizing evidence for pregalactic ^{6}Li production, which should be investigated further by studies of very metal-poor halo stars. On the other hand, the existence of ^{6}Li severely limits the allowed amount of stellar depletion of ^{7}Li in halo stars and thus better constrain Big Bang nucleosynthesis.

References

1. M. Asplund, *et al*, A&A **346**, L17 (1999).
2. M. Asplund, *et al*, A&A **357**, 729 (2000).
3. R. Cayrel, *et al*, A&A **343**, 923 (1999).
4. B.D. Fields and K.A. Olive, New Astronomy **4**, 255 (1999).
5. L.M. Hobbs and J.A. Thorburn, ApJ **491**, 772 (1997).
6. P.E. Nissen, *et al*, A&A **348**, 211 (1999).
7. P.E. Nissen, *et al*, A&A **357**, L49 (2000).
8. R. Ramaty, *et al*, ApJ **534**, 747 (2000).
9. V.V. Smith, *et al*, ApJ **408**, 262 (1993).
10. V.V. Smith, *et al*, ApJ **506**, 405 (1998).
11. E. Vangioni-Flam, *et al*, New Astronomy **4**, 245 (1999).

LITHIUM ALONG THE AGB OF LMC CLUSTERS

C. MACERONI, V. TESTA, F. D'ANTONA
Osservatorio Astronomico di Roma, via Frascati 33, I-00040 Monteporzio, Italy)
E-mail: maceroni,testa,dantona@coma.mporzio.astro.it

B. PLEZ
GRAAL cc72, Universite' de Montpellier II, 34095 Montpellier Cedex 5, France
E-mail: plez@graal.univ-montp2.fr

P. GARCIA LARIO
ISO Data Centre, Astrophysics Division, Space Science Department of ESA, Villafranca del Castillo, Apartado de Correos 50727, E-28080 Madrid, Spain
E-mail: pgarcia@iso.vilspa.esa.es

We present the first results of an exploration of Asymptotic Giant Branch (AGB) star properties in the young open clusters of the Magellanic Clouds. In accordance with theoretical predictions, Lithium lines of different strength are observed along the AGB, suggesting Li production by "Hot Bottom Burning" (HBB) in the brightest stars.

1 The AGB stars of NGC 1866

The evolution of Lithium in intermediate mass stars is a powerful probe of the convective envelope conditions. Lithium is first diluted, during the giant phase, then destroyed at the beginning of the HBB phase, and finally produced through the ^{3}He + ^{4}He $\rightarrow ^7$Be $\rightarrow ^7$Li chain. Therefore, in clusters with an AGB-evolving mass large enough for HBB to occur, the analysis of the surface abundances along the AGB should - in principle - reveal a complex behavior; the brightest stars should be Li-rich objects, while the early AGB stars should show a Li abundance lower than the original one, because of the preceding evolution that includes convective dilution during the giant phase. An intermediate state with no Li could also appear between the two, as a consequence of Li burning at the onset of HBB.

Our observing program is providing medium and high resolution spectra of the AGB stars of young LMC clusters, known for having one or more AGB members (NGC 1866, NGC 2031, NGC 2214). In Figure 1 we present the medium resolution (nominal R=5500) of the four brightest AGB stars of NGC 1866, obtained with EMMI at ESO NTT. The Doppler shifts were derived from Hα and other reference line positions yielding, for the whole sample of (nine) NGC 1866 AGB stars: V = 303 $\pm$ 17 km/sec, this is in

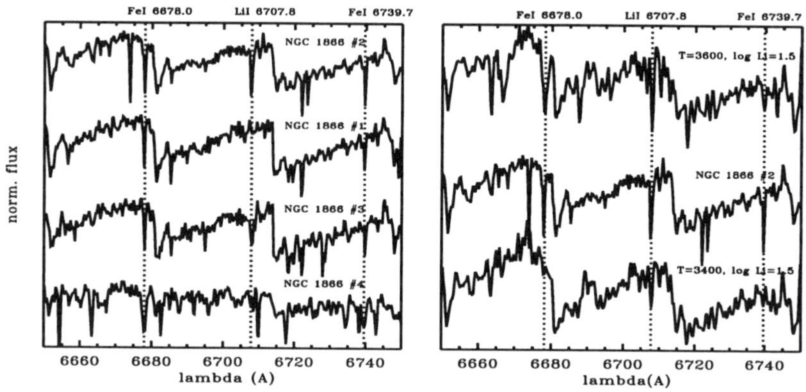

Figure 1. left panel: The spectra of the four most luminous stars in NGC 1866 in order (from top to bottom) of increasing R-magnitude at the epoch of observation. Right panel: comparison between the spectrum of NGC 1866 #2 and synthetic spectra with $T_e = 3400, 3600$ and $\log N(Li) = 3.0$.

agreement with dynamic studies (the systemic velocity, according to Fischer *et al.* [1], is 301 ± 1.2 km/s).

A strong Li line appears in star #2. This star turned out to be the brightest at the epoch of observations, in spite of the label related to increasing magnitude at another epoch (all these stars are long period variables). No Li line is detected in the spectrum of the star originally listed as #1 (Li burnt at the onset of HBB?), while a weak one is present in those of the hotter stars #3 and #4. The latter, at variance with the case of star #2, can be readily explained in terms of dilution, during the giant evolutionary phases, of the initial Li content. The figure shows as well a comparison between star #2 spectrum with two synthetic spectra (obtained by convolving theoretical spectra with a Gaussian of FWHM of 0.6 Å). The observed spectrum looks intermediate between the two models. A determination of Li abundance is not yet available at the time of writing, but it should be intermediate between $\log N(Li) = 1.5$ and 3.0. The strength of the last line in comparison with those found in early AGB stars suggests that we are witnessing *for the first time* Li manufacturing by Hot Bottom Burning in a cluster. An isochrone fit of the cluster HR diagram yields a value of the AGB evolving mass of $\sim 4.5\ M_\odot$.

References

1. P. Fischer *et al*, AJ **103**, 85 (1992)

LITHIUM ABUNDANCE : MORE ACCURACY REQUIRED

F. AND M. SPITE
Observatoire de Paris, Sect. de Meudon et UMR 8633 du CNRS, Place Janssen
92195 Meudon CEDEX, France,
E-mail: Francois.Spite@obspm.fr

The trends of lithium abundance in warm Pop II dwarfs are used as strong constraints in several important topics (Big Bang Nucleosynthesis, Cosmic Rays production, stellar structure ...). For example, a low scatter of the Li abundance in the stars of the plateau points towards a low lithium depletion, and a primordial abundance rather close to the value observed on the plateau. This scatter, and a small trend versus metallicity, are discussed, showing that firm conclusions need a still higher accuracy and therefore an urgent development of stellar physics, and of atomic and molecular physics.

1 First example : the controversial scatter of Li abundance

Lithium may be depleted in old stars by various processes (e. g. mixing, diffusion ...), which depend on mass, effective temperature T_{eff}, rotation, metallicity, age etc. so that any significant depletion should produce some scatter of the remaining lithium abundances A(Li) :

$$A(Li) = \log (N_{Li}/N_H) + 12.0$$

In the warm Pop II dwarfs, A(Li) is relatively constant (flat) versus T_{eff} (the temperature plateau), and also versus [Fe/H] (the metallicity plateau).

This fact is not well understood, diffusion or mixing or both can be lower than expected as shown by recent observations [1] (see also R. Gratton, this meeting). Also, a compensation (self-regulation ? see S. Vauclair this meeting) may occur in the processus of lithium depletion.

This scatter is difficult to determine, because it is masked by the errors of determination of A(Li), errors essentially due to the uncertainties of the effective temperature T_{eff}.

The difficulty has been judiciously circumvented (Ryan et al. [2]), by limiting the analysis to a sub-sample of very metal-poor stars, chosen in a limited range of colors (and therefore of temperature). It happens then that the disturbing masking effect of errors in A(Li) is kept to a minimum, and the remaining scatter is found to be extremely small. A rather similar result is reached for globular cluster stars (R. Gratton, this meeting).

2 A further example

Going a step further [2], a trend of A(Li) versus metallicity is found : Li is low in stars of low metallicity (see also [3]). The trend is not negligible, it has important consequences, it is not in agreement with previous studies, it deserves a careful analysis.

It happens that, by discarding 3 outliers, the slope dA(Li)/d[Fe/H] drops from 0.12 to 0.07 (a rather strong variation).

Moreover, it appears that the sample is not well balanced, the more metal-poor stars are cooler, the less metal-poor are warmer, so that there is some hidden (indirect) correlation between A(Li) and temperature through the ranking by metallicity. A different choice of a temperature scale should therefore change the numerical value of the trend. However, a simple calculation shows that the effect is negligible.

3 Conclusion

The trend [2] versus [Fe/H] is not strongly altered by a change in temperature scale, suggesting an early enrichment of Li (e. g. by cosmic rays : Vangioni-Flam et al. [4]) in spite of astration (e. g. Vangioni-Flam et al. [5]). However, this trend has not been found in previous studies of other authors (see R. Cayrel, this meeting). An obvious difficulty is the number of unrecognized binaries. The problem deserves further studies.

Anyhow, the primordial Li would be more safely derived from accurate abundances found in well observed VERY metal-poor stars, which in turn require a correct temperature determination, without systematic effect. Accurate atomic values (e. g. for excitation equilibrium), better model atmosphere (e. g. 3D convective models, see M. Asplund, this meeting) with accurate theoretical colors... are eagerly awaited.

References

1. B.V. Castilho, L. Pasquini, D. M. Allen, B. Barbuy, P. Molaro, A&A **361**, 92 (2000).
2. S. G. Ryan, J. E. Norris, T. C. Beers, ApJ **523**, 654 (1999).
3. S. G. Ryan, T. C. Beers, K. A. Olive, B. D. Fields, J. E. Norris, ApJ **530**, L57 (2000).
4. E. Vangioni-Flam, M. Cassé, R. Cayrel, J. Audouze, M. Spite, F. Spite, New Astr **4**, 245 (1999).
5. E. Vangioni-Flam, M. Cassé, Y. Oberto, ApJ **349**, 510 (1990).

LITHIUM ABUNDANCES IN HALO STARS

S. THÉADO, S. VAUCLAIR
Laboratoire d'Astrophysique, 14 av Ed. Belin, 31400 Toulouse, France
E-mail: stheado@ast.obs-mip.fr

The determination of the primordial lithium relies on the abundance observations in halo stars. The question of lithium depletion in stars is crucial in this respect. We claim that in case of rotation induced mixing the μ-gradient created by diffusion induces a feed-back effect on the meridional circulation and on the diffusion itself. This effect, which was not introduced in the previous computations, is able to account for the very small dispersion observed in the lithium plateau. The primordial value is derived as : Li/H = (3.16±0.21) 10^{-10}.

The lithium plateau observed in halo stars has long appeared as a paradox in the general context of the lithium abundance behavior in stellar outer layers. First the plateau is flat, second the lithium abundance dispersion is extremely small. This seems in contradiction with the large lithium variations observed in younger stars. It is also difficult to understand theoretically : as lithium nuclei are destroyed by nuclear reactions at a relatively low temperature ($\cong$ 2.5 million degrees), the occurrence of macroscopic motions in the stellar outer layers easily lead to lithium depletion at the surface. On the other hand, if no macroscopic motions occur in the stellar gas, lithium is subject to microscopic diffusion which, in the case of halo stars, should also lead to depletion, which is not observed.

Vauclair[1] pointed out that the effect of the μ-gradients induced by element settling was not introduced in previous computations of rotationally-induced mixing. The meridional circulation velocity in stars, in the presence of μ-gradients, is the sum of two terms, one due to the classical thermal imbalance (Ω-currents) and the other one due to the induced horizontal μ-gradients (μ-currents). In the most general cases, μ-currents are opposite to Ω-currents. When element settling occurs below the stellar outer convective zone in cool stars, a small helium gradient builds, even in the presence of circulation. The resulting μ-gradients are rapidly large enough to create μ-currents of the same order as Ω-currents. Then a self-regulating process may take place, in which both the circulation and the settling are strongly reduced.

Théado and Vauclair[2] have done precise computations of this process and its influence on the lithium abundance variations in halo stars.

Figure 1 displays the lithium variations with age for several rotation velocities for a $0.70M_{\odot}$ star with $[Fe/H] = -2.0$. We can see how the shape of the depletion curve varies with V_{rot} : for rotation velocities less than 6km.s^{-1},

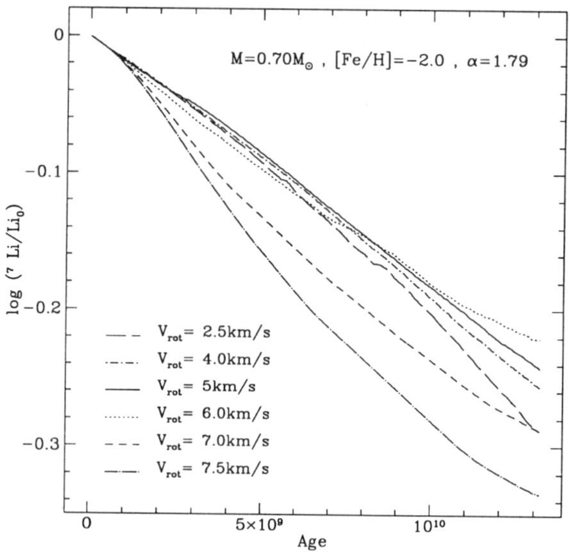

Figure 1: see text for details

settling is preponderant, although reduced in the layers below the convection zone. When V_{rot} increases, lithium is less and less depleted. Above 6km.s^{-1} however, the self-regulating process becomes more important and the lithium depletion curves are clearly bent. Increasing V_{rot} then leads to more and more lithium depletion as the time needed for the μ-currents to become equal to the Ω-currents increases. Altogether, for rotation velocities between 2.5 and 7.0 km.s^{-1}, the obtained lithium dispersion is less than $\cong$ 10%.

Similar computations have been done for $M = 0.75M_\odot$, $M = 0.65M_\odot$ and $M = 0.60M_\odot$ stars. The results show that the plateau is well reproduced with a very small dispersion as observed.

The depletion is (0.25 ± 0.03) dex, which leads to a primordial abundance $Li/H = (3.2 \pm 10^{-10})$, compatible with D/H and ^{4}He/H. The resulting value for the baryonic number is : $\eta_{10} = (5.5 \pm 1.0)$

References

1. Vauclair, S., 1999, A&A 351, 973
2. Théado, S., Vauclair, S., 2001, submitted A&A

EVOLUTION OF LITHIUM IN THE MILKY WAY

D. ROMANO
International School for Advanced Studies, SISSA/ISAS, Via Beirut 2–4, 34014 Trieste, Italy
E-mail: romano@sissa.it

F. MATTEUCCI
Dipartimento di Astronomia, Università di Trieste, Via G.B. Tiepolo 11, 34131 Trieste, Italy
E-mail: matteucci@ts.astro.it

P. VENTURA, F. D'ANTONA
Osservatorio Astronomico di Roma, Via Frascati 33, 00040 Monte Porzio Catone, Italy
E-mail: paolo, dantona@coma.mporzio.astro.it

We adopt up-to-date ^{7}Li yields from asymptotic giant branch (AGB) stars in order to study the temporal evolution of this element in the solar neighbourhood. Several lithium stellar sources are considered besides the AGBs: Type II supernovae (SNe), novae, low-mass giants. Galactic cosmic ray (GCR) nucleosynthesis is taken into account as well. We conclude that AGB stars do not substantially contribute to ^{7}Li enrichment on a Galactic scale. Therefore, a significant ^{7}Li production from novae and low-mass stars is needed to explain the late, steep rise of the ^{7}Li abundance in disk stars and the meteoritic ^{7}Li abundance.

1 Model Prescriptions and Results

The upper envelope of the observed $\log\epsilon(^7\mathrm{Li})$ vs. [Fe/H] diagram can constrain models of chemical evolution, provided that it actually traces the ^{7}Li enrichment history of the interstellar medium (ISM) in the solar neighbourhood. A nearly flat plateau at low metallicities (the *Spite Plateau*) is followed by a steep rise afterwards (big dots in Figure 1): these features are interpreted as due to ^{7}Li enhancement from a lower primordial value to a higher present one. The main purpose of this contribution is to show how the interplay between several categories of ^{7}Li producers may explain the rise from the Spite plateau and the meteoritic ^{7}Li abundance.

The adopted model assumes that the Galaxy forms out of two main episodes of accretion: the first one builds up the halo, the second one the disk. The Sun is located at a distance of 8 kpc from the Galactic center, the adopted Galactic age is 14 Gyr. Details on the model parameters and basic equations can be found in Chiappini *et al*[1 2]; detailed ^{7}Li synthesis prescriptions are given in Romano *et al*[3].

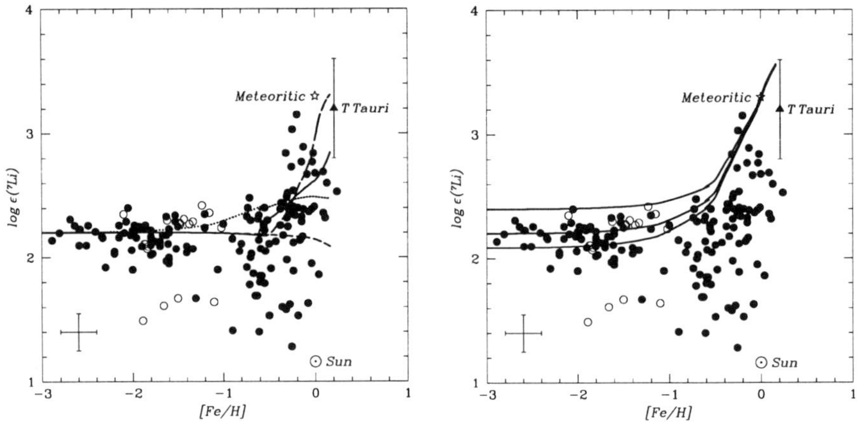

Figure 1: Model predictions compared to the data. The contribution from each category of stellar ^{7}Li producers is shown in the left panel (see text for details); the ^{7}Li evolution obtained by adding together in the model all the stellar sources and GCR nucleosynthesis is shown in the right panel for three different values of the primordial abundance. The results at [Fe/H] > -0.5 dex are nearly the same, owing to the consumption of ^{7}Li by stellar astration during the Galactic lifetime.

In Figure 1, left panel, the contributions to ^{7}Li enrichment from different stellar sources are shown: *i)* AGB stars (*short-dashed line*); *ii)* Type II SNe (*dotted line*); *iii)* low-mass giants (*long-dashed line*); and *iv)* novae (*continuous line*). Model predictions in the right panel of Figure 1 are obtained by taking into account in the model all these stellar sources plus *v)* GCRs.

The main conclusions are that AGBs are not substantial ^{7}Li producers on a Galactic scale (see also the contributions by P. Ventura and F. D'Antona, these proceedings) and that the meteoritic ^{7}Li abundance can be reproduced only if other stellar sources acting at late times (novae and low-mass stars) are contributing a substantial fraction ($\sim 60\%$ in our model) of the meteoritic ^{7}Li. However, it should be said that the fractional contribution of each of these two sources is still very uncertain, mainly due to uncertainties in the evolutionary histories of the progenitor stars and in the stellar yields.

References

1. C. Chiappini *et al*, *ApJ* **477**, 765 (1997).
2. C. Chiappini *et al*, *ApJ* , accepted (2001).
3. D. Romano *et al*, *A&A* , submitted (2001).

LITHIUM-BERYLLIUM-BORON EVOLUTION: FROM MENEGUZZI, AUDOUZE AND REEVES 1971 UP TO NOW

M. CASSÉ

Service d'Astrophysique, DSM/DAPNIA/ Orme des Merisiers, 91191 Saclay CEDEX, France and
Institut d'Astrophysique de Paris, CNRS, 98 bis bd Arago, 75014 Paris, France

E. VANGIONI-FLAM

Institut d'Astrophysique de Paris, CNRS, 98 bis bd Arago, 75014 Paris, France

J. AUDOUZE

Palais de la Découverte, avenue F. Roosevelt, 75008 Paris, France and
Institut d'Astrophysique de Paris, CNRS, 98 bis bd Arago 75014 Paris, France

We review the main sources of LiBeB production and show that a primary mechanism is at work in the early Galaxy involving both ejection and acceleration of He, C and O at moderate energy, which by nuclear interaction with H and He produce light isotopes. The precise measurement of the Be abundance at [Fe/H] = -3.3 and of ^6Li in halo stars find an explanation in this framework. Thus, the preservation of ^6Li in the atmosphere of metal poor stars implied, points toward the fact the Spite plateau reflects the primordial value of Li. Consequently, it can be used as a baryodensitometer.

1 Basic facts and historical perspective

LiBeB are exceptional since they are both simple and rare. In the Solar System, their abundances amount to Li/H = 2.10^{-9}, B/H = 7.610^{-10} and Be/H = 2.510^{-11}, respectively, whereas istopic ratios of lithium and boron are $^7Li/^6Li = 12.6$, $^{11}B/^{10}B = 4$. Indeed they are rare because their nuclei are fragile. The fact that nuclear structures with A = 5 and 8 are violently unstable has limited primordial nucleosynthesis by thermonuclear fusion to ^7Li. Effectively Standard BBN is hoplessly ineffective in generating ^6Li , ^9Be, ^{10}B, ^{11}B [1]. Stellar nucleosynthesis (quiescent or explosive) produces nuclei from C to U but destroy LiBeB, since their burning temperature is especially low.

Another mechanism (different from fusion) is thus necessary to yield LiBeB. This one is the break-up of CNO by spallation. It is worth noting that among light isotopes, Li, is special, since it owes its origin to numerous sources, involving processes as different than BBN ($^7Li/H$ about 10^{-10}), nuclear spallation ($< 2.10^{-10}$), neutrino spallation in SNII ($< 2.10^{-10}$), and stellar nucleosynthesis through low mass stars and novae (about 10^{-9}).

The origin of LiBeB is an old problem. Initially, the debate has opposed supporters of a Stellar/Circumstellar origin [4] to those of an interstellar origin related to Galactic Cosmic Rays (Reeves and collaborators). In a seminal work Meneguzzi, Audouze and Reeves (1971), MAR [5], have given credance to the second explanation. They were able to device a consistent scenario linking measurements of local abundances, spallation cross sections and the measured GCR flux. The merit of their work is the clear identification of the production process i.e. the Galactic Cosmic Rays / ISM interaction, essentially via fast (p,α) impacting on CNO. Furthermore, MAR offered a quantitative estimate of the present production rate of each isotope, and integrating over time (assuming a constant rate) they got local (cumulated) abundances that compare favorably with that of ^6Li , ^9Be and ^{10}B in the solar system.

The beauty of the spallative interpretation is that the abundance hierarchy: ^{11}B, ^{10}B, ^9Be is reflected by the production cross sections. This is a gratifying proof that nature follows nuclear physics! It is worth emphasizing that alpha induced reaction have lower energy threshold than the corresponding proton reactions. This is of importance for the postulated low energy component associated to superbubbles (see below). Despite the success of the MAR approach, some problems remained however, concerning the $^7Li/^6Li$ ratio (the calculated production ratio is 1.2 and the observed one in meteorites is 12.6). Thus other sources of ^7Li were required, namely stars that can produce it through the fusion of ^4He and ^3He. Also the $^{11}B/^{10}B$ ratio was discrepant (2.5 against 4 in meteorites). Thus an ad-hoc hypothesis was proposed: the production of ^{11}B by a low energy flux, not observable in the solar cavity.

2 Production processes and ratios

In the 70', LiBeB observations were limited to the Solar System and closeby stars. Since the 90's they have been continuously pushed down to lower and lower metallicities. The great surprise was to find linear correlations between both Be, B and Fe up to [Fe/H] =-1, whereas a quadratic relation was expected from the GCR theory.

The production rate of species L is a function of target composition and of the projectile characteristics (composition and energy spectrum):

$dN(L)/dt = N_T < \sigma > \phi$

One distinguishes two cases: If p and alphas are the projectiles and interstellar CNO, the targets, one deals with a Secondary Production mechanism.

The fast protons and alphas of the standard GCR react on CNO at rest in the ISM, giving rise to LiBeB nuclei. Then:

$dN(L)/dt = N_{CNO} < \sigma > \phi_{\alpha p}$

Making the reasonable assumption that SNII both produce and eject O (essentially) and accelerate GCR (through the shock waves they produce), one gets:

$\phi(t)\ \alpha\ SN(t)\ \alpha\ dO/dt$. The flux decreases as time increases.

$N_{CNO}(t)\ \alpha\ SN(t)\ \alpha\ O$. The target abundance increases as time increases. Then,

$dBe/dt\ \alpha\ dO/dt.O$ and $Be/H\ \alpha\ (O/H)^2$

Conversely, if C and O are the projectiles, and H and He the targets, the case is different because H and He do not evolve significantly in the course of the galactic lifetime, their abundance being essentially fixed by the big-bang nuclesosynthesis. One deals, in this case with a Primary Production process.

$dN'(L)/dt = N_{H-He} < \sigma > \phi_{CO}$

$\phi(t)\ \alpha\ SN(t)\ \alpha\ dO/dt$. Then,

$dBe/dt\ \alpha\ dO/dt$ and $Be/H\ \alpha\ O/H$

This primary process as been astrophysically associated with galactic superbubbles (SB) which are interstellar cavities produced by the collective effect of SN and WR. In these low density regions fresh products of nucleosynthesis (essentially O in our case) accumulate and are accelerated by shock waves induced by successive supernovae[7], [8]. In the following we will call Superbubble Accelerated Particles (SAP) the fast component related to these objects. Standard GCR and SAP have in common the acceleration agent: type II Supernovae. But, they could differ by i) the source composition, SAP being enriched w.r.t GCR in O and specifically He (by a factor 3), and this has a strong bearing on the ^6Li problem (see below) ii) the energy spectrum : the mean energy of GCR is presumably higher than that of SAP, 1GeV/n typically, against say 100 MeV/n. It is expected that SAP, due to their low energy do not escape easily from the SB regions, and even if they escape, they cannot enter the solar cavity due to the repelling effect of the solar wind. Thus, in this case there is some chance that SAP interacting with the dense shells enclosing superbubbles could produce C and O gamma ray lines and the LiBe feature (a combination of two broad lines around 450 keV). An existence test of this low energy He and O rich component is the gamma ray emission of local OB associations/superbubbles (Cyg OB2, PerOB2, Orion, Vela). This is a good prospect for the european INTEGRAL mission (J. Paul, this meeting). Above all the two mechanisms in question (standard GCR and SAP) differ by the type of production process, secondary in the case of standard GCR, ($BeB\ \alpha\ O^2$) and primary in the case of SAP ($BeB\ \alpha\ O$). Before closing this section on physical production processes of light elements, it is worth mentionning another primary mechanism, neutrino spallation, which takes place within supernovae. It is expected to generate significant amounts of ^{11}B and ^{7}Li but not Be. However, Nadyozhin (this conference) has warned us that the discussion is not closed. Indeed many uncertainties are yet attached to this mechanism.

3 Empirical O-Fe correlations and evolutionary models

The [O/Fe] behaviour at low metallicity is a central issue, since the physical correlation is between BeB and O, the products and the progenitor, whereas the observable one is between BeB and Fe. According to the lines analysed (IR triplet, [OI] or permitted OI), the estimates of the O abundance differ.

Over a long period, work largely based on the forbidden OI doublet observed in giants has established that [O/Fe] levels off below [Fe/H]= -1 at about 0.5 dex . Consequently, the observed proportionality between Be and Fe leads naturally to the proportionality between Be and O. Thus, this implies a clear primary origin of BeB.

Subsequently data were released based on UV OH and IR triplet leading to [O/Fe] = 0.35 [Fe/H][9 10]. The proportionality between Be and Fe leads in this case to the correlation: log(Be/H) proportional to about 1.6 [O/H]. Thus the slope of the correlation is between 1 and 2, a quite ambiguous situation, in which a purely secondary origin linked to GCR is still possible ([11]).

To summarize the situation, three mechanisms operate: i) Standard GCR (secondary) ii) SAP (primary) iii)Neutrino Spallation (primary).

Different models have been proposed to explain the evolution of LiBeB. i) A pure secondary Standard GCR mode based on variable [O/Fe] [11]. This one has to confront the following difficulties: is [O/Fe] really increasing at low metallicities? Many observers would be reluctant to accept this variation. Why O nucleosynthesis should be special compared to other alpha elements? Also the energetics of the BeB production in this case is quite demanding in the early Galaxy (Ramaty this conference).

ii) A pure primary component arising from SB assimilated to GCR. Galactic Cosmic Rays in this view becomes primary [13]. But it seems that their source composition cannot be reconciled with the SAP/SN one, [12]. Moreover this kind of model suffers from an underproduction of ^{6}Li in the halo due to lack of incident alphas.

iii) An hybrid model giving consecutive roles to SAP (acting in the halo) and standard GCR (active in the disk) seems to fulfil the essential observational constraints [2,3].

Accurate oxygen abundances of old, metal poor stars are required if one desires to reach a satisfactory solution of the halo BeB isssue. Indeed, a lot of efforts have been put recently on observations and atmosphere modelisation and a special session of the IAU at Manchester 2000 has been devoted to the O-Fe correlation. The general feeling is that the flat (or slightly increasing) [O/Fe] vs [Fe/H] at decreasing Fe/H is favoured. The need for a primary component in the early Galaxy is thus strenghthened. Moreover, the recent observation of Primas et al (2000) [15] indicating a flatenning of the Be-Fe relation at very low metallicity, reinforces a primary origin of Be through the most massive stars in agreement with the prediction of Vangioni-Flam et al (1998) [2]. Thus a big-bang origin of Be is not required to explain these data.

4 The case of Lithium 6 and 7

Recently much work has been devoted to the non thermal production of Li in the early Galaxy, especially via the reaction $\alpha + \alpha$. An important condition has to be fulfiled: lithium should not be overproduced at low metallicity. In other words, the Spite plateau should not be traversed by the rising Li. Recent observations on ^6Li in halo stars , amounting to $^6Li/^7Li$ about 2 to 510^{-2} (Cayrel and Asplund, this conference) set new constraints on spallative processes in the early Galaxy. Indeed GCR (either standard as proposed by Fields and Olive or of primary type as proposed by Ramaty et al) are not sufficient to explain the observed ^6Li abundances due to their low alpha content contrary to SAP, enriched in alphas [14]. In the SAP context, it is probable that ^6Li is almost intact in the atmosphere of halo stars, meaning that ^7Li, much less fragile, has essentially its big-bang value. Accordingly, Li becomes an interesting baryon density indicator (Coc et al, Olive, this conference).

5 Conclusion

Spallation is confirmed as the production process of ^6Li, Be, B. The present observational trend is that Fe-O is quite flat at low metallicities indicating a that primary production mechanism (SAP) is predominant in halo phase. ^6Li observations strengthen this assumption since they require an alpha-rich fast component in the early Galaxy. In the disk phase, standard GCR take increasing importance. The agent of acceleration common to both SAP and GCR is SNII. The favored sources of low energy nuclei (SAP) are massive stars gathered in OB associations and exploding in superbubbles.

For the future, the needs are the following:

concerning the observational aspect, simultaneous observations O, Fe, Be,

B, ^6Li and ^7Li in halo stars would be the most welcome. In this context VLT missions are very promising. A detection of the O* gammay-ray lines and the LiBe feature by INTEGRAL (to be launched in 2002) would be a clue to the existence of a low energy He-O rich non thermal component. Finally, we would like to draw attention on an intriguing possibility, offering a new nucleosynthetic site for LiBeB, namely hypernovae. Nomoto's velocity and composition profiles of exploding CO cores associated to Nadyozhin's analytic treatement of outer layers, allow to determine the number and energy spectrum of C and O particles ejected and the LiBeB they produce by impacting on the surrounding medium. the O + α reactions are favoured due to their low energy threshold. A simple rescaling of the work of Fields et al (1996) [16] shows a generous production of LiBeB.

MAR children and cousins express their aknowledgements to Jean Audouze for his major contribution to light element nucleosynthesis and evolution.

References

1. E. Vangioni-Flam, A. Coc and M. Cassé, *A&A* **360**, 15 (2000)
2. E. Vangioni-Flam et al, *A&A* **337**, 714 (1998)
3. E. Vangioni-Flam, M. Cassé, B. Fields and K. Olive, *ApJ* **468**, 199 (1996)
4. W. Fowler, J.E. Greenstein and F. Hoyle, *Geoph. J. R.A.S.* **6**, 148 (1962)
5. M. Meneguzzi, J. Audouze and H. Reeves, *A&A* **15**, 337 (1971)
6. E. Vangioni-Flam, M. Cassé and J. Audouze, *Phys. Rep.* **333 - 334**, 365 (2000)
7. A. Bykov, *Space Sci. Rev.* **74**, 397 (1995)
8. E. Parizot, M. Cassé and E. Vangioni-Flam, *A&A* **328**, 107 (1997)
9. G. Israelian, R.F. Garcia-Lopez and R. Rebolo, *ApJ* **507**, 357 (1998)
10. A.M. Boesgaard, J.R. King, C.P. Deliyannis and S.S. Vogt *Astron. Journ.* **117**, 492 (1999)
11. B. Fields and K. Olive, *ApJ* **516**, 797 (1999)
12. J.P. Meyer, LOC Drury and D.C. Ellison, *ApJ* **488**, 730 (1997)
13. R. Ramaty, B. Kozlovsky and R.E. Lingenfelter, *ApJ* **456**, 525 (1996)
14. E. Vangioni-Flam et al, *New Astron.* **4**, 245 (1999)
15. F. Primas et al, *A&A* in press, astro-ph/0009482 (2000)
16. B. Fields, M. Cassé, E. Vangioni-Flam and K. Nomoto, *ApJ* **462**, 276 (1996)

THE OBSERVED TREND OF BORON AND OXYGEN IN DISK STARS

KATIA CUNHA

Observatório Nacional, Rua General José Cristino 77, São Cristóvão, RJ, Brazil, 20921-400; katia@on.br

VERNE V. SMITH

Department of Physics, University of Texas at El Paso, El Paso, TX 79968; verne@barium.physics.utep.edu

Boron abundances have recently been derived for a sample of roughly solar metallicity stars from Hubble Space Telescope UV spectra. Using the weak oxygen lines as the primary abundance indicators, oxygen abundances were derived for a sub-sample of these stars and a slope of 1.5 $\pm$0.2 is determined for the relation of A(B) versus A(O) in the 'metallicity' range of A(O)=8.3-9.0. Literature data on boron and oxygen abundances for halo stars agree quite well, in the region of overlapping abundances, with the abundances presented here. When taken together, the 'metal-poor' and 'metal-rich' samples exhibit similar B-O slopes: the metal-poor sample, by themselves, yields a slope of 1.3 $\pm$0.1 and, if combined, the overall slope in the sample is 1.4 $\pm$0.1. The boron-oxygen relation seems to be intermediate between pure-primary and pure-secondary production.

1 Introduction

It was only in the 1990's that boron abundances could be systematically derived by means of UV spectra observed with the Hubble Space Telecope. The derived B abundances for a sample of metal-poor stars indicated the surprising result that boron behaved as a primary element relative to iron (Duncan, Lambert & Lemke 1992; Duncan et al. 1997). Such behavior of boron with iron was not expected as the theoretical predictions for boron production in the early Galaxy indicated a secondary behavior of boron with metallicity (Meneguzzi, Audouze & Reeves 1971). More recently, boron abundances were measured in stars with roughly solar metallicity ([Fe/H]$\geq$-1.0; Cunha et al. 2000) and the observed behavior of boron versus [Fe/H] could be defined in more metal-rich stars. These results indicated that the behavior of boron versus iron remained roughly unchanged from the halo to the disk of the Galaxy.

However, oxygen and not iron is more directly related to boron production in the Galaxy: boron is probably produced by the ν-process and it should scale with oxygen, as both the ν-process and O-production result from SN II. Spallation reactions (that are responsible for the main production of B) result, to a large degree from O, either as energetic protons spallating ISM oxygen, or accelerated oxygen being spalled from ISM H and He. Oxygen, however, is more difficult to observe and analyse than iron. The choice of available lines is more limited and reliability of these lines (and consequently the derived O abundances) remains a topic of active research and controversy. Here, we discuss O abundance results that were obtained for a sub-sample of stars with disk metallicity in order to investigate the observed trend of B and O in the disk.

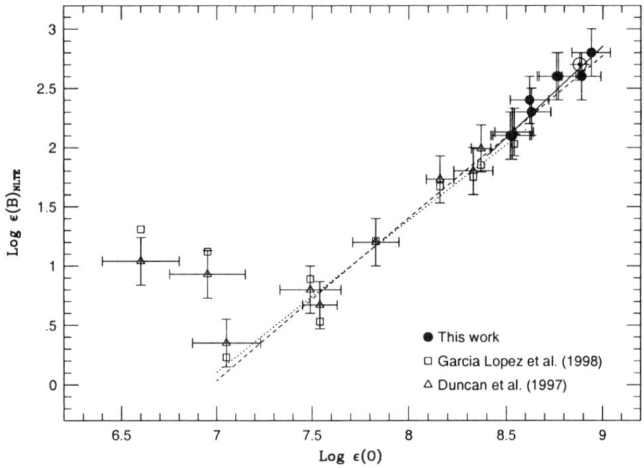

Figure 1. Results from boron versus oxygen for disk-metallicity stars compared to literature results for more metal-poor stars from Duncan et al. (1997) and García López et al. (1998). The solid line through the metal-rich sample of stars is the slope 1.54± 0.17 relation. The short-dashed line is a 'linear' fit (on this log-log plot) to the Duncan et al. (1997) and García López et al. (1998) results, which has a slope of 1.28±0.07 (and a good correlation coefficient of 0.985). We note that the two most metal-poor stars are considered to have boron upper limits and were not included in the fit. There is no clear evidence of any significant change of slope in logB versus logO over two orders of magnitude in oxygen abundance. The overlap region of the metal-rich and the metal-poor studies, near A(O)∼ 8.4-8.6 is excellent. The short-dashed line is a linear fit to all the data, metal-rich and metal-poor, which has a slope of 1.37±0.04 (and a correlation coefficient of 0.989).

2 Boron and Oxygen Abundances and Trends

Oxygen abundances have been derived using exclusively weak O I features: a combination of the [O I] λ6300Å line, and the permitted excited lines at λ6157Å and 6158Å. The weak O I lines, by all accounts, seem to provide reliable oxygen abundances, even in an LTE analysis, so the resulting O abundances presented here, when combined with the non-LTE boron abundances from Cunha et al. (2000), yield an accurately defined relation between B and O at the metal-rich end of the Galactic metallicity distribution. The boron and oxygen abundances are represented by filled circles in Figure 1. The derived slope between A(B) and A(O) is 1.54± 0.1. Given the accuracy of the derived abundances, this slope should best be characterized by 1.5± 0.2.

When the results for the metal-rich stars are compared to those published previously for the metal-poor halo dwarfs by Duncan et al. (1997) and García López et al. (1998), the comparison is good. The metal-poor data span A(O) ~ 7.0-8.6 while the disk results go from A(O) ~ 8.4-9.0. Plotted together, they agree quite well in the overlap region. In addition, linear fits (in log-log space) to the Duncan et al. (1997) and García López et al. (1998) abundances yield a B-O slope of 1.3± 0.1, while the combined data sets provide a slope of 1.37± 0.04. The different ways of treating the various abundance sets yield consistent slopes of 1.3-1.5; there may be a slight change of slope at A(O) ~ 8.0-8.4 from 1.3 to 1.5, but this possibility is not significant.

References

1. Cunha, K., & Smith, V.V., Boesgaard, A.M., & Lambert, D.L. 2000, ApJ, 530, 9 39
2. Duncan, D.K., Lambert, D.L. & Lemke, M. 1992, ApJ, 401, 584
3. Duncan D.K., Primas, F., Rebull, L.M., Boesgaard, A.M., Deliyannis, C.P, Hobbs , L.M., King, J.R., & Ryan, S.G. 1997, ApJ, 488, 338
4. García López, R.J., Lambert, D.L., Edvardsson, B., Gustafsson, B., Kiselman, D., & Rebolo, R. 1998, ApJ., 500, 241
5. Meneguzzi, M., Audouze, J., & Reeves, H. 1971, A&A, 15, 337

OBSERVING BERYLLIUM WITH UVES AT THE VLT

FRANCESCA PRIMAS
European Southern Observatory, Garching, Germany
E-mail: fprimas@eso.org

MARTIN ASPLUND
Uppsala Astronomical Observatory, Uppsala, Sweden
E-mail: martin@astro.uu.se

POUL ERIK NISSEN
Institute of Physics and Astronomy, Aarhus University, Denmark
E-mail: pen@ifa.au.dk

We present preliminary results of a new analysis of Be abundances in a large sample of stars, spanning the metallicity range $-2.4 \leq$[Fe/H]≤ -0.7. The observations were taken with the Ultraviolet and Visible Echelle Spectrograph (UVES) at the ESO Very Large Telescope (VLT), which made it possible to obtain very high data quality and measurements accuracy. The aim of the project is to further investigate the trend of Be with metallicity, and the possible presence of dispersion. Our recent detections of very high Be abundances in two of the most metal-poor stars suggest either a Be plateau or significant scatter at the lowest metallicities.

1 Introduction

Among the light elements, beryllium has the advantage of being a pure product of cosmic-ray (CR) spallation nucleosynthesis. It was suggested to originate primarily from the bombardment of C and O by protons and α-particles (Reeves et al[13]; Meneguzzi et al[6]). From this scenario, a slope of two in the logarithmic plane Be vs. O is then expected. If [O/Fe] in metal-poor halo stars is constant (where [A/H] = log(A/H)$_*$ / log(A/H)$_\odot$), then this same slope holds vs. Fe as well. However, recent studies of Be in halo stars (e.g. Molaro et al[7], Boesgaard et al[3]) have detected a linear trend of [Be/H] vs. [Fe/H] with a slope of about one, which cannot be easily reproduced by the above mentioned spallation reactions. Type II supernovae (SN) have been proposed as possible accelerators of freshly synthesized C and O nuclei which subsequently fragment into Be and B (Vangioni-Flam et al[14]).

Although a complete and successful scenario has not yet been proposed, it is clear that oxygen is a key element, being directly involved in the production of Be. But the uncertainties currently plaguing its measurement in stars (different results from different abundance indicators, cf Lambert 2001[5]) also affect the interpretation of the entire Be evolution in the Galaxy. A combined effort aimed at solving the oxygen question and gathering a high quality sample of Be abundances is needed. Here, we report on the latter one.

2 Observations and Analysis

Measuring beryllium in stars, especially in the most metal-poor ones, is challenging. The resonant doublet normally used for this purpose falls very close to the atmospheric cut-off (3130Å), and in a highly populated spectral region. Very efficient UV detectors and high resolution are thus needed. Our targets were observed with UVES at the VLT in dichroic mode. The large spectral range thus covered allow us to determine the abundance of several other key elements, like Fe, O, and Li (e.g. Nissen et al[9]; Asplund et al[2]). All the spectra are characterized by a resolving power of R$\sim$ 45,000 (in the blue), and S/N $\geq$ 100 at the Be doublet.

The Be abundances were measured via spectrum synthesis. After having determined the basic stellar parameters ($T_{\rm eff}$ from the $(b\text{-}y)_0$ calibration of Alonso et al[1], log g from the Hipparcos parallaxes (e.g. Nissen et al[8]), and [Fe/H] from Fe II lines for most of the stars), synthetic spectra were computed with the Kurucz suite of codes and LTE model atmospheres (without approximate overshooting). All the Be abundances were then corrected for NLTE effects, based on the recent computations by Garcìa Pèrez et al[4].

3 Preliminary Results

Our results are summarized in Figure 1 (filled circles and triangles - the latter representing the two most metal-poor stars observed so far with UVES at the VLT for which a Be detection has been claimed, cf Primas et al[11,12]), together with the abundances derived from Keck spectra of similar quality (Boesgaard et al[3]; open circles). The filled square represents the Sun.

Although preliminary, this new set of data immediately draws our attention on two main issues: the presence of dispersion at given metallicity, and the apparently high Be abundances measured in two of the most metal-poor stars. This could imply that in the early days of the Galaxy, when the interstellar medium (ISM) was low in the abundance of CNO target nuclei, light elements still were formed with high efficiency. Otherwise, the light elements may have been formed with different efficiencies throughout the entire galactic history depending on e.g. if the star formed within an OB association or in the vicinity of an isolated SN, as recently proposed by Parizot and Drury[10]. These are indeed important issues that need further investigation: they will

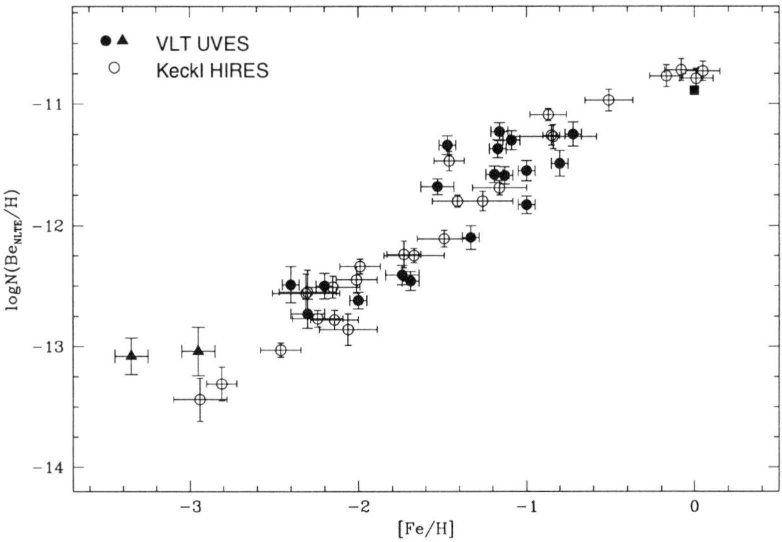

Figure 1. Be as a function of metallicity for the highest quality data currently available. See text for symbols explanation.

be fully addressed in the final analysis of this new set of data.

References

1. A. Alonso *et al*, A&AS **117**, 227 (1996).
2. M. Asplund *et al*, *this volume.*
3. A.M. Boesgaard *et al*, AJ **117**, 1549 (1999).
4. A.E. Garcìa Pèrez *et al*, *this volume.*
5. D.L. Lambert, in *Proc. IAU JD #8*, ed. Elsevier, in press.
6. M. Meneguzzi *et al*, A&A **15**, 337 (1971).
7. P. Molaro *et al*, A&A **319**, 593 (1997).
8. P.E. Nissen *et al*, in *Proc. ESA Hipparcos Symp.*, ESA SP-402, 1997.
9. P.E. Nissen *et al*, in *Proc. IAU JD #8*, ed. Elsevier, in press.
10. E. Parizot and L. Drury, A&A **356L**, 66 (2000).
11. F. Primas *et al*, A&A **362**, 666 (2000a).
12. F. Primas *et al*, A&A **364L**, 42 (2000b).
13. H. Reeves *et al*, Nature **226**, 727 (1970).
14. E. Vangioni-Flam *et al*, ApJ **337**, 714 (1998).

Evolution of Light Elements in an Inhomogeneous Galactic Halo

Takeru K. Suzuki
Theoretical Astrophysics Division, National Astronomical Observatory, Mitaka, Tokyo, 181-8588 Japan; stakeru@th.nao.ac.jp;
Department of Astronomy, School of Science, University of Tokyo (UT)

Yuzuru Yoshii
Institute of Astronomy, School of Science, UT, Mitaka, Tokyo, 181-8588 Japan;
Research Center for the Early Universe, Faculty of Science, UT

We present the two interesting results of evolution of light elements based on a model of SN-induced chemical evolution: (1)Light elements as cosmic clocks (2)Issues on energetics of cosmic rays.

1 Introduction and Our Model of SN-induced Chemical Evolution

Recent analysis [1] and observations [2 3] uncovered evidence that the chemical evolution of the early Galactic halo was quite inhomogeneous. A model fully incorporating this inhomogeneity have been constructed by our group for investigation of both heavy elements [4] and light elements (LiBeB) [5 6]. An essential point of this model is that all the stars of the subsequent generations are born from individual supernova (SN) remnant shells of the previous generations. Therefore, we can precisely take into account inhomogeneous nature of the non-well mixed gas arising from the stochastic process of star formation induced by the SN explosions at early epochs. The light elements are produced by nuclear reactions of collisions between cosmic rays (CRs) and the ambient atoms. The CRs are assumed to originate from the SN explosions. The results [4 5 6 7] derived from the model explain the observed large scatters [2 3] in metal-poor stars very well. In this contribution, we would like to focus on evolution of light elements based on this model.

2 Light Elements as Cosmic Clocks

Fig.1 shows distribution of stellar abundance of each element as a function of time elapsed from the formation of the first star. Due to the inhomogeneous nature of the gas making up the early Galactic halo, the correlation between Fe and the time is considerably poor at early epochs. In other words, the age-metallicity relation is completely broken down in the very early phase [4]. However, stellar abundances of the light elements produced by the nuclear reactions of cosmic rays are better-correlated with the cosmic time than that

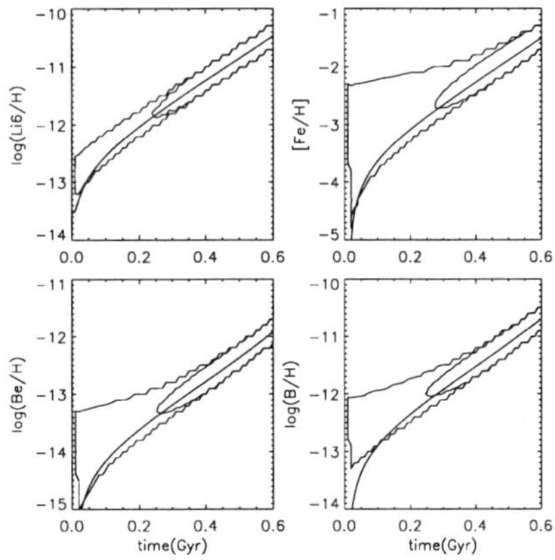

Figure 1: Predicted frequency distribution of long-lived stars in the t-$\log(X_i/\mathrm{H})$ planes as a function of time. Two contour lines, from the inside to the outside, correspond to those of constant probability density 10^{-3}, and 10^{-5} in unit area of $\Delta t = 10(\mathrm{Myr}) \times \Delta\log(X_i/\mathrm{H})=0.1$.

of Fe, because the transported speed of the cosmic rays is much faster than the speed of hydrodynamical mixing of the gas so that the light elements are expected to distribute more uniformly in the Galactic halo than the heavy elements. Among the three light elements, the best correlation is achieved for ^{6}Li, because ^{6}Li is mainly produced by the fusion of α-particles, which are the BBN products, and are distributed globally throughout the entire halo.

3 Cosmic Ray Energetics and Acceleration of Cosmic Rays

3.1 Energetics

In order to explain the present observation of the light elements, two distinct features should be accomplished for the past activities of CRs if the SN explosion is assumed to be the only accelerator: (1)Heavy CRs (e.g. CNO) should be effectively accelerated [8]. (2)One SN has to give more energy to accelerate CRs than today (See fig.2 in Suzuki & Yoshii [5]). So as to achieve these conditions superbubble was introduced by more than one group [9 10].

3.2 Acceleration of Cosmic Rays by Individual SNe

In the very early epochs individual SNe are generally believed to give less contribution to acceleration of heavy CRs than superbubbles, since the forward shock cannot easily accelerate metal-rich SN-ejecta. However, we would like

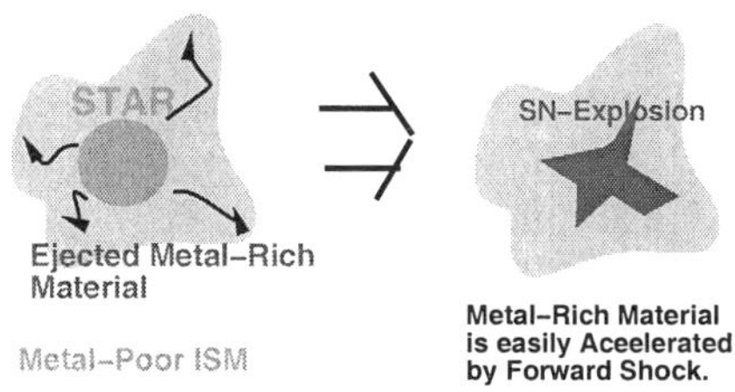

Figure 2: Mass loss before the SN explosion makes it possible that individual SN accelerates heavy CRs very efficiently!!!

to point out that individual SNe can give a significant contribution to heavy CRs because metal-rich circumstellar envelopes formed due to mass loss of the progenitor star *before* the SN explosion can be quite effectively accelerated by the forward shock[5] (See Fig. 2).

3.3 Acceleration by Large Scale Shocks

Shocks with large scale are expected to exist at early epochs originating from gravitational collapse of objects through the formation of structure. These shocks would accelerate CRs besides those due to SN explosions (both individual SNe and superbubbles) to compensate the energetic deficiency of CRs.

References

1. Audouze, J. & Silk, J. 1995, ApJ, 451, L49
2. McWilliam,A., Preston,W., Sneden,C., & Searle,L. 1995, AJ, 109, 2757
3. Ryan, S. G., Norris, J. E., & Beers, T. C. 1996, ApJ, 471, 254
4. Tsujimoto, T., Shigeyama, T., & Yoshii, Y. 1999, ApJ, 519, L63
5. Suzuki, T. K. & Yoshii, Y. 2001, ApJ, in press (vol.548)
6. Suzuki, T. K., Yoshii, Y., & Kajino, T. 1999, ApJ, 522, L125
7. Tsujimoto, T., Shigeyama, T., & Yoshii, Y. 1999, ApJ, 531, L33
8. Ramaty, R., Scully, S. T., Lingenfelter, R. E., & Kozlovsky, B. 2000, ApJ, 534, 747
9. Higdon, J. C., Lingenfelter, R. E., & Ramaty, R. 1998 ApJ, 509, L33
10. Parizot, E. & Drury, L. 1999, A&A, 349, 673

NEUTRINO NUCLEOSYNTHESIS (modern status)

D. K. NADYOZHIN
Institute of Theoretical and Experimental Physics
25 B.Cheremushkinskaya Street, Moscow 117259, Russia
E-mail: nadezhin@vxitep.itep.ru

The most important issues of the neutrino nucleosynthesis are the contribution to the cosmic abundances of some specific isotopes and the traces left by the neutrino-nuclear interaction in matter surrounding the stellar core collapsed into a neutron star or a black hole. The first issue deals with the neutrino-induced production of such isotopes as ^{7}Li, ^{11}B, ^{19}F, ^{49}V, p-nuclei, and some others. The second issue is of especial importance as a tool to detect the so-called silent collapses which are not accompanied by the violent supernova-like outburst of stellar envelope. Modern status of the two issues is briefly discussed with special emphasis placed on the possibility to observe the "fingerprints" of the neutrino flux passed through the material around collapsed objects.

1 General Scheme

The neutrino flux generated by collapsing stellar core penetrates through the onion-like presupernova structure (Fig. 1) to produce various nuclear transmutations in matter to be expelled by the supernova shock wave. Figure 2 shows the typical normalized neutrino bolometric luminosity $l_\nu(t)$ (the sum of all the neutrino-antineutrino flavors) and a fraction of the total energy $\mathcal{E}_{\nu\bar{\nu}}$, radiated in form of neutrino by the time t, $E_{\nu\,\mathrm{tot}}(t)$ $[E_{\nu\,\mathrm{tot}}(\infty) = 1]$. To obtain the actual neutrino luminosity and radiated energy at given instant t, one has to multiply $l_\nu(t)$ and $E_{\nu\,\mathrm{tot}}(t)$ by $\mathcal{E}_{\nu\bar{\nu}} = (3-5)\cdot 10^{53}\,\mathrm{erg}$ for the mass of collapsed stellar core limited within $(1.4-2)\,\mathrm{M}_\odot$. The neutrino-induced transformations in stellar matter begin with the following reactions

$$
\begin{aligned}
&\nu_e + (A,Z) \Longrightarrow (A,Z+1)^* + e^-\,, \\
&\tilde{\nu}_e + (A,Z) \Longrightarrow (A,Z-1)^* + e^+\,, \\
&\nu_i + (A,Z) \Longrightarrow (A,Z)^* + \nu_i'\,, \qquad\qquad (\nu_i = \nu_e, \widetilde{\nu}_e, \nu_\mu, \widetilde{\nu}_\mu, \nu_\tau, \widetilde{\nu}_\tau)\,, \\
&(A,Z_i)^* \Longrightarrow \begin{cases} (A-1, Z_i) & +\ n \\ (A-1, Z_i-1) & +\ p \\ (A-4, Z_i-2) & +\ \alpha \end{cases} \qquad (Z_i = Z,\ Z+1,\ Z-1)\,.
\end{aligned}
$$

which give rise to a long chain of thermonuclear reactions between the products of the above key reactions and background nuclei. During already more than 20 years[2,3,4,5], the neutrino nucleosynthesis has been studied in the framework of such a scheme.

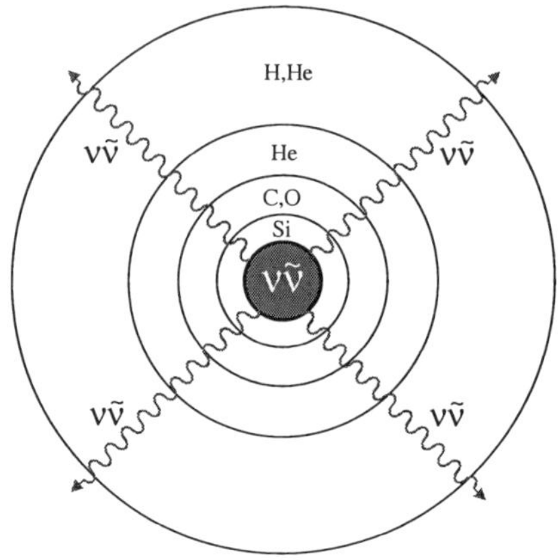

Figure 1: The onion-like structure of a presupernova

It was shown that the neutrino yields depend on the physical conditions in the presupernova envelope (temperature, density and radii of the "onion" shells prior to the explosion) and on the properties of the supernova shock wave. The dependence on the neutrino luminosity parameters seems to be less pronounced as long as the thermal nature, usually assumed for the bulk of the neutrino flux, remains true.

2 Main Achievements

The main results obtained so far can be listed as follows

- The neutrino contribution to the cosmic abundances of ^{7}Li and ^{11}B gives an elegant explanation of the observed high cosmic ratios ^{7}Li/^{6}Li and ^{11}B/^{10}B if one assumes that the cosmic ray spallation process produces these isotopes in comparable quantities while the differences (^{7}Li – ^{6}Li) and (^{11}B – ^{10}B) are mainly of the neutrino origin.

- The neutrino nucleosynthesis can contribute to the production of various rare intermediate mass isotopes such as ^{15}N, ^{19}F, ^{26}Al, ^{31}P, ^{35}Cl, ^{39,40,41}K, ^{50}V, and some others.

- The neutrino process affords an opportunity to explain the whole set of p-nuclei (from ^{74}Se right up to ^{196}Hg including ^{138}La and ^{180}Ta) within one and the same environment, rather than to play with different physical conditions for different small groups of these nuclei.

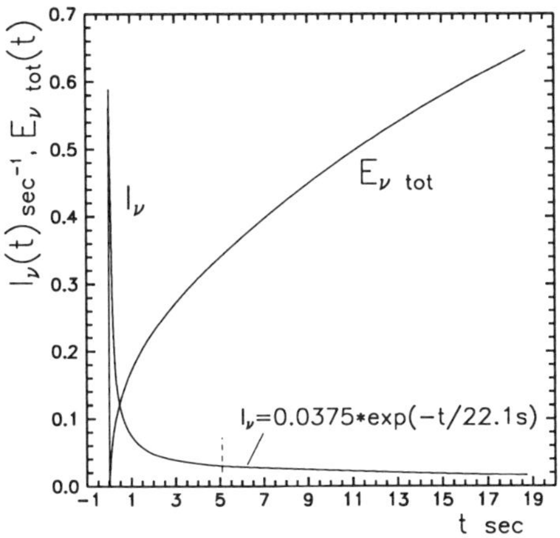

Figure 2: The normalized neutrino-antineutrino light curve [1]

- The neutrino breakup of ^{4}He in the helium shell of the metal-deficient stars seems to be a good neutron source for the predicted[6] weak r-process component to explain the origin of the r-process nuclei for the mass numbers $A \lesssim 130 - 140$ [5,7].

3 Neutrino Diagnostics of Collapses

Since the neutrino-nucleosynthesis yields are sensitive to the supernova structure and hydrodynamics, they can serve as a useful tool for diagnostics of the gravitational collapses, especially the "silent" ones. Even in those cases when the neutrino yields do not contribute appreciably to the cosmic abundances they represent a specific signature left by a collapsed object in the surrounding medium. It would be a challenging project to try to discover the neutrino "fingerprints" in debris of supernova 1987 A, in the filaments of the Crab Nebula and other young supernova remnants — e.g., Cas A, G11.2-0.3 (historical supernova 386 AD). Of special interest would be to explore spectroscopically the material nearby a lonely neutron star discovered by ROSAT (RX J1856.5-3754) and observed optically by the HST in 1997. This neutron star might be the product of a silent collapse. It should be also emphasized that any boundaries imposed on the neutrino-nucleosynthesis yields [8] can be, in principle, transformed into the constraints on the theoretical supernova models.

Acknowledgments

It is a pleasure to thank the Organizing Committee of this Conference for the financial support and hospitality. The financial supports of Russian Foundation for Basic Research (Grant 00-02-17230) and Swiss National Science Foundation are also gratefully acknowledged.

References

1. D.K. Nadyozhin *Ap. Space Sci.* **53**, 131 (1978).
2. G.V. Domogatsky and D.K.Nadyozhin *MNRAS* **178**, 33P (1977).
3. G.V. Domogatsky, Eramzhyan R.A., and Nadyozhin D.K. *Ap. Space Sci.* **58**, 273 (1978).
4. S.E. Woosley et al. *ApJ* **356**, 272 (1990).
5. D.K. Nadyozhin, I.V. Panov, and S.I. Blinnikov *A&A* **335**, 207 (1998).
6. G.J. Wasserburg, M. Busso, and R. Gallino *ApJ* **466**, L109 (1996).
7. D.K. Nadyozhin and I.V. Panov in *Proceedings NIC-2000*, Aarhus, Denmark. *Nuc. Phys.* A, in press.
8. E. Vangioni-Flam et al. *ApJ* **468**, 199 (1996).

GALACTIC CHEMICAL EVOLUTION AND THE ABUNDANCES OF LITHIUM, BERYLLIUM AND BORON

ANDREU ALIBÉS, JAVIER LABAY AND RAMON CANAL

Departament d'Astronomia i Meteorologia, Universitat de Barcelona, Martí i Franquès 1, 08028 Barcelona, Spain. E-mail: aalibes, javier, ramon@am.ub.es

A LiBeB evolution model including Galactic Cosmic Ray nucleosynthesis, the ν-process, novae, AGB and C-stars is presented.

We have included Galactic Cosmic Ray Nucleosynthesis (GCRN) in a complete Chemical Evolution Model that takes into account 76 stable isotopes from hydrogen to zinc. Any successful LiBeB evolution model should also be compatible with other observational constraints like the age-metallicity relation, the G-dwarf distribution or the evolution of other elements. At the same time, we have checked how different would be a model that took into account the last observations by Wakker et al. (1999) of metal-enriched clouds falling onto the disk, from a primordial infall model.

We have integrated the standard evolution equations, using a SFR proportional to $\sigma^m \sigma_g^n$ and a double exponential infall (Chiappini et al. 1997) with $\tau_h = 1$ Gyr, $\tau_d = 7$ Gyr and $t_{max} = 1$ Gyr. We have used two kinds of infall material: in the primordial model (PM), the galaxy acretes only primordial material during all its lifetime (13 Gyr), while the enriched model (EM) considers a first Gyr of primordial infall, followed by 12 Gyr of enriched infall (0.1 $Z_{\odot}$, indicated by recent observations of clouds falling onto the disk by Wakker et al. (1999)).

We use the Kroupa et al. (1993) IMF and different prescriptions for the nucleosynthesis in the various mass ranges: low and intermediate-mass stars (van den Hoek and Groenewegen, 1997), type II SNe (Woosley and Weaver, 1995) and type Ia SNe (model W7 of Nomoto et al., 1997). Finally, novae have also been included, by means of the José and Hernanz (1998) yields.

The main source of ^{6}Li, ^{9}Be and ^{10}B is the **GCRN**. In the superbubble (SB) scenario, we have calculated the production rate by this mechanism, taking a GCR composition that comes from inside the SB, where newly synthesized material ejected by a SN is accelerated by the shock wave of other SN and mixed with the ISM at that epoch. The energy spectrum of the GCR is: $q(E) \propto \frac{p^{-2.2}}{\beta} e^{-\frac{E}{E_0}}$, with $E_0 = 10$ GeV/n. For **the ν-process** we have fixed the contribution of WW95 yields (about 25%) by means of $(^{11}\mathrm{B}/^{10}\mathrm{B})_{\odot}$. **AGB**

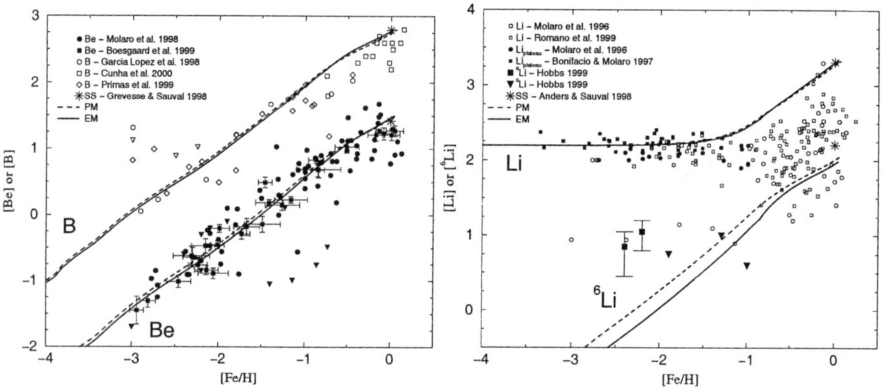

stars and C-stars are also producers of ^{7}Li. We use the time-dependent production rate suggested by Abia et al. (1993) and a current production rate of $1.5{\cdot}10^{-8}$ $M_\odot pc^{-2} Gyr^{-1}$. Finally, we have taken an average yield per **nova outburst** of $1.03{\cdot}10^{-10}$ $M_\odot$, considering that 30% of the outbursts come from ONe WD and 70% from CO ones.

Our two-infall model, both primordial and enriched, is able to reproduce the main solar neighborhood data; in particular, the G-dwarf distribution, which is the main constraint, and the evolution of the majority of chemical elements, specially the CNO ones. Together with the superbubble scenario for the acceleration of GCR is also able to reproduce the LiBeB evolution, specially the linear relationships of Be vs Fe and B vs Fe, when taking a GCR composition as a mixture of ejecta of SNe (20%) and ISM material (80%). All the ^{7}Li sources considered are necessary to reproduce the Li evolution.

References

1. Abia, C., Isern, J., Canal, R. 1993 A&A **275**, 96
2. Chiappini, C., Matteucci, F., & Gratton, R. 1997 ApJ **477**, 765
3. José, J., & Hernanz, M. 1998 ApJ **494**, 680
4. Kroupa, P., Tout, C., & Gilmore, G. 1993 MNRAS **262**, 545
5. Nomoto, et al. 1997 Nucl. Phys. A **621**, 467c
6. van den Hoek, L.B., & Groenewegen, M.A.T. 1997 A&ASS **123**, 305
7. Wakker, B.P. et al. 1999 Nature **402**, 338
8. Woosley, S.E., & Weaver, T.A. 1995 ApJS **101**, 181

DEPARTURES FROM LTE IN Be LINE FORMATION

A. E. GARCÍA PÉREZ, M. ASPLUND
Uppsala Astronomiska Observatorium, Box 515, SE-751 20 Uppsala, Sweden
E-mail: aegp@astro.uu.se, martin@astro.uu.se

D. KISELMAN
Kungliga Vetenskapsakademien, Stockholms Observatorium, SE-133 36 Saltsjöbaden, Sweden
E-mail: dan@astro.su.se

Stellar abundances of Be can be affected by NLTE effects. We have calculated the NLTE corrections to the LTE abundances obtained from the Be II doublet at 313.0 and 313.1 nm for a range of stellar parameters. These lines are not formed in LTE due to over-ionization and over-excitation but the NLTE corrections are in general small (~ 0.1 dex).

1 Introduction

Stellar abundances of Be give important clues to the Galactic chemical evolution, cosmic ray physics and perhaps even Big Bang Nucleosynthesis. Furthermore, due to its fragile nature, Be is a sensitive probe of mixing events during the lifetime of stars. Although much work has recently been devoted to determine Be abundances in metal-poor halo stars, no consensus has yet been reached whether Be is of primary or secondary origin, partly due to the still uncertain O abundances at low metallicities (Asplund & García Pérez [1]).

2 NLTE calculations

In order to obtain accurate Be abundances, allowance must be made for departures from LTE (García López et al.[4]). Here we present such calculations based on the Be II doublet lines at 313.0 and 313.1 nm, for which a new extended Be atom consisting of 71 levels and 243 bound-bound radiative transitions has been constructed. At this stage, no collisional excitation by hydrogen atoms have been included and the photo-ionization rates have been computed with UV radiation fields provided by 1D MARCS model atmospheres (Asplund et al. [2]) for the grid of models under study ($T_{\rm eff} = 5000, 5500, 6000, 6500$ K; $\log g = 3.5, 4.0, 4.5$ [cgs]; [Fe/H] $= 0, -1, -2, -3$). The observed Be abundances of Boesgaard *et al.*[3] have been adopted as input abundances.

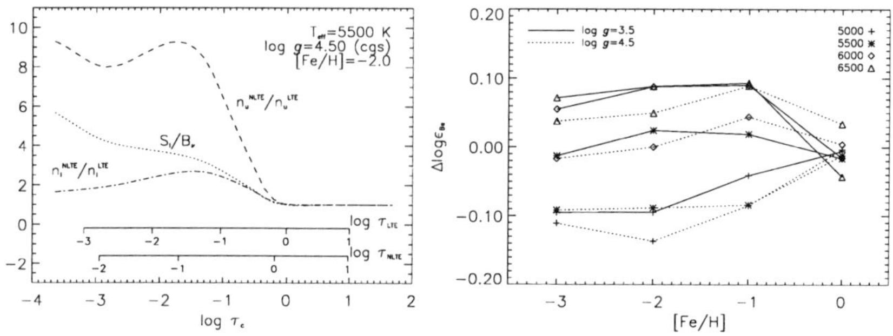

Figure 1: *Left panel*: Departures from LTE for the populations of the upper (n_u) and the lower (n_l) level and for the line source function S_l. *Right panel*: The mean NLTE Be abundance corrections for the Be II 313.0 and 313.1 nm lines for four different T_{eff} and two different $\log g$.

3 Results

It is clear that the Be II resonance lines are <u>not</u> formed in LTE (Kiselman & Carlsson [5]). However, the two main NLTE effects – over-excitation and over-ionization – partly balance each other (Fig. 1). Over-ionization increases the line opacity and thus makes the lines stronger while over-excitation makes it weaker because the line source function S_l exceeds its LTE Planck-value B_ν.

The NLTE abundance corrections are small but significant (Fig. 1). At solar metallicities, the abundance corrections practically vanish, while for metal-poor stars the departures from LTE can be either positive or negative depending on which of the two NLTE effects dominate for the different stellar parameters. For higher T_{eff}, over-excitation dominates while for higher $\log g$ the balance is shifted towards over-ionization. The largest NLTE corrections are found for the coolest, most metal-poor models, $\simeq -0.15$ dex. Typically the NLTE corrections are on the order of 0.1 dex.

References

1. Asplund M., García Pérez A.E., (2000), submitted to *A&A*
2. Asplund M., Gustafsson B., Kiselman D., *et al*, *A&A* **318**, 521 (1997)
3. Boesgaard A.M., Deliyannis C.P., King J.R., *et al.*, *AJ* **117**, 1549 (1997)
4. García López R.J., Severino G., Gomez M.T., *A&A* **297**, 787 (1995)
5. Kiselman D., Carlsson M., in *The Light Element Abundances*, ESO, ed. P. Crane, Springer, p.372 (1994)

ISSUES OF LIBEB, OXYGEN AND FE EVOLUTION

REUVEN RAMATY
NASA/GSFC, Greenbelt, MD, USA, ramaty@gsfc.nasa.gov

RICHARD E. LINGENFELTER
University of California, San Diego, USA, RICHL@cass01.ucsd.edu

BENZION KOZLOVSKY
Tel Aviv University, Israel, benz@wise1.tau.ac.il

We discuss the highlights of our recent research, specifically the refractory vs. volatile interpretation of the rise of [O/Fe] with decreasing [Fe/H], and the issue of primary vs. secondary evolution of Be.

1 Introduction

The evolution of the light elements, Li, Be, B (LiBeB), is an important ingredient in studies of cosmic evolution. Closely related are the evolutions of O and Fe, because [LiBeB/H] are measured as functions of either [Fe/H] or [O/H]. In several recent papers we treated in detail LiBeB production[1] by cosmic rays, and LiBeB, O and Fe evolution[2,3]. We emphasized a model in which the cosmic ray source composition is independent of [Fe/H], based on cosmic ray acceleration in metal enriched superbubbles[4,5]. In the evolutionary calculations, we employed a Monte Carlo approach in which supernova explosions are treated as discrete events in time. This approach allowed us to introduce delays between the time of the explosion and the deposition time of the nucleosynthetic products into star formation regions. For the LiBeB the delays are due to the transport of the cosmic rays. For refractory supernova synthesized elements, e.g. Fe, the delay is due to the slowing down time of fast moving supernova grains, provided that the bulk of the ejecta are in such grains, suggested by observation of the width of the 1.809 MeV gamma ray line from ^{26}Al decay. Below we briefly summarize our most important results.

2 Discussion

We showed that the rise of [O/Fe] with decreasing [Fe/H], recently confirmed[6], could be due to the delayed deposition of refractory Fe relative to the mostly volatile O. This model has the advantage over several other models, e.g.[7], because it provides a simple explanation for the observation that the abundance ratios of the refractory α-nuclei Mg, Si, Ca and Ti relative to Fe do not increase with decreasing [Fe/H] below [Fe/H] $= -1$. The reason being that the bulk of these refractories should also be in grains and thus behave like Fe. On the other hand, the model predicts[3] that volatile S should behave like O, a trend which appears to be confirmed[8].

We showed[2] that secondary Be evolution, due to production by cosmic rays accelerated out of the average metal poor ISM in the early Galaxy, is energetically untenable, the increase of the O abundance at low [Fe/H] notwithstanding. On the other hand, by employing a cosmic ray origin model[5], in which the cosmic rays are accelerated in superbubbles out of a suprathermal population generated from the sputtering of, and scattering by fast supernova grains, we demonstrated[2,3] that primary Be evolution, entirely consistent with cosmic ray energetics, can well account for the observations, e.g.[9] The fact that the slope of log(Be/H) was found[9] to be steeper vs. [O/H] than vs. [Fe/H] (1.45 ± 0.04 and 0.96 ± 0.04, respectively) appears to suggest a combination of primary and secondary evolutions. However, we have shown[3] that if the delay of Be deposition due to cosmic ray transport is taken into account, the observed trend vs. [O/H] is consistent with primary evolution. The often mentioned notion that the trend of Be vs. O is more relevant than vs. Fe is a misconception, because at low [Fe/H] Be is produced by cosmic ray O which is not accelerated out of the average ISM.

References

1. R. Ramaty, B. Kozlovsky, R.E. Lingenfelter and H. Reeves, *Astrophys. J.* **488**, 730 (1997).
2. R. Ramaty, S. Scully, R.E. Lingenfelter and B. Kozlovsky, *Astrophys. J.* **534**, 747 (2000).
3. R. Ramaty, R.E. Lingenfelter and B. Kozlovsky, in *The Light Elements and their Evolution*, (IAU Symp. 198., Astron. Soc. Pacific, 2000), ed. L. de Silva et al., p. 51, astro-ph/0001466.
4. H.C. Higdon, R.E. Lingenfelter and R. Ramaty *Astrophys. J.* **500**, L33 (1998).
5. R.E Lingenfelter, J.C. Higdon and R. Ramaty, in *Acceleration and Transport of Energetic Particles Observed in the Heliosphere*, (AIP Conf. Proc. 528, 2000), eds. R. Mewaldt et al., p. 375, astro-ph/0004166.
6. G. Israelian et al., *Astrophys. J.*, in press, astro-ph/0101032.
7. A. Goswami and N. Prantzos, *Astron. Astrophys.* **359**, 191 (2000).
8. Y. Takeda et al. *Astrophys. J.* (2000) submitted.
9. A. Boesgaard et al. *Astron. J.* **117**, 1549 (1999).

Part 3. Abundances and Nucleosynthesis

INTEGRAL AND NUCLEAR ASTROPHYSICS

JACQUES PAUL, MICHEL CASSÉ

SAp/DAPNIA/DSM/CEA, Orme des Merisiers, 91 191 Gif/Yvette CEDEX, France and

Institut d'Astrophysique de Paris, CNRS, 98 bis bd Arago, 75014 Paris, France

ELISABETH VANGIONI-FLAM

Institut d'Astrophysique de Paris, CNRS, 98 bis bd Arago, 75014 Paris, France

We briefly review the fundamentals of nuclear gamma-ray line astronomy (radioactive astronomy), focusing on its role to decipher the intimate physics of supernovae, either immediatly (via ^{56}Co) or after a time delay (via ^{44}Ti). All kinds of supernovae can be in principle tested through their radioactivities and their associated gamma-ray lines. Dedicated to the spectroscopy and imaging of celestial sources in the 15 keV to 10 MeV band, the ESA scientific observatory INTEGRAL will open a golden age of nuclear astrophysics in Europe.

1 Why a new gamma-ray astronomy mission?

Gamma-ray astronomy explores the most energetic phenomena that occur in nature and addresses some of the most fundamental problems in astrophysics. It embraces a great variety of gamma-ray continuum and gamma-ray line production processes: nuclear excitation, radioactivity, positron annihilation and Compton scattering; and an even a greater diversity of astrophysical objects and phenomena: nucleosynthesis, nova and supernova explosions, the interstellar medium, cosmic-ray interactions and sources, neutron stars, black holes, gamma-ray bursts, active galactic nuclei and the cosmic gamma-ray background. Not only do gamma rays allow us to see deeper into these objects, but the bulk of the power radiated by them is often at gamma-ray energies.

In the low-energy gamma-ray band, line-forming processes such as nuclear excitation, radioactivity, positron annihilation, cyclotron emission and absorption become important, and when used as astrophysical tools. Unique astrophysical information is contained in the spectral shift, line width, and line profiles. Detailed studies of these processes require the resolving power of a germanium spectrometer. Lower-resolution spectrometers (e.g. SIGMA, OSSE, COMPTEL) did not have sufficient energy resolution to permit a study of the parameters of these lines. The last high-resolution space instrument, that on HEAO-3 in 1979-80, was 100 times less sensitive than required to tackle the scientific subjects outlined below. Solid observational and theoretical grounds already exist for predicting detectable line emission from such varied celestial objects as the Central region of the Galaxy, the interstellar medium, compact objects, novae and supernovae and a variety of active galactic nuclei.

2 Nuclear astrophysics and gamma-ray astronomy

2.1 *Generalities*

The observability of gamma-ray lines is conditionned by three factors:

a. Thermonuclear synthesis in stellar furnaces of fresh radioisotopes in significant abundance (N).

b. Quick removal (before decay) from the dense medium of birth to avoid burning and allow transfer to low density regions, transparent to gamma rays.

c. Lifetime sufficiently short (radioactive constant, λ, high enough) in order to get measurable activity (λN), i.e. enough photons to be detected at galactic/cosmic distances.

The list of candidates consequently is rather brief; it includes principally ^{56}Co, ^{44}Ti and ^{26}Al (the lifetimes are respectively 112 days, 87 years, 10^6 years).

The observation of nuclides through the gamma-ray lines emitted in the course of their natural decay in the circumstellar/interstellar media reveals the recent nucleosynthesis activities on time scales comensurate to their lifetime. According to the species involved, one can probe individual events, like supernovae and novae or collective effects of these objects at galactic scale. The two limiting cases are the following:

i) if the lifetime is short(^{56}Co) compared to the frequency of the relevant event (SN explosion in this case) one has the possibility of observing individual events.

ii) if the lifetime is long compared to the frequency of fertile events (^{26}Al) one should observe the collective effect of many events under the form of a diffuse galactic emission. For lifetimes of the order of the recurrence time (rate) , i.e. ^{22}Na and ^{44}Ti, the case is intermediate. One should then observe diffuse extended sources.

The line profiles offer a wealth of information on the detailed physics/ dynamics of the astrophysical sources of unstable isotopes [1 2]. Gamma ray lines are indeed ideal diagnostics of nucleosynthesis. Stringent constraints on stellar and supernova yields can be set by specific line intensity or even upper limits on these. Gamma-ray line astronomy, in principle should allow to test both quiescent (hydrostatic) nucleosynthesis (in AGB and Wolf- Rayet stars) and explosive nucleosynthesis (novae and supernovae). We focus here on supernovae (for the other objects see J. Knodlseder and M. Hernanz, this conference). We are concerned here by the high potential of these exploding objects in both the synthesis and acceleration of nuclei.

2.2 Gamma-ray lines and supernova physics

A highligt of gamma-ray line astronomy has been the observation of signatures of the radioactive decay (weak interaction) of $^{56,57}Co$ isotopes, implied in the $^{56,57}Ni$ chain ending at $^{56,57}Fe$ from the supernova SN1987A in the Large Magellanic Cloud. Radioactive decay proceeds, in all cases considered, through the weak interaction (β^+ decay or electron capture) transforming inside a nucleus a proton into neutron. Indeed the synthesized species, due to the physics involved, are too proton rich to be stable. This proton richness is itself related to the fact that most of the reactions that built complex nuclei are induced by protons and alphas. This is especially true for explosive nucleosynthesis in supernovae, where the symetrical ^{44}Ti (22 protons and 22 neutrons) and ^{56}Ni (28 protons, 28 neutrons) are formed by successive alpha captures on ^{28}Si.

All kinds of supernovae can be tested (Ia,b,c and II) both through their radioactivity and also their (collective) particle acceleration effect.

i) individual SNII: There is very little chance to capture the cobalt-iron gamma-ray line emission of a new core collapse supernova during the lifetime of the present generation of gamma-ray satellite. However young supernova remnants could show up through the ^{44}Ti lines and through superbubbles excavated in the interstellar medium by the combined effect of strong stellar winds of massive stars and explosions. Due to its lifetime, of the order of a century, ^{44}Ti is well suited to discover young SNR hidden in the galactic dust. ^{44}Ti decay lines from recent SNR (Cas A and Vela junior) without optical counterpart have already been detected. Thus we have at least two cases of hidden supernovae. How many others will the INTEGRAL mission discover? The estimates are rather uncertain [3]. The detailed exploration of the known sources, associated to X ray data from Chandra and XMM, should allow to derive the distribution of ^{44}Ti and its mass (taking into account the fact that its lifetime against electron capture is possibly modified by very high state of ionisation of the emitting regions), setting important constraints on the physics of the explosion.

ii) collective effects of SNII in Superbubbles: Non thermal nucleosynthesis of Lithium-Beryllium-Boron (see M. Cassé, this conference) could also be accompanied by specific (rather wide) gamma ray lines, specially those of C*, O* and the LiBe feature arising from the alpha + alpha reaction [4] [5].

Indeed, superbubbles seem to be the most favourable sites of production of non thermal gamma-ray lines.

iii) SNIa: A direct observation of the gamma ray lines associated to ^{56}Co from a SNIa (closer than 15 Mpc) would be a very happy event. It could serve to clarify the thorny physics of these objects, which serve, on the other hand, as distance indicators for cosmology.

3 The INTEGRAL mission

3.1 Overall presentation

Dedicated to the spectroscopy and imaging of celestial sources in the 15 keV to 10 MeV band, the ESA (European Space Agency) scientific mission INTEGRAL (International Gamma-Ray Astrophysics Laboratory) will address the scientific objectives defined above through the simultaneous use of two main gamma-ray instruments, the high resolution spectrometer SPI and the fine imaging telescope IBIS, with concurrent source monitoring in the X-ray and visible bands. High resolution spectroscopy with fine imaging and accurate positioning of celestial sources over the entire energy range are mandatory to reach the scientific goals of the mission. Fine spectroscopy over the entire energy range will permit spectral features to be uniquely identified and line profiles to be determined for physical studies of the source region. Fine imaging capability within a large field of view will permit the accurate location and hence identification of the gamma-ray emitting objects with counterpart at other wavelengths. It will also enable extended regions to be distinguished from point sources and provide considerable serendipitous science. Explicit references on INTEGRAL science can be found in the proceedings of past INTEGRAL Workshops held in les Diablerets [6], Saint Malo [7], Taormina [8], and Alicante [9].

The INTEGRAL spacecraft (Figure 1) consists of a service module, commonly designed with the service module of the ESA XMM-Newton mission containing all spacecraft subsystem and a payload module containing the scientific instruments. INTEGRAL, with a payload mass of approximatly 2000 kg and a total launch mass of about 4000 kg will be launched in April 2002 by a Russian PROTON launcher into a highly eccentric 72-hour orbit (initial perigee height 10 000 km, initial apogee height 153 000 km). The particle induced background affects the performance of high-energy detectors, and scientific observations will therefore be carried out while the spacecraft is above an altitude of nominally 40 000 km, implying that 90% of the time spent on the orbit can be used for scientific observations. However, when taking into account several in-orbit activities (as e.g. slew and instrument calibration), the average observation efficiency becomes about 85 % per year. The spacecraft employs fixed solar arrays: this means, that the target pointing of the spacecraft (at any point in time) will remain outside an avoidance cone around the sun and anti-sun. This leads to a minimum angle between any celestial source and the sun/anti-sun of 50 degrees during the nominal mission life.

INTEGRAL will be an observatory-type mission with a nominal lifetime of 2 years, an extension up to 5 years is technically possible. Most of the observing time (65% during year 1, 70 % year 2 and 75 % after) will be awarded to the scientific community at large as the General Program. Typical observations will last up two weeks. Proposals following a standard AO process, will be selected on their scientific merit only by a single Time Allocation Committee. The first call for observation proposals, released in November 2000, has been closed on February 16, 2001. The remaining fraction of the observing time will be reserved, as guaranteed time, for the INTEGRAL Science Working Team for its contribution to the program. This fraction, the Core Program, will be devoted a Galactic plane survey, a deep exposure of the central radian of the Galaxy and pointed observations of selected regions (such as the Vela region) and targets of opportunity (TOO). In accordance with ESA's policy on data rights, all scientific data will be made available to the scientific community at large one year after they have been released to the observer. For more information see the INTEGRAL World Wide Web pages[10].

The INTEGRAL Science Data Center (ISDC) located in Versoix, Switzerland, close to the Geneva Observatory, will be the center in which the INTEGRAL payload telemetry data will be processed to a level at which all users can pursue the scientific interpretation of the data. The data will be corrected for instrumental signature and some standard scientific processing and analysis will be performed. The ISDC is the place where the archive and derived products will be built and made accessible to the world wide astronomical community. The ISDC will routinely monitor the instrument science performance and conduct a quick-look science analysis. Most of the TOO showing up during the lifetime of INTEGRAL will be detected at the ISDC during the routine scrutiny of the data. Scientific data products obtained by standard analysis tools will be distributed to the observers and archived for later use by the science community.

3.2 The spectrometer SPI

One of the two main INTEGRAL instruments, the spectrometer SPI (Figure 1), has been designed to perform spectral analysis of localized and extended

Figure 1. (Left) The INTEGRAL spacecraft. The cylindrically shaped SPI is next to the larger rectangular payload module (PLM) structure housing the IBIS and JEM-X detectors inside. The top of the PLM carries the coded mask for IBIS (squared) and the two coded masks for the two JEM-X detectors. The OMC is located at the left top of the PLM. (Right) Schematic view of the spectrometer SPI.

regions with performances at 1 MeV such as an energy resolution $E/\Delta E \sim 500$ and a 3σ line sensitivity of 5 10^{-6} photons s^{-1} cm^{-2} (10^6 s exposure). The instrument features an array of 19 hexagonal high purity germanium detectors cooled by an active device to an operating temperature of 85 K. An hexagonal coded aperture mask is located 1.7 m above the detection plane in order to image large regions of the sky (fully coded field of view $\sim$ 16 degrees) with an angular resolution of 2 degrees. In order to reduce background radiation, the detector assembly is shielded by a veto system which extends around the bottom and side of the detector almost completely up to the coded mask. A plastic veto is provided below the mask to further reduce the 511 keV background. The spectrometer SPI is being developed under prime contractor responsibility of CNES (the French Space Agency) by a consortium of institutes in France (CESR Toulouse, CEA Saclay), Germany (MPE Garching), Italy (IFCTR Milano), Spain (U Valencia), Belgium (U Louvain), UK (U Birmingham), USA (UC San Diego, LBL Berkeley, NASA/GSFC Greenbelt). Principal Investigators are G. Vedrenne (CESR Toulouse), and V. Schoenfelder (MPE Garching).

4 Conclusion

With the launch in the near future of the INTEGRAL spacecraft, a golden age of nuclear astrophysics will open in Europe, at the point of convergence of nuclear physics and astrophysics. Time is ripe to join this effort, especially having in mind the fact that INTEGRAL is an observatory-type mission with most of the total observing time being awarded as the general program to the scientific community at large.

References

1. R. Diehl, in "Astronomy with Radioactivities", Workshop Proceedings, Ringberg, sept 1999, MPE Report, 274 3 1999
2. J. Knodlseder, in "The Interplay Between Massive Stars and the ISM" parallel session of JENAM99, sept 1999, Toulouse, Edts D. Schaerer and R. Gonzalez-Delgado, astro-ph/9912131
3. L.S. The, in "Astronomy with Radioactivities", Workshop Proceedings, Ringberg, sept 1999, MPE Report, 274, p.77, 1999
4. E. Parizot, M. Cassé and E. Vangioni-Flam *A&A* **328**, 107 (1997)
5. V. Tatischeff et al, in the 4th INTEGRAL Workshop "Exploring the Gamma-Ray Universe", Alicante, sept. 2000, in press
6. Proc. 1st INTEGRAL Workshop, The Multiwavelength Approach to Gamma-Ray Astronomy, ApJS 92, No. 2, 1994
7. Proc. 2nd INTEGRAL Workshop, The Transparent Universe, Eds: C. Winkler, T. J.-L. Courvoisier, Ph. Durouchoux, ESA SP-382, 1997
8. Proc. 3rd INTEGRAL Workshop, The Extreme Universe, Astrophys. Let. & Communications, 38 (Part I) & 39 (Part II), 1999
9. Proc. 4th INTEGRAL Workshop, Exploring the Gamma-Ray Universe, 2001 in press
10. http://astro.estec.esa.nl/SA-general/Projects/Integral/integral.html

IMPACT OF ROTATION AND MAGNETIC FIELDS ON INTERNAL STELLAR COMPOSITION

S. TURCK-CHIÈZE
Service d'Astrophysique, CEA/DSM/DAPNIA, CE Saclay, Gif sur Yvette, 91191, FRANCE
E-mail: cturck@cea.fr

For decades, the post BBNS history of the elements in different media has been studied mainly theoretically. Here, we focus on the helioseismic SOHO satellite results. The precision of the used methods allows nowadays a direct insight of the solar interior, revealing the internal composition with a remarkable accuracy. Therefore, the role of the rotation and magnetic field begins to be studied and unsolved problems as lithium destruction and neutrino puzzle progress consequently. This method will be extended to other stars for progressing on dynamical processes of stellar evolution which contribute to the enrichment of the interstellar medium.

The history of elements in the Univers is rich and full of informative processes. Through stars, it connects also the Univers to the human existence. In this field, Jean Audouze has been one of the first guides in France, he has largely contributed to develop this area in very different directions with all his students and friends. It is a pleasure for me to offer him a new piece of the puzzle, looking directly inside the stars to discover missing or incorrectly described processes improving our vision of elements in the Univers.

1 Helioseismology and Stellar evolution

Acoustic modes are suitable probes to check the internal stellar structure. They must bring interesting constraints on turbulence regions, mixing in stellar interiors and may be magnetic field. The satellite SOHO, through the seismic instruments GOLF, MDI and VIRGO, has significantly improved the knowledge of these modes owing to an excellent duty cycle (greater than 90%) and the capability to detect very low amplitude modes, down to 3mm/s after five years of measurement. This step garantees a first real view of the solar interior from the surface to the energy-generating core. Effectively the modes of low frequency (low signal/noise) are less dependent on the turbulent surface than previous measurements. Three quantities are nowadays accessible: the rotation profile down to 0.4 $R_\odot$, the sound speed and density profiles down to 0.07 $R_\odot$. The sound speed is the natural quantity that we can deduce from acoustic modes, but this quantity is rather smooth and is weakly varying during evolution: 9% in 4.6 Gyr for the Sun. Hopefully, we extract the sound speed profile from the frequency of the modes (or the acoustic travel time of

Table 1. Along solar radius, some element can be used to check the presence of physical processes which are accessible through an helioseismic observable

radius ($R_\odot$)	element	observables	physical processes	solved ?
0.98	^4He	Γ_1	diffusion	yes
0.71	^{16}O	κ	transition R/C	yes
0.70	^7Li	c^2, rotation	NP,turbulence	yes
			magnetic tubes	no
0.57	^9Be	c^2	NP, turbulence	yes
0.25	^3He	c^2, density	mixing ?	yes
0.05	^7Be	c^2, density	mixing?	yes
0- 0.1	^{56}Fe	κ	central temp.?	correct

the modes), so the high accuracy is directly determined by the resolution in the frequency domain, which depends on the integration duration. After one (or several) year of integration, we get frequencies at the level of several 10^{-5}. With the three instruments aboard, we are able to study all the bias due to the used methods, or to the variation with the solar cycle (Toutain et al, 1997; Basu et al. 2000; Turck-Chièze et al. 2000; Garcia et al, 2001). Consequently we are quite confident today to determine the sound speed at the level of about 10^{-4}. Consequently a variation of the sound speed between the Sun and the model at a level of 10^{-3} is significant and allows a check of the physical processes at the level of several % (see table 2 of Turck-Chièze et al. 2001a). This is exactly what we need to check the hypotheses of stellar evolution and add the missing processes necessary to introduce dynamical effects in stellar evolution which contribute to the enrichment of the Univers. Using these variables, Table 1 illustrates this point, showing that in different locations, elements can be used to check different processes (R/C: radiation/convection, NP: nuclear processes) through a variable accessible by acoustic modes, also noted if the problem is already solved (from Turck-Chièze et al, 2001a).

2 Dynamics in the solar interior

In connecting the different quantities we have already get the following answers with the SOHO satellite:

- mixing is favoured at the base of the convection zone in a very narrow band of about 0.02 $R_\odot$, corresponding to the region where differential latitudinal rotation vanishes. The introduction of an (magneto) hydrodynamical instability is sufficient to find the good order of magnitude for lithium burning in the Sun with an absence of ^9Be burning (Brun, Turck-Chièze and Zahn 1999 (and references therein); Piau and Turck-Chièze 2001 for PMS).

- turbulent mixing is not favoured in the region of the ^{3}He peak (except very small turbulence), and in the very central part of the Sun but if there are some instabilities produced by g modes, they are not yet accessible to the present acoustic mode study (Turck-Chièze et al. 2001b),

- helioseismology is extremely sensitive to the fundamental (p,p) reaction rate, an increase of only 1% coupled to +3-5 % on heavy element composition gives a well description of the solar core (Turck-Chièze et al., 2001b),

- dynamical screening or any deviation from Maxwellian distribution on $(^{3}He,^{3}He)$ and $(^{3}He,^{4}He)$ and heavier elements is not really favoured by the present study, but the sound speed is not really sensitive to these processes, the density profile could be a way to progress (Turck-Chièze et al., 2001b).

These last ten years, we have reduced the difference between the sound speed profile deduced from acoustic modes and the model one by a factor 10, in improving the physics of the stellar representation. This is a great success of atomic physics, nuclear physics and astrophysics. Now the residual is so small that we can build a seimic representation of the Sun which lets no place for important misunderstanding of the physics of the core, in the hypothesis that the Sun is spherically symmetrical and in hydrostatic and thermal balance. These hypotheses will be verified in looking for time variabilities and in detecting g modes which will increase the spatial resolution in the whole nuclear core. We deduce from this seismic model, the emitted neutrino fluxes which are not compatible with earth detection, *demonstrating experimentally* that the deficit cannot be supported by the solar mean internal structure. The limitations of such results and the way to progress on the neutrino enigma, as far as the solar medium, is mentionned just above.

Magnetic field has been looked for in all helioseismic data, if it is clearly present at the surface, the induced effect on the solar structure is so small that we have no direct signature of its presence in the solar interior. Today we have only indirect manifestation of its presence (dynamo effect) and limits of its strength along the solar radius.

Complementary informations on other stars will be extremely useful.

Bibliography:

Brun, S., Turck-Chièze, S. & Zahn, JP., ApJ, 1999, 525, 1032.

Basu, S. et al., ApJ, 2000,535, 1078.

Garcia, R. et al., Sol. Phys, 2001, 200, in press.

Piau, L. & Turck-Chièze, S., ApJ, 2001, submitted.

Toutain, T. et al., Sol. Phys, 1997, 175, 311.

Turck-Chièze, S. et al., in SOHO10/GONG 2000, ESA SP464, in press.

Turck-Chèze, S., et al., 2001a, Sol. Phys, 200, in press;

Turck-Chèze, S., et al., 2001b, submitted ApJ lett.

NUCLEOSYNTHESIS IN CLASSICAL NOVAE

MARGARITA HERNANZ

Institut d'Estudis Espacials de Catalunya, IEEC, and Instituto de Ciencias del Espacio (CSIC), Edifici Nexus, C/Gran Capità, 2-4, E-08034 Barcelona, Spain

JORDI JOSÉ

Institut d'Estudis Espacials de Catalunya, IEEC, and Departament de Física i Enginyeria Nuclear (UPC), Avda. Víctor Balaguer, s/n, E-08800 Vilanova i la Geltrú (Barcelona), Spain

ALAIN COC

Centre de Spectrométrie Nucléaire et de Spectrométrie de Masse, IN2P3-CNRS, Université de Paris Sud, Bâtiment 104, F-91405 Orsay Cedex, France

A general review of the relevance of classical novae for the chemical evolution of the Galaxy, as well as for Galactic radioactivity is presented. A special emphasis is put on the pioneering work done by Jim Truran in this field of research. The impact of recent developments in nuclear astrophysics on nova nucleosynthesis, together with the prospects for observability of novae radioactivities with the INTEGRAL satellite are discussed.

1 Introduction

Classical novae are explosions occurring on the top of white dwarfs accreting hydrogen-rich matter in close binary systems, of the cataclysmic variable type. The transfer of matter results from Roche lobe overflow from the main sequence companion star. The accreted hydrogen is compressed up to degenerate conditions, leading to a thermonuclear runaway. Explosive hydrogen burning occurs via the CNO cycle, out of equilibrium because radioactive nuclei with timescales longer than the evolutionary timescale are produced. These nuclei are transported by convection to the outer envelope, where they are preserved from destruction until they decay. This late decay is the final responsible of the envelope expansion and increase in luminosity characteristic of nova outbursts.

It is important to stress that Jim Truran, with Sumner Starrfield and collaborators, was a pioneer in the study of thermonuclear explosions. Starrfield, Truran, Sparks and Kutter wrote a series of papers, starting in the early 70's, which presented the first hydrodynamic computations of classical novae [1,2,3,4,5]. Models had the accreted envelope "in place", i.e, the accretion phase was not followed. One of the main conclusions was that the envelope should be enhanced in CNO nuclei in order to power the nova outburst. This is still a valid fact, and it poses severe problems to the current theory, because there isn't a clear understanding of the mixing mechanism responsible for that enhancement. In 1978, another group also developed a hydro code to study nova explosions [6], with similar limitations concerning the accretion phase. It is also worth mentioning that Jean Audouze also contributed with Jim Truran to the progress in the nova field, since he studied the properties of the hot CNO burning [7,8], which (although not "so hot") is crucial for the explosion.

Classical nova explosions are very common: ~35 per year in our Galaxy, from which only 3 to 5 are discovered. They release large amounts of energy, $\sim 10^{45}$ erg, but not large enough to have an impact on the dynamics of the interstellar medium, like supernovae. The small mass ejected (in the range $10^{-5} - 10^{-4}$ $M_\odot$) implies that novae scarcely contribute to the chemical evolution of the Galaxy. If we adopt an ejected mass of $2 \times 10^{-5} M_\odot$, a Galactic nova rate of 35 yr^{-1} and an age of the Galaxy of 10^{10} yr, novae can account only for $\sim 1/3000$ of the gas and dust in the Galactic disk; this number is somehow a lower limit, since the ejected mass deduced from observations of some novae is larger than the typical theoretical value adopted here. However, novae are the main producers of some particular isotopes (^{13}C, ^{15}N and ^{17}O), they can help to explain the ^{7}Li versus metallicity relation in the Galaxy and they contribute to the radioactivity of the Galaxy, through the ejection of medium and long-lived radioactive isotopes (such as ^{7}Be, ^{22}Na and ^{26}Al). In addition, novae often form dust, which manifests through their IR emission; therefore they could explain the anomalous isotopic signatures measured in some presolar grains [9,10] (see also the excellent recent review [11]).

2 Novae and the chemical evolution of the Galaxy

Novae can be classified in two types, according to their observed abundances. The often called "standard" novae are objects where enrichments in CNO nuclei have been reported, whereas the so-called "neon" novae show also enrichments in neon. This class of novae was first studied hydrodynamically by Starrfield et al. [12], including the follow-up of the accretion phase. The two nova types are interpreted as the result of explosions on accreting carbon-oxygen (CO) and oxygen-neon (ONe) white dwarfs[a], respectively, because the chemical composition of the ejecta reflects that of the underlying white dwarf.

[a] Until recently, massive white dwarfs were called ONeMg white dwarfs, because they were thought to be made of these elements. But now it is known that these stars are almost devoid of magnesium and, consequently, should be called ONe white dwarfs

The reason is that some mixing between the accreted envelope (presumed to have close to solar composition) and the underlying core is required, in order both to power the explosion and to explain the observed enrichments in CNO and/or Ne. The question of the frequency of occurrence of ONe white dwarfs in classical novae was addressed by Truran and Livio [13,14], who concluded that these white dwarfs can account for $\sim 1/3$ of all observed outbursts.

The mechanism responsible of the mixing has not been determined yet (shear mixing, diffusion, convection). Recent two and three-dimensional calculations [15,16,17] address the convective mixing during the hydrodynamic phase following degenerate ignition of hydrogen. There isn't yet a clear way to mix and probably the mixing, or some part of it, should proceed in the previous hydrostatic accretion phase. Diffusion during a number of succesive flashes has been also simulated [18,19]. In this case, moderate enhancement can be obtained for some combinations of initial conditions (mass, luminosity, accretion rate); the remaining problem of this scenario is the handling of mass-loss during succesive flashes.

Until some clear mechanism is identified and understood, we [20] adopt a parametrization of the mixing with the underlying core (ranging from 25 to 75 %, in order to explain the range of observed metallicities in novae ejecta; other authors [21,22] adopt only 50%). In table 1 we show the main nucleosynthetic products of a handful of nova models, representative of both classes of novae (see José and Hernanz [20] for a larger sample). CO white dwarfs have masses up to ~ 1.15 $M_\odot$, whereas ONe have larger masses (the exact limit between both types of degenerate cores, related to the previous evolution during the AGB phase, is still an open issue). The accretion rate adopted is 2×10^{-10} $M_\odot.yr^{-1}$ and the initial luminosity 10^{-2} $L_\odot$ (initial mixing 50 %). We show the ejected masses of the most overproduced isotopes (^{13}C, ^{15}N and ^{17}O), together with ejected masses of medium and long-lived radioactive nuclei. In figure 1 we show the overproduction factors, with respect to solar, of all the isotopes contained in the ejecta of a CO and an ONe novae, of the same mass, in order to illustrate the impact of the chemical composition of the white dwarf. ^{13}C, ^{15}N and ^{17}O are overproduced in both nova types by similar factors. ONe novae also eject Ne, required to fit the observations. The radioactive isotopes ^{22}Na and ^{26}Al are almost only produced in ONe novae, because of the presence of seed nuclei in the accreted envelope mixed with the core; this presence is necessary to power the NeNa-MgAl reaction rate cycles responsible of the synthesis of ^{22}Na and ^{26}Al synthesis. On the other hand, ^{7}Li (coming from the radioactive ^{7}Be) is produced in larger amounts in CO novae, because of the shorter duration of the accretion phase (with T smaller than $\sim 10^8$ K), which is the critical one for ^{3}He survival, with ^{3}He being essential for ^{7}Be synthesis (see Hernanz et al. [23] for details). The production of ^{7}Li in novae was already predicted by the computations of Starrfield et al. [24] with their hydro code more than 20 years ago, but at the epoch they didn't follow the accretion phase, since they had the accreted envelope in place; therefore, a crucial phase for ^{7}Li synthesis was missed. The relevance of novae for the Galactic content of ^{7}Li is still an open issue. A rough estimate of their average contribution gives a maximum of the 20%, but the important fact is at which stage of the Galactic evolution they contribute. In a recent paper [25], it has been shown that ^{7}Li contribution from novae is required in order to reproduce the shape of ^{7}Li versus metallicity relationship, which grows at low metallicities.

Comparison between theoretical abundances in the ejecta and observed abundances in particular novae gives a quite good agreement [20] both for CO and ONe novae [22], as well as a global comparison of average properties in CO novae [19].

3 Novae and Galactic radioactivity

The content of CO and ONe novae ejecta in radioactive isotopes is shown in table 1 for some significative models. As mentioned before, CO novae are more successful in producing ^{7}Be, whereas ONe novae produce larger quantities of ^{22}Na and ^{26}Al than CO novae. This fact has important consequences for the γ-ray signatures expected for both types of novae, as shown in detail in our recent papers [26,27]: individual CO novae will preferentially emit photons of 478 keV (with τ=77 days), whereas ONe will emit 1275 keV photons (τ=3.75 yr). Concerning the cumulative emission, ONe novae can contribute to the Galactic content of ^{26}Al ($\tau = 10^6$ yr), although their contribution is not the most important one: according to the observations of the Galactic 1809 keV emission, made with the CGRO/COMPTEL instrument, this emission seems to be preferentially associated with a massive star population.

In the recent years, we have studied in detail the impact that uncertainties in nuclear reactions have on nova yields, specially for the radioactive ones [28,29], in order to make predictions of detectability of classical novae by the future ESA satellite INTEGRAL, to be launched in April 2002. Novae (ONe) up to distances of $\sim$ 2kpc could be detected through the ^{22}Na line at 1275 keV, whereas CO novae should be at shorter distances to be detected through the ^{7}Be line at 478 keV [26]. The most powerful emission of γ-rays from novae is the line emission at 511 keV and the continuum at energies between 20 and 511 keV, related to e^--e^+ annihilation, with the positrons coming from the very short-lived isotopes ^{13}N and ^{18}F. This emission could be detected up to 3

Table 1. Nucleosynthesis in CO and ONe novae

Nova	$M_{wd}(M_\odot)$	^{13}C ($M_\odot$)	^{15}N ($M_\odot$)	^{17}O ($M_\odot$)	^{7}Be ($M_\odot$)	^{22}Na ($M_\odot$)	^{26}Al ($M_\odot$)
CO	0.8	7.7×10^{-6}	9.9×10^{-8}	2.6×10^{-7}	6.0×10^{-11}	7.4×10^{-11}	1.7×10^{-10}
CO	1.15	1.3×10^{-6}	5.4×10^{-7}	2.7×10^{-7}	1.1×10^{-11}	1.1×10^{-10}	6.1×10^{-10}
ONe	1.15	8.0×10^{-7}	6.8×10^{-7}	7.9×10^{-7}	1.6×10^{-11}	6.4×10^{-9}	2.1×10^{-8}
ONe	1.25	6.0×10^{-7}	7.7×10^{-7}	6.9×10^{-7}	1.2×10^{-11}	5.9×10^{-9}	1.1×10^{-8}

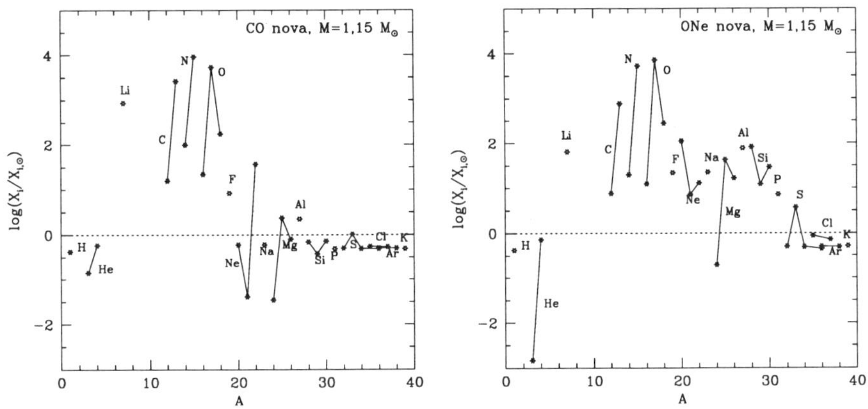

Figure 1. Overproduction factors relative to solar abundances, versus mass number for: (left) CO nova with $M=1.15M_\odot$, (right) ONe nova with $M=1.15M_\odot$

kpc with INTEGRAL, but it has very short duration and appears very early, even before optical detection. Detection of novae with INTEGRAL would provide a direct proof of the thermonuclear runaway model.

Acknowledgments

Research partially supported by the CICYT (ESP98-1348, PB98-1183-C03-02 and PB98-1183-C03-03) and by the AIHF1999-0140.

References

1. S. Starrfield, J.W. Truran, W.M. Sparks and G.S. Kutter, *ApJ* **176**, 169 (1972).
2. S. Starrfield, W.M. Sparks and J.W. Truran, *ApJS* **28**, 247 (1974).
3. S. Starrfield, W.M. Sparks and J.W. Truran, *ApJ* **192**, 647 (1974).
4. W.M. Sparks, S. Starrfield and J.W. Truran, *ApJ* **208**, 819 (1976).
5. S. Starrfield, J.W. Truran and W.M. Sparks, *ApJ* **226**, 186 (1978).
6. D. Prialnik, M.M. Shara and G. Shaviv, *A&A* **61**, 363 (1978).
7. J. Audouze, J.W. Truran and B.A. Zimmerman, *ApJ* **184**, 493 (1973).
8. B. Lazareff, J. Audouze, S. Starrfield and J.W. Truran, *ApJ* **228**, 875 (1979).
9. S. Amari, X. Gao, L. Nittler, E. Zinner, J. José, M. Hernanz and R. Lewis, *ApJ* **in press**, (astro-ph/0012465) (2001).
10. J. José, M. Hernanz, S. Amari and E. Zinner, *these proceedings*.
11. R.D. Gehrz, J.W. Truran, R.E. Williams and S. Starrfield, *Publ. Astr. Soc. Pac.* **110**, 3 (1998).
12. S. Starrfield, W.M. Sparks and J.W. Truran, *ApJ* **303**, L5 (1986).
13. J.W. Truran and M. Livio, *ApJ* **308**, 721 (1986).
14. M. Livio and J.W. Truran, *ApJ* **425**, 797 (1994).
15. S.A. Glasner, E. Livne and J.W. Truran, *ApJ* **475**, 754 (1997).
16. A. Kercek, W. Hillebrandt and J.W. Truran, *A&A* **337**, 379 (1998).
17. A. Kercek, W. Hillebrandt and J.W. Truran, *A&A* **345**, 831 (1999).
18. D. Prialnik and A. Kovetz, *ApJ* **445**, 789 (1995).
19. A. Kovetz and D. Prialnik, *ApJ* **477**, 356 (1997).
20. J. José and M. Hernanz, *ApJ* **494**, 680 (1998).
21. M. Politano, S. Starrfield, J.W. Truran, A. Weiss and W.M. Sparks, *ApJ* **448**, 807 (1995).
22. S. Starrfield, J.W. Truran, M.C. Wiescher and W.M. Sparks, *MNRAS* **296**, 502 (1998).
23. M. Hernanz, J. José, A. Coc and J. Isern, *ApJ* **465**, L27 (1996).
24. S. Starrfield, J.W. Truran, W.M. Sparks and M. Arnould, *ApJ* **222**, 600 (1978).
25. D. Romano, F. Matteucci, P. Molaro, P. Bonifacio, *A&A* **352**, 117 (1999)
26. J. Gómez-Gomar, M. Hernanz, J. José and J. Isern, *MNRAS* **296**, 913 (1998).
27. M. Hernanz, J. José, A. Coc, J. Gómez-Gomar and J. Isern, *ApJ* **526**, L97 (1999).
28. J. José, A.Coc and M. Hernanz, *ApJ* **520**, 347 (1999).
29. A. Coc, M. Hernanz, J. José and J.P. Thibaud, *A&A* **357**, 561 (2000).

THE WINDS OF NOVAE

MICHAEL FRIEDJUNG

Institut d'Astrophysique (CNRS), 98 bis Boulevard Arago, 75014 Paris, France

the main properties of the winds of novae during outburst are reviewed both from what is indicated by observations and from what is expected according to present theory. Two types of models (or two points of view) are compared.

1 Introduction

There is very good evidence that novae have strong winds during their activity, following the start of their outbursts. Though most specialists agree about this, there are strong disagreements about the exact nature of these winds. There are at present two main points of view, which are as follows:

(a)— The wind is optically thick in the continuum and is mainly accelerated probably by radiation pressure at great depths below an apparent photosphere. [1,10,7]. The outer layers of the envelope have lower velocities than that of the wind at a particular time.

(b)— Most acceleration of the wind occurs above the apparent photosphere; this acceleration could be due to radiation pressure acting in the lines as for more classical winds of hot stars. The point of view (b) has been assumed in recent attempts to treat rigorously radiation transfer in the wind[5,6]; such theories do not include hydrodynamics.

2 Observational considerations

Multiwavelength observations have shown that after optical maximum, the spectral energy distribution of radiation emitted shifts towards shorter wavelengths, the total luminosity decreasing much more slowly than that emitted only in the optical. Such observations suggest at first sight that radiaion is emitted by a kind of photosphere, which becomes hotter and shrinks with time. In addition there are indications that luminosities are sometimes above the Thomson scattering Eddington limit, equal to $1.44\ 10^{38}$ erg s^{-1} for a solar mass star with solar element abundances. A very striking case is that of nova LMC 1988 no. 1 [11]; a fit of the observed energy distribution, to that predicted by a theoretical model, suggested a bolometric luminosity of $3\ 10^{38}$ erg s^{-1} for 45 days after optical maximum.

Complex line profiles are seen in outburst nova spectra, which can be understood as produced by layers expanding at different velocities. Mc Laughlin

[8] showed that velocities were larger in inner regions than in outer regions of a nova envelope after maximum visual brightness, that is the more blueshifted absorption components were formed in more central regions.

However recent studies of line profiles have indicated that the width of the emission part of profiles decreases in time after maximum visual brightness. This might be understood as indicating a different kind of velocity stratification. If lower velocity material was in deeper layers, line emission from such layers, which would be less broadened by the Doppler effect, would only be able to escape at later times after a decrease of the continuum optical thickness of the expanding envelope. However the interpretation of such an effect is far from unambiguous. For instance the emission line narrowing observed in the spectrum of nova V1500 Cyg was preceded by emission line widening shortly after optical maximum[4]. One can suppose that after the time its line emission became visible, due to a decrease of the continuous absorption, the high velocity wind became less intense, so the proportion of line emission produced by it became smaller compared with the line emission of lower velocity material ejected earlier. In addition one needs to take account of the outwards motion of zones with a certain level of ionization, as the photosphere shrinks and becomes hotter [2]. When expansion velocities are less in layers, which are further from the centre, the outwards motion of a region with a certain level of ionization will bring it to layers, ejected earlier, where the expansion velocity is less, so leading to an apparent decrease of line width.

3 Theoretical considerations

It is tempting to explain the existence of apparently discrete layers, with different expansion velocities in nova envelopes, by collisions between ejecta and winds, whose velocities are not the same. I proposed [3] that a shell, containing most of the ejected mass, is formed by the collision of a high velocity wind with slower moving material ejected before optical maximum. The slower moving material would appear to be at least often instantaniously ejected by an initial shock at the beginning of a nova outburst. Such processes can occur when slower moving material is further from the centre of the envelope than higher velocity material (point of view (a)). The observed hard X ray emission is also explained as a result of this type of collision [9]. The complex structures of envelopes, seen on direct images, could also be at least partly due to instabilities produced by such collisions.

The models of Kato and Hachisu [7] for the acceleration of winds at large optical depths, have been criticized. Their mechanism of wind acceleration is not that involving radiation pressure acting in spectral lines, but that of the action of radiation pressure in the continuum. There is in fact no inconsistency, as temperatures are very large near the base of the wind, while the luminosity is only slightly below the Eddington limit there. The scale height of the atmosphere of the white dwarf is large and a thermal wind is at first accelerated by gas pressure. The calculations of Kato and Hachisu indicate that, in the layers just above where the wind passes the critical point, the OPAL opacity increases considerably, leading to the luminosity becoming locally above the Eddington limit and strong acceleration of the wind by radiation pressure. The layers where this occurs are far below the apparent photosphere, where the wind becomes optically thin in the continuum. However, the theory does not account for luminosities above the Thomson scattering Eddington limit. More exotic physical effects might have to be invoked, such as that of most of the radiation being advected at large optical depths.

Spectral syntheses of nova spectra using good radiative transfer theory have been described in a number of papers,[5,6]. The wind was assumed to be accelerated above the photosphere (point of view (b)). Relatively good fits were claimed. However more recently Short et al (2001) [12] were only able to well fit low resolution ultraviolet spectra of nova V1974 Cyg, discrepancies appearing when fits were attempted to high resolution spectra. It is clear that there is something wrong in the velocity law of the wind, assumed by the authors.

References

1. M. Friedjung, *MNRAS* **132**, 317 (1966a)
2. M. Friedjung, *MNRAS* **133**, 401 (1966b)
3. M. Friedjung, *A&A* **180**, 157 (1987)
4. M. Friedjung, J. Mikolajewska, M. Mikolajewski, *A&A* **348**, 475 (1999)
5. P.H. Hauschildt ,R. Wehrse, S. Starrfield, G. Shaviv, *ApJ* **393**, 307 (1992)
6. P.H. Hauschildt, S. Starrfield, S. Austin, R.M. Wagner, S.N. Shore, G. Sonnenborn, *ApJ* **422**, 831 (1994)
7. M. Kato, I. Hachisu, *ApJ* **377**, 802 (1994)
8. D.B. McLaughlin, *PASP* **59**, 244 (1947)
9. T.J. O'Brien, H.M. Lloyd, M.F. Bode, *MNRAS* **271**, 155 (1994)
10. C.L.N. Ruggles, G.T. Bath, *A&A* **80**, 97 (1979)
11. G.J. Schwarz, P.H. Hauschildt, S. Starrfield, P.A. Whitelock, E. Baron, G. Sonnenborn, *MNRAS* **300**, 931 (1998)
12. C.I. Short, P.H. Hauschildt, S. Starrfield, E. Baron, ApJ 2000, in press

NOVA OUTBURSTS AND DUST FORMATION

JORDI JOSÉ
Dept. Física i Enginyeria Nuclear, UPC, and
Institut d'Estudis Espacials de Catalunya (IEEC/UPC), Barcelona, Spain

MARGARITA HERNANZ
Institut d'Estudis Espacials de Catalunya (IEEC/CSIC), Barcelona, Spain

SACHIKO AMARI, ERNST ZINNER
Lab. for Space Sciences and Physics Dept., Washington Univ., St. Louis, USA

IR observations of nova light curves reveal that novae form grains in their expanding shells, ejected into the interstellar medium by an explosion powered by a thermonuclear runaway. Quite different condensates are expected to form in CO and ONe nova ejecta. In this paper, we outline different isotopic signatures expected for nova grains on the basis of detailed nucleosynthesis calculations for both CO and ONe novae.

1 Introduction

About 10^{-5} $M_{\odot}$ of nuclear processed material (largely enhanced in ^{13}C, ^{15}N and ^{17}O) are ejected into the interstellar medium during classical nova outbursts. The evolution of such systems, which includes a long-lasting accretion phase, followed by a thermonuclear runaway that heats the envelope's base up to several hundred million degrees, imprints a unique chemical pattern onto the ejecta. Up to now, the identification of presolar nova grains, presumably condensed in the shells ejected by the explosion, relied only on high ^{22}Ne/^{20}Ne ratios (attributed to ^{22}Na decay), but quite recently five silicon carbide grains and one graphite grain with C, N, Al, and Si isotopic ratios that point towards a nova origin have been finally discovered[1]. In this paper, we present a summary of signatures expected for nova grains on the basis of detailed nucleosynthesis results from recent hydrodynamic calculations of nova outbursts[2,3,4].

2 Theoretical isotopic ratios in nova outbursts

Explosions in CO novae result in higher ^{14}N/^{15}N and lower ^{12}C/^{13}C ratios than in ONe novae. This is a direct consequence of a higher initial ^{12}C content in CO nova envelopes, which favors ^{13}C synthesis through $^{12}\mathrm{C}(\mathrm{p},\gamma)^{13}\mathrm{N}(\beta^+)^{13}\mathrm{C}$. On the other hand, the higher peak temperatures achieved in ONe models allow (p,γ) reactions on ^{13}C to proceed up to ^{15}N through $^{13}\mathrm{C}(\mathrm{p},\gamma)^{14}\mathrm{N}(\mathrm{p},\gamma)^{15}\mathrm{O}(\beta^+)^{15}\mathrm{N}$ (this also suggests that the ^{14}N/^{15}N ratio decreases as the white dwarf mass increases, since peak temperatures increase with the mass of the star). Similar ^{26}Al/^{27}Al ratios are obtained for both CO and ONe nova models: although ^{26}Al is efficiently synthesized in ONe novae, the larger initial amount of ^{27}Al in such novae (which is not strongly modified during the explosion) results in similar ratios to those obtained for CO novae. Therefore, the ^{26}Al/^{27}Al ratio does not provide a robust criterion to distinguish between CO and ONe novae. Both CO and ONe nova models yield low ^{26}Mg/^{25}Mg and ^{24}Mg/^{25}Mg ratios, a consequence of the efficient destruction of ^{24,26}Mg in all nova models.

CO novae show very limited nuclear activity beyond the CNO cycle, because of the moderate peak temperatures attained in the explosion and because of the low initial abundance of seed nuclei above this mass range. Therefore, calculations of CO novae yield close-to-solar Si contents in the ejecta. In contrast, the higher temperatures achieved in ONe novae (together with a higher initial ^{27}Al abundance) allow significant synthesis of ^{29,30}Si. This requires peak temperatures in excess of 2×10^8 K and prior production of ^{28}Si (first through ^{27}Al(p,γ)^{28}Si, followed by ^{26}Al^{m,g}(p,γ)^{27}Si(p,γ)^{28}P(β^+)^{28}Si, which dominate destruction through (p,γ) reactions).

Oxygen isotopic ratios strongly depend on white dwarf mass and composition (i.e., CO or ONe). Since ^{17,18}O synthesis proceeds through (p,γ) reactions on initial ^{16}O, which require high enough temperatures, lower ^{16}O/^{17}O and ^{16}O/^{18}O ratios are expected for ONe novae than for CO ones. Moreover, we expect significantly larger ^{20}Ne/^{22}Ne ratios in ONe than in CO novae, as a result of the larger initial ^{20}Ne content in the former. In fact, ^{20}Ne is scarcely modified in most nova models, whereas ^{21}Ne, a very fragile isotope, is almost completely destroyed, which accounts for the large ^{20}Ne/^{21}Ne ratios obtained for both nova types.

Comparison of our results with theoretical predictions from other hydrodynamic models [5,6] yields a general agreement.

References

1. S. Amari, X. Gao, L. Nittler, E. Zinner, J. José, M. Hernanz and R. Lewis, *ApJ* , in press (astro-ph/0012465).
2. M. Hernanz, J. José, A. Coc, J. Gómez-Gomar and J. Isern, *ApJ* **526**, L97 (1999).
3. J. José, A. Coc and M. Hernanz, *ApJ* **520**, 347 (1999).
4. J. José and M. Hernanz, *ApJ* **494**, 680 (1998).
5. A. Kovetz and D. Prialnik, *ApJ* **477**, 356 (1997).
6. S. Starrfield, J. Truran, M. Wiescher and W. Sparks, *MNRAS* **296**, 502 (1998).

SUPERNOVAE AND GALACTIC EVOLUTION INDICATORS OF THEIR NUCLEOSYNTHESIS

F.-K. THIELEMANN, D. ARGAST, F. BRACHWITZ, O.E. GERHARD, G. MARTINEZ-PINEDO AND M. SAMLAND

Department of Physics & Astronomy, Univ. of Basel, Klingelbergstrasse 82, CH-4056 Basel, Switzerland
E-mail: fkt@quasar.physik.unibas.ch

W.R. HIX, M. LIEBENDÖRFER, A. MEZZACAPPA

Physics Division, Oak Ridge National Laboratory, Oak Ridge, TN 37831-6371, USA
E-mail: mezz@mail.phy.ornl.gov

K. NOMOTO, T. NAKAMURA, AND H. UMEDA

Department of Astronomy, University of Tokyo, Tokyo 113-0033, Japan
E-mail: nomoto@astron.s.u-tokyo.ac.jp

K. IWAMOTO

Department of Astronomy, Nihon University, Tokyo, Japan
E-mail: iwamoto@etoile.phys.cst.nihon-u.ac.jp

Supernovae of type Ia and type II/Ib/Ic are the main contributors to nucleosynthesis in galaxies. Despite many efforts, a full and self-consistent understanding of their explosion mechanisms is not existing, yet. However, they leave fingerprints in spectra, lightcurves, radioactivities/gamma-rays or in galactic evolution. The aim of the present paper is to highlight how model uncertainties enter abundance predictions and how clues can be obtained from abundance observations.

1 Introduction

The evolution of galaxies is dominated by its main contributors SNe II, SNe Ia, and planetary nebulae. The latter do not contribute to the element abundances in the range O through Ni (although a few specific minor isotopes can be produced in the s-process). Thus, for this element range we have to explain galactic evolution and also solar abundances by the combined action of SNe II and Ia. The surface abundances of stars, indicating the interstellar medium abundances at the time of their birth are a clear indicator of galactic evolution as a function of metallicity [Fe/H]. Observational data for [x/Fe] at low metallicities ($-2 <$[Fe/H]$- < 1$), x standing for elements from O through

Ca, show an enhancement of the alpha elements (O through Ca) by a factor of 2-3 (0.3 to 0.5 dex in [x/Fe]) in comparison to Fe [10,29,2]. This is the clear fingerprint of the exclusive early contribution of fast evolving massive stars, i.e. SNe II. The ratio of Fe-group elements to Si-Ca in SNe Ia is important to compensate for these overabundance in SNe II in order to obtain solar abundance ratios for the combined nucleosynthesis products at solar metallicity [Fe/H]=0. Such global tests can constrain the average features of SNe II and SNe Ia mixed in frequency ratios of $N_{\rm Ia}/N_{\rm II} = 0.15 - 0.27$ [48,5] and require an Fe contribution of SNe Ia from 50-60%, if average ejected Fe masses of 0.1 $M_\odot$ and 0.6 $M_\odot$ are taken for SN II and SN Ia.

An interesting feature is that also the Fe-group ejecta of both types of supernovae have to differ. For Ti, Sc, Cr, Mn, Co, Ni one finds at low metallicities average SN II values [x/Fe] of 0.25, 0, -0.1, -0.3, -0.1, -0.1 [18]. Taken the typical uncertainty of 0.1 dex, this leaves Ti and Mn as elements with clear signatures for a SN II behavior different from solar, which asks for the opposite SN Ia behavior.

The features discussed so far apply to average SNe II and Ia compositions. A more complete understanding, however, would require observational clues for the range of individual SNe II and Ia which can directly constrain theoretical models. After presenting shortly the present status of SNe II and Ia nucleosynthesis predictions, we want to give examples how that might be done.

2 Type II Supernovae

The present situation is that self-consistent spherically-symmetric calculations (with the presently known microphysics) do not yield successful explosions [25,20,36]. This seems to be the same for multi-D calculations, which however lack good neutrino transport schemes and do not yet consider the combined action of rotation and magnetic fields. However, typical kinetic energies of 10^{51} erg are observed in supernova remnants. Thus, light curve as well as explosive nucleosynthesis calculations have been performed by introducing a shock of appropriate energy in the pre-collapse stellar model [50,41,17,26,47,37]. Such induced calculations lack self-consistency and cannot predict the ejected ^{56}Ni-masses from the innermost explosive Si-burning layers (powering supernova light curves by the decay chain ^{56}Ni-^{56}Co-^{56}Fe) due to missing knowledge about the mass cut between the neutron star and supernova ejecta.

A correct prediction of the amount of Fe-group nuclei ejected (which includes also one of the so-called alpha elements, i.e. Ti) and their relative composition depends directly on the explosion mechanism and the size of the collapsing Fe-core. Three types of uncertainties are inherent in the Fe-group ejecta, related to (i) the total amount of Fe(group) nuclei ejected and the mass cut between neutron star and ejecta, mostly measured by ^{56}Ni decaying to ^{56}Fe, (ii) the total explosion energy which influences the entropy of the ejecta and with it the amount of radioactive ^{44}Ti as well as ^{48}Cr, the latter decaying later to ^{48}Ti and being responsible for elemental Ti, and (iii) finally the neutron richness or $Y_e = < Z/A >$ of the ejecta, dependent on stellar structure, electron captures and neutrino interactions. Y_e influences strongly the ratios of isotopes 57/56 in Ni(Co,Fe) and the overall elemental Ni/Fe ratio. The latter being dominated by ^{58}Ni and ^{56}Fe. The pending understanding of the explosion mechanism also affects possible r-process yields for SNe II [40,52,35,8,24].

The intermediate mass elements Si-Ca provide information about the explosion energy and the stellar structure of the progenitor star, while elements like O and Mg are essentially determined by the stellar progenitor evolution.

3 Type Ia Supernovae

There are strong observational and theoretical indications that SNe Ia are thermonuclear explosions of accreting white dwarfs in binary stellar systems [15,33,32,21] with carbon ignition and a thermonuclear runway causing a complete explosive disruption of the white dwarf [31,49]. High accretion rates favor lower ignition densities. A flame front then propagates at a subsonic speed as a deflagration wave due to heat transport across the front [13]. The averaged spherical flame speed depends on the development of instabilities of various scales at the flame front. Multi-dimensional hydro simulations suggest a speed $v_{\rm def}$ as slow as a few percent of the sound speed $v_{\rm s}$ in the central region of the white dwarf. Electron capture affects the central electron fraction Y_e and depends on (i) the electron capture rates of nuclei, (ii) $v_{\rm def}$, influencing the time duration of matter at high temperatures (and with it the availability of free protons for electron captures), and (iii) the central density of the white dwarf ρ_{ign} (increasing the electron chemical potential i.e. their Fermi energy) [18,3,19]. After an initial deflagration in the central layers, the deflagration might turn into a detonation (supersonic burning front) at lower densities [28].

Nucleosynthesis constraints can help to find the "average" SN Ia conditions responsible for their contribution to galactic evolution, i.e. especially the Fe-group composition. While ignition densities ρ_{ign} determine the very central amount of electron capture and thus Y_e, the deflagration speed $v_{\rm def}$ determines the resulting Y_e -gradient as a function of radius [18]. Y_e values of 0.47-0.485 lead to dominant abundances of ^{54}Fe and ^{58}Ni, values between 0.46

and 0.47 produce dominantly ^{56}Fe, values in the range of 0.45 and below are responsible for ^{58}Fe, ^{54}Cr, ^{50}Ti, ^{64}Ni, and values below 0.43-0.42 are responsible for ^{48}Ca. The intermediate Y_e-values 0.47-0.485 exist in all cases, but the masses encountered which experience these conditions depend on the Y_e-gradient and thus v_{def}. Whether the lower vales with Y_e<0.45 are attained, depends on the central ignition density ρ_{ign}. Therefore, ^{54}Fe and ^{58}Ni are indicators of v_{def} while ^{58}Fe, ^{54}Cr, ^{50}Ti, ^{64}Ni, and ^{48}Ca are a measure of ρ_{ign}. These are the (hydrodynamic) model parameters. An additional uncertainty is the central C/O ratio of the exploding white dwarf [14]. Nuclear uncertainties based on electron capture rates enter as well [3,4]. Conclusions from these results are: (i) a v_{def} in the range 1.5–3% of the sound speed is preferred [18], and (ii) the change in electron capture rates [19] made it possible to have ignition densities as high as ρ_{ign}=2 × 10^9 g cm^{-3} without distroying the agreement with solar abundances of very neutron-rich species [18,3]. It seems, however, hard to produce amounts of ^{48}Ca sufficient to explain solar abundances from SNe Ia when applying more realistic electron capture rates and C/O ratios [51,4].

4 Observational Constraints and Galactic Evolution

There have been detailed discussions [42,42,18,41] on observational constraints from supernova spectra, lightcurves, and X-ray and gamma-ray observations of remnants. They refer mainly to ^{56,57}Ni, and ^{44}Ti abundances and possibly stable Ni/Fe ratios in SNe II, giving insight into the details of the explosion mechanism with respect to the mass cut between the neutron star and the SN ejecta, the total energy of the explosion, the entropy and the Y_e in the innermost ejecta (see section 2). The intermediate mass elements Si-Ca provide information about the explosion energy and the stellar structure of the progenitor star, while elements like O and Mg are essentially determined by the stellar progenitor evolution.

In SNe Ia the ^{56}Ni production and the Si-Ca/Fe ratio are related to the total explosion energy and burning front speed in layers of the exploding white dwarf extending as far as $M(r)$=1M$_\odot$ and beyond. Constraints on the ignition density and burning front speed in the central regions (as discussed in section 3) are reflected e.g. in minor isotopic abundances like ^{50}Ti and ^{54}Cr, where direct supernova or remnant observations cannot be used. Here one can only make use of global abundance constraints from galactic evolution, which therefore permit only statements about an "average" SN Ia. The application of present day electron capture rates makes it hard to account for our solar system ^{48}Ca [4]. On the other hand, observations of varying stable Ni/Fe ratios during the evolution of a supernova (by spectral means) related to ^{54}Fe and ^{58}Ni, might give clues to the metallicity of the exploding white dwarf [16].

In the present paper we want to focus on more recent promising trends in galactic evolution modeling, which might provide constraints on individual supernova models rather than only global properties of SNe II and SNe Ia. The reason for this possibility is the fact that there is no instantaneous mixing of ejecta with the interstellar medium, and therefore early phases of galactic evolution can present a connection between low metallicity star observations and a single supernova event. On average, each supernova pollutes a volume of the interstellar medium containing $\approx (3-5) \times 10^4 M_\odot$. After a supernova polluted such a previously pristine environment mass it results in values for [x/Fe] and [Fe/H] in the remnant and a scatter in [x/Fe] is expected for the same [Fe/H]. The amount of the polluted volume depends on the explosion energy, and if there is a strong variation of explosion energies with progenitor mass this could affect the relation between [x/Fe] and the metallicity [Fe/H] [26]. The scatter in [x/Fe] expected for the same [Fe/H] is observed up to metallicities of [Fe/H]=-2 where it vanishes because overlapping contributions from mant SNe II which behave like a well mixed medium [1] (each volume of the interstellar medium containing $\approx 3 \times 10^4 M_\odot$ needs to be enriched by $\approx 10^3$ SNe in order to obtain solar metallicities).

Some results from such an approach [1,2] are displayed in Fig.1 which shows O/Fe, Mg/Fe, Si/Fe and Ca/Fe ratios. Three types of symbols are given, circles, squares and dots. The squares show observations of low metallicity stars with a scatter at very low metallicities, as expected if mixing of ejecta with the ISM is not instantaneous and contributions from individual supernovae with different progenitor masses give their different signatures. This scatter apparently becomes as small as the observational uncertainties at [Fe/H]=-2, where everywhere in the ISM the average signature of IMF integrated SNe II yields emerges. The open circles indicate the x/Fe and Fe/H ratios in a volume of $5 \times 10^4 M_\odot$ of an initially pristine (big bang) composition, polluted by a single supernova with our ejecta compositions [41,30]. The numbers inside the circles refer to the progenitor mass of a supernova whose ejecta composition was mixed with $5 \times 10^4 M_\odot$ of prestine ISM. Such a remnant would therefore show directly the [Fe/H] metallicity caused by a single SN event. However, we compare with observations from low metallicity stars rather than remnants. The model stars are indicated by dots [1,2]. They assume a stochastic star formation at random positions in the ISM and with random progenitor masses (however a statistical distribution according to a Salpeter IMF). One sees on the one hand that much smaller contributions to a stellar progenitor cloud can be possible than expected from a single remnant (leading to Fe/H ratios much

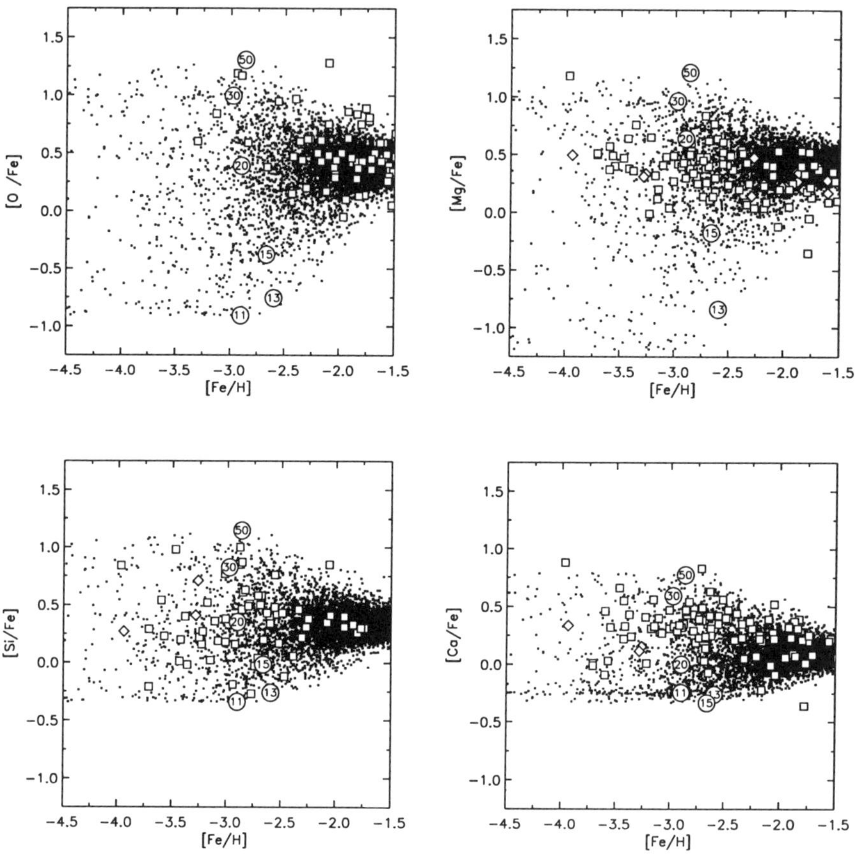

Figure 1. Comparison of low metallicity observations with SN II yields and a galactic evolution model, see text for details.

smaller than given in the circles). On the other hand, successive enrichment of many (and finally overlapping) remnants leads to a metallicity evolution which approaches at [Fe/H]=-2 the IMF-averaged SNe II yields. These are apparently in agreement with the observations.

The features (at [Fe/H]=-2) agree nicely with previous galactic evolution calculations which applied our yields [46,44,22,6], assuming instantaneous mixing of yields with the ISM. However, the observed and predicted scatter at very low metallicities bears information which was previously unavailable. The scatter for Si and Ca/Fe seems essentially correct, possibly only slightly too large, which could indicate that the lower mass SNe II (11 and 13 $M_\odot$) produce slightly too small ratios and the high mas end (50 $M_\odot$) slightly too large ones. This can be related to the unertainty of the mass cut and the assumptions made in our previous papers on the Fe-yields as a function of progenitor mass. Thus, while the average x/Fe is correct, the progenior mass dependence of the Fe-yields could be slightly too strong. This is, however, a small effect for the elements Si and Ca, produced in explosive burning. Much larger deviations can be seen for O and Mg, which are dominated by uncertainties in stellar evolution (not the explosion). Here the lower mass stars (11, 13, 15 $M_\odot$) predict clearly much too small ratios, far beyond the obervational scatter, while the 50 $M_\odot$ model predicts slightly too large ones. This indicates a problem in the stellar models which might improve with more recent calculations [7,47,11,12] and has to be tested.

The main conclusion we can draw from these results is that such investigations can also test individual stellar yields rather than only IMF integrated samples. This is a large advantage over the very few data points we have from individual supernova observations. Such tests seem also very useful for other applications, where one is not certain about the stellar site of a nucleosynthesis product or possible contributions from objects with different evolution timescales. Hypernova contributions should be considered in a similar way [27].

The r-process is an example, where the alternative site to supernovae, neutron star megers, have a much smaller occurance frequency. In that case, the mixed phase (orrurring for SNe II at about [Fe/H]=-2) should be delayed to larger metallicities. In addition, one expects with the large amounts of r-process ejecta from each occasionally occurring neutron star merger [9] a much larger scatter than for the smoothly changing supernova yields as a function of stellar mass. Both effects are seen in the r/Fe observations (a scatter of more than a factor of 100 [39] at low metallicities, which still amounts to about a factor of 10 at [Fe/H]=-1, see Fig. 2 in [45]. Combined with the suppression of abundances below A=130 in low-metallicity stars, these could be taken as supportive features for a fission cycling r-process from a low frequency source in galactic evolution. At least it argues for a very rare event, even if some type of supernovae are responsible for it. Similar galactic evolution calculations as presented here are needed in order to test the expected amount of scatter as a function of metallicity to give clues on the r-process site. A discussion of the advantages and disadvantages of both possible r-process sources (SNe II vs. neutron star mergers) is given in refs. [34,38].

Acknowledgement

We want to thank Jean Audouze and Jim Truran, who inspired us to our first nucleosynthesis calculations and made clear what one can learn from galactic chemical evolution.

References

1. Argast, D., Samland, M., Gerhard, O.E., Thielemann, F.-K. 2000, A & A 356, 873
2. Argast, D., Samland, M., Gerhard, O.E., Thielemann, F.-K. 2001, to be submitted to A & A
3. Brachwitz, F. et al. 2000, Ap. J., 536, 934
4. Brachwitz, F. et al. 2001, Ph. D. thesis, Univ. of Basel, unpublished
5. Cappellaro, E. et al. 1997, A&A 322, 431
6. Chiappini, C., Matteucci, F., Beers, T. C., Nomoto, K. 1999, ApJ 515, 226
7. Chieffi, A., Limongi, M., Straniero, O. 1998, Ap. J. 502, 737
8. Freiburghaus, C. et al. 1999, Ap. J. 516, 381
9. Freiburghaus, C., Rosswog, S., Thielemann, F.-K. 1999, Ap. J. 525, L121
10. Gratton, R.G., Sneden, C. 1991, A & A 241, 501
11. Heger, A., Langer, N., Woosley, S.E. 2000, Ap. J., 528, 368
12. Heger, A., Langanke, K., Martinez-Pinedo, G., Woosley, S.E. 2001, Phys. Rev. Lett 86, 1768
13. Hillebrandt, W., Niemeyer, J.C. 2000, Ann. Rev. of Astron. Astrophys. 38, 191
14. Höflich, P. 2001, in 1st KIAS Astrophysics Workshop, Seoul/Corea, IAP-Publishing, ed. I. Yi, in press
15. Höflich, P., Khokhlov, A. 1996, Ap. J., 457, 500
16. Höflich, P., Wheeler, J.C., & Thielemann, F.-K. 1998, ApJ 495, 617
17. Hoffman, R.D. et al. 1999, Ap. J. 521, 735
18. Iwamoto, K. et al. 1999, Ap. J. Suppl. 125, 439
19. Langanke, K., Martinez-Pinedo, G. 2000, Nucl. Phys. A673, 481
20. Liebendörfer, M. et al. 2001, astro-ph/0006418, Phys. Rev. D, in press
21. Livio, M. 2000 in *Type Ia Supernovae: Theory and Cosmology*, Cambridge Univ. Press, in press
22. Matteucci, F., Romano, D., Molaro, P. 1999, A&A 341, 458
23. Martinez-Pinedo, G. et al. 2000, ApJS, 126, 493
24. McLaughlin, G.C. et al. 1999, Phys. Rev. C59, 2873
25. Mezzacappa, A. et al. 2001, astro-ph/0004059, PRL, in press

26. Nakamura, T. et al. 1999, Ap. J. 517, 193
27. Nakamura, T. et al. 2001, Ap, J., in press (astro-ph/0011184)
28. Niemeyer, J.C. 1999, Ap. J. 523, L57
29. Nissen, P.E., Gustafsson, B., Edvardsson, B., & Gilmore, G. 1994, A & A 285, 440
30. Nomoto, K. et al. 1997, Nucl. Phys., A161, 79c
31. Nomoto, K., Thielemann, F.-K., Yokoi, K. 1984, Ap. J. 286, 644
32. Nomoto, K. et al. 2000, in *Type Ia Supernovae: Theory and Cosmology*, Cambridge Univ. Press, eds. J. Niemeyer & J.W. Truran, in press (astro-ph/9907386)
33. Nugent, P. et al. 1997, Ap. J. 485, 812
34. Qian, Y.-Z., Ap. J. 534, L67
35. Qian, Y.-Z., Woosley, S. E. 1996, Ap. J., 471, 331
36. Rampp, M., Janka, H.T. 2000, Ap. J. 539, L33
37. Rauscher, T., Heger, A., Hoffman, R.D., Woosley, S.E. 2001, Nucl. Phys. A, in press (astro-ph/0010021)
38. Rosswog, S.K., Freiburghaus, C., Thielemann, F.-K. 2001, Nucl. Phys. A, in press
39. Sneden, C. et al. 2000, Ap. J. 533, 139
40. Takahashi, K., Witti, J., Janka, H.-T. 1994, A&A, 286, 857
41. Thielemann, F.-K., Nomoto, K., Hashimoto, M. 1996, Ap. J. 460, 408
42. Thielemann, F.-K. 2000, in *Astronomy with Radioactivities*, eds. R. Diehl, D. Hartmann, MPE Report, p. 123
43. Thielemann, F.-K. et al. 2001, in *The Largest Explosions Since the Big Bang: Supernovae and Gamma-Ray Bursts*, eds. M. Livio, K. Sahu, N. Panagia, Cambridge University Press
44. Thomas, D., Greggio, L., Bender, R. 1998, MNRAS 296, 119
45. Truran, J.W., Cowan, J.J., Sneden, C., Burris, D.L., Pilacowski, C.A., in *The First Stars*, Springer, eds. A. Weiss et al.
46. Tsujimoto, T. et al. 1995, MNRAS 277, 945
47. Umeda, H., Nomoto, K., Nakamura, T. 2000, in *The First Stars*, Springer, eds. A. Weiss et al. (astro-ph/9912248)
48. van den Bergh, S., & Tammann, G. 1991, ARA&A 29, 363
49. Woosley, S.E., Weaver, T.A. 1994, in *Les Houches, Session LIV, Supernovae*, eds. S.R. Bludman, R. Mochkovitch, J. Zinn-Justin, Elsevier Science Publ., p. 63
50. Woosley, S.E., Weaver, T.A. 1995, Ap. J. Suppl. 101, 181
51. Woosley, S.E. 1997, Ap. J. 476, 801
52. Woosley, S.E. et al. 1994, Ap. J. 433, 229

HYPERNOVA NUCLEOSYNTHESIS AND IMPLICATIONS FOR COSMIC CHEMICAL EVOLUTION

TAKAYOSHI NAKAMURA[1], HIDEYUKI UMEDA[1], KOICHI IWAMOTO[2], KEN'ICHI NOMOTO[1], MASA-AKI HASHIMOTO[3], W. RAPHAEL HIX[4], AND FRIEDRICH-KARL THIELEMANN[5]

[1]*Department of Astronomy, School of Science, University of Tokyo, Tokyo, Japan*
[2]*Dept. of Physics, College of Science and Technology, Nihon Univ., Tokyo, Japan*
[3]*Department of Physics, School of Sciences, University of Kyushu, Fukuoka, Japan*
[4]*Department of Physics and Astronomy, University of Tennessee, Knoxville, TN and Physics Division, Oak Ridge National Laboratory, Oak Ridge, TN, USA*
[5]*Department für Physik und Astronomie, Universität Basel, Switzerland*

We examine the characteristics of nucleosynthesis in 'hypernovae', i.e., supernovae with very large explosion energies ($\gtrsim 10^{52}$ ergs). Implications for the cosmic chemical evolution and the abundances in M82 are discussed.

1 Enhancement of [Fe/O], [Ti/O], and [Si, S/O]

The explosion energy of SN1998bw is estimated to be $E = 3$ - 6×10^{52} ergs, which is about thirty times larger than that of a normal supernova (Iwamoto et al. 1998; Woosley et al. 1999; Nakamura et al. 2001). We explore nucleosynthesis in such energetic core-collapse supernovae, called 'hypernovae' (Nakamura et al. 2000), and find the following characteristics. In hypernovae, both complete and incomplete Si-burning take place in more extended, and hence, lower density regions, so that the α-rich freezeout is enhanced in comparison with normal supernova nucleosynthesis. Thus ^{44}Ca ($\leftarrow$ ^{44}Ti) and ^{48}Ti are produced more abundantly than in canonical supernovae. Oxygen and carbon burning also takes place in more extended regions for the larger explosion energy. Therefore, the fuel elements O, C, Al are less abundant while a larger amount of burning products such as Si, S, and Ar are synthesized by oxygen burning. In short, [Ti/O] and [Si, S, Ar/O] as well as [Fe/O] are enhanced in hypernovae.

2 Comparison with abundances in metal-poor stars

Ti is known to be deficient in Galactic chemical evolution models that use supernova yields currently available (e.g., Timmes et al. 1995; Thielemann et al. 1996), especially at [Fe/H] $\lesssim -1$, when Type Ia supernovae (SNe Ia) have not contributed to the Galactic chemical evolution. However, if the contribution from hypernovae is relatively large, this problem could be relaxed, because, as we have seen, the α-rich freezeout is enhanced in hypernovae and Ti is produced more abundantly than normal supernovae.

Another feature of hypernova nucleosynthesis is a large amount of Fe. One hypernova can produce 2 - 10 times more Fe than normal core-collapse supernovae. This large iron production leads to small ratios of α-elements over iron in hypernovae. In this connection, the abundance pattern of the very metal-poor binary CS22873-139 ([Fe/H] $= -3.4$) is interesting. This binary has only an upper limit to [Sr/Fe] < -1.5, and therefore was suggested to be a second generation star (Nordström et al. 2000; Spite et al. 2000). The interesting pattern is that this binary shows almost solar Mg/Fe and Ca/Fe ratios, as is the case with hypernovae. Another feature of CS22873-139 is enhanced Ti/Fe ([Ti/Fe] $\sim +0.6$; Nordström et al. 2000; Spite et al. 2000), which could be explained by a hypernova explosion.

3 Abundances in the starburst galaxy M82

X-ray emissions from the starburst galaxy M82 were observed with ASCA and the abundances of several heavy elements were measured (Tsuru et al. 1997). Tsuru et al. (1997) found that the overall metallicity of M82 is quite low, i.e., O/H and Fe/H are only 0.06 - 0.05 times solar, while Si/H and S/H are $\sim$ 0.40 - 0.47 times solar. This implies that the abundance ratios are peculiar, i.e., the ratio O/Fe is about solar, while the ratios of Si and S relative to O and Fe are as high as $\sim$ 6 - 8. These ratios are very different from those ratios in Type II supernovae (SNe II). The age of M82 is estimated to be $\lesssim 10^8$ years, which is too young for SNe Ia to contribute to enhance Fe relative to O. Tsuru et al. (1997) also estimated that the explosion energy required to produce the observed amount of hot plasma per oxygen mass is significantly larger than that of normal SNe II (here the oxygen mass dominates the total mass of the heavy elements). Tsuru et al. (1997) thus concluded that neither SN Ia nor SN II can reproduce the observed abundance pattern of M82.

Compared with normal SNe II, the important characteristic of hypernova nucleosynthesis is the large Si/O, S/O, and Fe/O ratios. Figure 1 shows the good agreement between the hypernova model and the observed abundances in M82 (Umeda et al. 2001). Hypernovae could also produce larger E per oxygen mass than normal SNe II. We therefore suggest that hypernova explosions may make important contributions to the metal enrichment and energy input to the interstellar matter in M82. If the IMF of the star burst is relatively flat compared with Salpeter IMF, the contribution of very massive stars and thus hypernovae could be much larger than in our Galaxy.

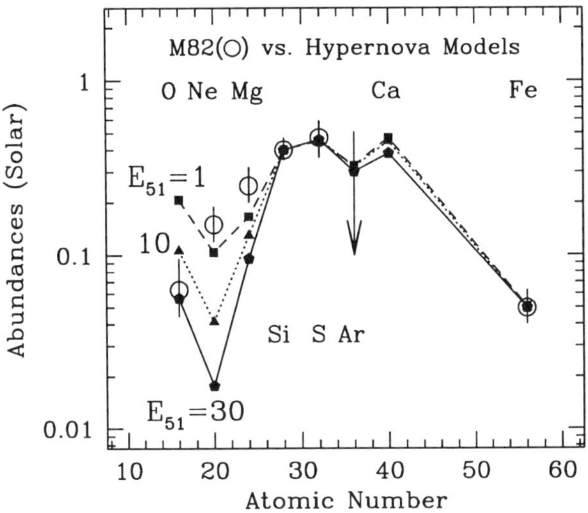

Figure 1. Abundance patterns in the ejecta of $25M_\odot$ metal-free SN II and hypernova models compared with abundances (relative to the solar values) of M82 observed with ASCA (Tsuru et al. 1997). Here, the open circles with error bars show the M82 data. The filled square, triangle, and pentagons represent E_{51}=1, 10, and 30 models, respectively, where E_{51} is the explosion energy in 10^{51} ergs. Theoretical abundances are normalized to the observed Si data, and the mass cuts are chosen to eject 0.07, 0.095, and 0.12 ($M_\odot$) Fe for E_{51} = 1, 10, and 30, respectively (Umeda et al. 2001).

References

1. Iwamoto, K. et al. Nature **395**, 672 (1998).
2. Nakamura, T., Mazzali, P. A., Nomoto, K., & Iwamoto, K. ApJ, 550, in press (2001), (asrto-ph/0007010).
3. Nakamura, T., Umeda, H., Iwamoto, K., Nomoto, K., Hashimoto, M., Hix, R., & Thielemann, F.-K. ApJ, submitted (2000), (astro-ph/0011184).
4. Nordström, B. et al., in The First Stars (MPA/ESO Astrophysics Symposia), eds. A. Weiss, T. Abel, & V. Hill (Berlin: Springer), 76, (2000).
5. Spite et al. A&A **360**, 1077 (2000).
6. Thielemann, F.-K., Nomoto, K., & Hashimoto, M. ApJ **460**, 408 (1996).
7. Timmes, F. X., Woosley, S. E., Weaver, T. A. ApJS **98**, 617 (1995).
8. Tsuru, T. G., Awaki, H., Koyama K., Ptak, A. PASJ **49**, 619 (1997).
9. Umeda, H. et al., in preparation (2001).
10. Woosley, S. E., Eastman, R. G., & Schmidt, B. P. ApJ **516**, 788 (1999).

NUCLEOSYNTHESIS WITH SPI: THE CASE OF ^{26}Al

JÜRGEN KNÖDLSEDER

Centre d'Etude Spatiale des Rayonnements, B.P. 4346, 31028 Toulouse, FRANCE
E-mail: knodlseder@cesr.fr

This article reviews our knowledge about galactic ^{26}Al production and traces the science perspectives of ^{26}Al observations with the telescope SPI aboard the future INTEGRAL observatory.

The COMPTEL telescope aboard the Compton Gamma-Ray Observatory has provided the first all-sky map[9] of the radioactive decay line of ^{26}Al at 1.809 MeV, confirming that nucleosynthesis is still active in our Galaxy. The map shows the galactic disk as the most prominent emission feature. The observed intensity profile along the galactic plane shows asymmetries and localised emission enhancements, characteristic for a massive star population that follows the galactic spiral structure. Thus, ^{26}Al production seems related to the massive star population. Correlation studies using tracer maps for various source candidate populations strongly support this suggestion. Knödlseder et al. (1999a) demonstrated that the 1.809 MeV gamma-ray line emission follows closely the distribution of galactic free-free emission which is powered by the ionising radiation of stars with initial masses $> 20\ M_{\odot}$. This suggests that explosive ^{26}Al production in supernovae may be less important than previously thought (e.g. Timmes et al., 1995), and hydrostatic nucleosynthesis in massive mass-losing stars may possibly be the primary production channel for galactic ^{26}Al.

Having established the correlation between 1.809 MeV emission and massive star populations, ^{26}Al becomes an excellent tracer of recent galactic star formation. By refining the knowledge about the 1.809 MeV emission distribution, the spectrometer SPI aboard the future INTEGRAL observatory (see Paul, these proceedings) will provide a unique view on the star formation activity in our Galaxy. To illustrate this potential, the radial ^{26}Al mass density distribution as derived by COMPTEL is shown in Fig. 1. The bulk of galactic star formation occurs at distances of less than 6 kpc from the galactic centre. Star formation is also present within the central 3 kpc of the Galaxy, although at a poorly determined rate. There are indications for enhanced star formation between 3 – 6 kpc, coinciding with the molecular ring structure. Enhanced star formation is also seen in the solar neighbourhood (8 – 9 kpc) which probably corresponds to activity in the local spiral arm.

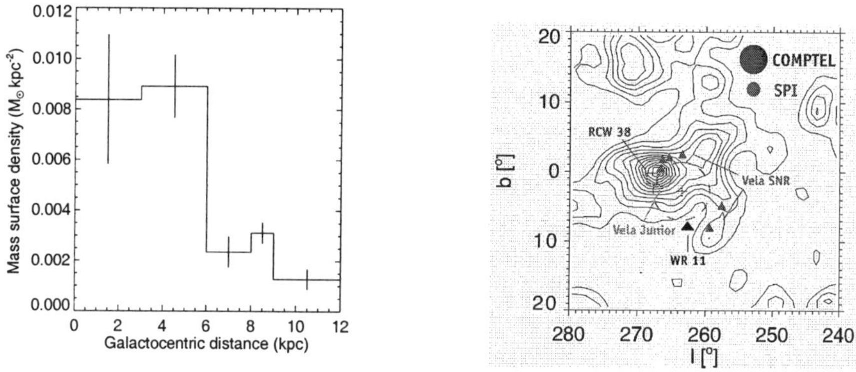

Figure 1. Radial ^{26}Al density profile (left) and 1.809 MeV image of Vela (right).

The ^{26}Al profile is probably not strictly proportional to the star formation rate since ^{26}Al nucleosynthesis may depend on metallicity.[6] Valuable information about the metallicity dependence will come from a precise determination of the 1.809 MeV longitude profile by SPI, and the comparison of this profile to other tracers of star formation activity, such as galactic free-free emission. A precise determination of the 1.809 MeV latitude profile by SPI will provide important information about the dynamics and the mixing of ^{26}Al ejecta within the interstellar medium. High velocity ^{26}Al has been suggested by measurements of a broadened 1.809 MeV line by the *GRIS* spectrometer[7], although this observation is at some point inconsistent with the earlier observation of a narrow line by *HEAO 3*.[5] In any case, the propagation of ^{26}Al away from its origin should lead to a latitude broadening with respect to the scale height of the source population, and the observation of this broadening may allow the study of galactic outflows and the mass transfer between disk and halo of the Galaxy. The excellent energy resolution of SPI will easily allow to decide whether the 1.809 MeV line is broadened or not, and the improved angular resolution and sensitivity with respect to COMPTEL will allow to determine the scale height of the galactic ^{26}Al distribution much more precisely.

The expected energy resolution of SPI of ~ 2.5 keV at 1.809 MeV converts into a velocity resolution of ~ 400 km s^{-1}, allowing for line centroid determinations of the order of 50 km s^{-1} for bright emission features. Thus, in the case of an intrinsically narrow 1.809 MeV line, line shifts due to galactic rotation should be measurable by SPI. Although this objective figures certainly about the most ambitious goals of SPI observations, a coarse distance determination of 1.809 MeV emission features based on the galactic rotation curve seems in principle possible.

Complementary to the study of the large-scale distribution will be observations of nearby, localised 1.809 MeV emission regions, such as Vela, Cygnus, Carina, or Orion. The aim of these observations will be the identification of emission counterparts at other wavelengths in order to associate the nucleosynthesis activity to individual objects or specific groups such as OB associations or young open clusters. Already with COMPTEL data, such studies have proven to provide important insights in the nature of the ^{26}Al progenitor sources, and in constraining nucleosynthetic yields for individual objects.

To illustrate the potential of SPI, a contour map of the 1.809 MeV emission in the Vela region obtained by COMPTEL is shown in Fig. 1. There is a wealth of potential ^{26}Al sources in this field, but the limited angular resolution and sensitivity of COMPTEL does not allow for a clear identification of the dominant contributors. SPI will help to overcome this problem. Deep exposures of localised emission features will sufficiently constrain the 1.809 MeV morphology to associate the structure with candidate sources in the field. In the Vela region a detection of the Wolf-Rayet star γ-Vel is awaited, and the contributions of the Vela SNR, the new RX J0852.0-4622 supernova remnant, and OB star associations should be measurable. In the Cygnus region, the young globular cluster Cyg OB2 should be detectable by SPI, allowing the study of nucleosynthesis in an individual massive star association.[1] In Carina, the point-like 1.809 MeV emission feature detected by COMPTEL[4] may be resolved, and the contributions from individual young massive star clusters identified. And in Orion, hints[8] for 1.809 MeV emission from a nearby OB association may be confirmed, enabling the study of nucleosynthesis in one of the closest and best studied star forming regions in the Milky Way.

References

1. Knödlseder, J., et al. 2001, in: 4th INTEGRAL workshop, in press
2. Knödlseder, J., Bennett, K., Bloemen, H., et al. 1999a, A&A, 344, 68
3. Knödlseder, J., Dixon, D., Bennett, K., et al. 1999b, A&A,
4. Knödlseder, J., Bennett, K., Bloemen, H., et al. 1996, A&A, 120, 327C
5. Mahoney, W.A., Ling, J.C., Wheaton, A., et al. 1984, ApJ, 286, 578
6. Meynet, G. 1994, ApJS, 92, 441
7. Naya, J.E., Barthelmy, S.D., Bartlett, L.M., et al. 1996, Natur, 384, 44
8. Oberlack, U. 1997, PhD Thesis, Technische Universität München
9. Oberlack, U., Bennett, K., Bloemen, H., et al. 1996, A&AS, 120C, 311
10. Timmes, F.X., Woosley, S.E., Hartmann, D.H., et al. 1995, ApJ, 449, 204

NEUTRON-CAPTURE ELEMENT TRENDS IN THE HALO

C. SNEDEN
Department of Astronomy, University of Texas, Austin, TX 78712, USA
E-mail chris@verdi.as.utexas.edu

J. J. COWAN
Department of Physics and Astronomy, University of Oklahoma, Norman, OK 73019, USA
E-mail: cowan@physast.nhn.ou.edu

J. W. TRURAN
Department of Astronomy & Astrophysics, University of Chicago, Chicago, IL 60637, USA
E-mail: truran@nova.uchicago.edu

In a brief review of abundances neutron-capture elements (Z > 30) in metal-poor halo stars, attention is called to their star-to-star scatter, the dominance of *r*-process synthesis at lowest metallicities, the puzzle of the lighter members of this element group, and the possibility of a better *s*-/*s*-process discriminant.

1 Introduction

Most isotopes of elements with atomic numbers Z > 30 are synthesized via neutron capture reactions. These "n-capture" elements are the majority of the periodic table. In the so-called *s*-process, neutron fluxes are small enough to allow β-decays to occur between successive neutron captures, and element buildup proceeds along the valley of β-stability. In the *r*-process, huge but short-lived neutron fluxes overwhelm β-decays, creating very neutron-rich isotopes out to the neutron drip line. Then multiple β-decays drive the nuclei back to the valley of β-stability. The final isotopic mixes will be very different in *r*- and *s*-process synthesis episodes, as will the elemental abundances summed over the isotopes. The *s*-process mainly occurs in the helium shell burning phases of low-intermediate mass AGB stars, while the *r*-process is probably associated with the explosive deaths of high mass stars. Thus, *r*- and *s*-process elemental abundance variations with metallicity should trace the contributions of different mass ranges of stars over the Galaxy's history.

Detailed comparisons of solar system meteoritic abundances of n-capture isotopes (Cameron[1], Käppeler *et al.*[2]) have yielded accurate breakdowns into *r*- and *s*-process parts of each isotope. But stellar spectroscopy generally cannot resolve the finely split isotopic absorption components of atomic transitions, so elemental r- and s-process abundances summed over the isotopes have been computed for solar system material by Burris *et al.*[6], and references therein. No n-capture element with $Z \leq 83$ can be identified solely with the r- or s-process but some have been clearly dominated by a single synthesis mode. For example, Ba and Ce have s-process fractions, and Eu, Gd, and Dy have r-process fractions greater than 80% in solar system material. Such elements are usually labeled as r-process or s-process, regardless of their synthesis history in non-solar Galactic material.

Here we will comment on several aspects of observed n-capture abundance distributions of metal-poor stars: (a) the bulk n-capture abundance levels; (b) the relative s-/r-process dominance among heavier n-capture elements in the lowest metallicity stars, (c) the best spectroscopic indicator of those ratios, and (d) the difficulty in ascribing a nucleosynthetic origin for the lighter n-capture elements. Theoretical interpretation of these observational results will be considered in a companion paper by Cowan *et al.* in this volume.

2 Variations in Bulk n-Capture Abundance Levels

Early large-sample surveys of n-capture elements in metal-poor stars (*e.g.* Luck and Bond[3], Gilroy *et al.*[4]) discovered apparently significant star-to-star scatter in the ratios [n-capture/Fe], where the n-capture elements usually considered were Sr, Y, Zr, Ba, La, Nd, and Eu. But the observational data were usually of modest resolution (R < 30,000) and sometimes low S/N, raising questions about the reality of the n-capture scatter. Recent analyses of much better spectroscopic data have decided the issue unambiguously in favor of σ[n-capture/Fe] > 1 at metallicities [Fe/H] < –2. The reality of these large star-to-star variations can be demonstrated through inspection of spectra of stars with similar atmospheric parameters ($T_{eff}/\log g/[Fe/H]/v_t$) but enormous line strength differences of some n-capture elements. For example, compare the spectra of HD 115444 and HD 122563 (Figure 1 of Westin *et al.*[5]; Δ[Fe/H] $\simeq$ –0.2, Δ[Eu/H] $\simeq$ +1.0) and those of HD 6268 and BD +9 2870 (Figure 3 of Burris *et al.*[6]; Δ[Fe/A] $\simeq$ 0.0, Δ[Eu/H] $\simeq$ 0.9). The analyses of these stars and others (*e.g.* McWilliam *et al.*[7]; Ryan *et al.*[8]) quantify these impressions; clearly σ[n-capture/Fe] grows with decreasing [Fe/H]. To date there have been few stars with [Fe/H] $\sim$ –3 (in which n-capture lines are weak) fully analyzed with the highest quality spectra, and in stars with [Fe/H]>-2 many n-capture lines are saturated, so abundances derived from them are less reliable. New VLT data should help at the low metallicity end of the scale; future n-capture surveys at intermediate metallicities would also help.

3 Dominance of the r-Process at Lowest Metallicities

Perhaps the first indication of non-solar abundance ratios of n-capture elements in halo stars was the Wallerstein *et al.*[9] assertion that the bright very metal poor ([Fe/H] $\simeq$ –2.7) giant HD 122563 has very small amounts of these elements compared to Fe. Re-analysis of their data by Pagel[10], and further studies of in succeeding decades have shown that Ba is much more deficient than Eu in HD 122563: [Ba/Eu] $\simeq$ –0.7. But attention to s-/r-process abundance mixes was really first brought by Spite and Spite[11], who found persistent deficiencies of Ba with respect to Eu in their sample of 11 halo stars. These abundances (as well as those of Y) suggested to Truran[12] that "...the observed trends follow in a natural and straightforward manner from the assumption that the Y and Ba in the most extreme metal-poor stars represent products of r-process nucleosynthesis." Since then the supporting data have improved but the basic conclusion has not. The observational attack has been on two fronts: (a) determination a few key abundance ratios in many stars, and (b) mapping the detailed abundance pattern in a few n-capture-rich stars.

Metal-poor stars with [n-capture/Fe] $\gg$ 0 have ideal spectra for studying the entire range of n-capture elements, because the ubiquitous Fe-peak element transitions weaken substantially relative to those of n-capture elements, allowing rarely-detected elements (*e.g.* Tb, Ho, Hf) to be detected. Two prominent examples are CS 22892-052 (Sneden *et al.*[13]) and HD 115444 (Westin *et al.*[5]). These stars have n-capture abundance patterns for elements with Z $\geq$ 56 that are near-perfect matches to a scaled solar r-process abundance distribution. The presence of s-process synthesis cannot be detected. These stars' abundances suggest that regardless of the site(s) that are responsible for the r-process nuclei, they release their products into the ISM in a remarkably uniform pattern.[a]

Most observers simply estimate s-/r-process influence from a few abundance ratios of elements whose syntheses are dominated by one or the other mechanism in the solar system. In Figure 1 we show likely candidate elements, through comparison of their solar number densities $\log_{10}\epsilon$. Clearly there should be a major abundance shift between elements Ba$\leftrightarrow$Ce with respect to Eu$\leftrightarrow$Tm if the synthesis shifts from the solar (combined s- and r-) mix to a pure r-process. In practice, this has come down to derivations of [Ba/Eu] ratios, since these two elements have the strongest transitions in most routinely accessible spectral regions. Nearly all large-sample n-capture abundance studies of low metallicity stars have correlated [Ba/Eu] with [Fe/H],

[a] S-process-rich stars do exist at very low metallicities, such as the "CH stars" discussed by Norris *et al.*[14] and McWilliam *et al.*[7], but these appear to be in the minority.

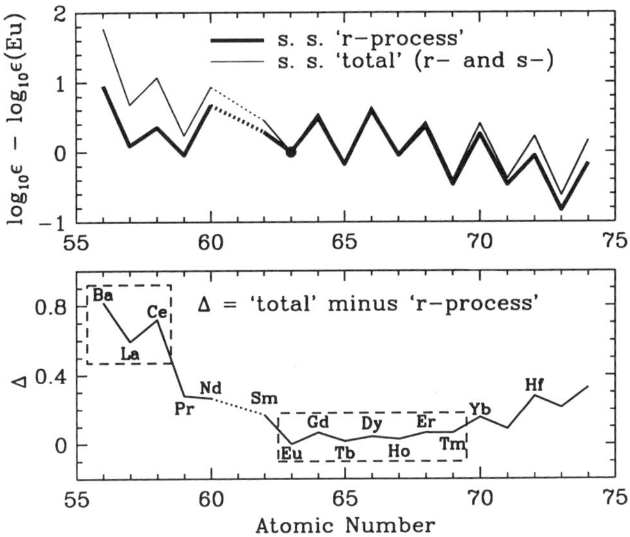

Figure 1. Top panel: total solar-system abundances and r-process parts only of n-capture elements in the range $56 \leq Z \leq 74$, normalized to the Eu abundance. Bottom panel: the subtraction of the two curves from the top panel. Ratios of any element within the upper left dashed box to any element within the lower middle dashed box should in principle be useful indicators of the s-/r-process ratios in n-capture material.

and a consensus has arisen that $<$[Ba/Eu]$>$ $\sim$ 0 for $-2 \leq$ [Fe/H] $\leq$ 0, and then the mean ratio declines to ~ -0.9 as [Fe/H] ~ -3. Little is known about $<$[Ba/Eu]$>$ at even lower metallicities because the Eu transitions usually become extremely weak; it would be useful to detect this element even in a few stars with [Fe/H] < -3.

Burris *et al.*[6] have estimated the s-process component of Ba in their sample of halo stars by first assuming that Eu is a 100% r-process product and that the r-process part of Ba is fixed by the solar system r-process Ba/Eu ratio. They then subtract the inferred $Ba_{r-process}$ from Ba_{total}. The result (see their Figure 7) suggests a complete absence of the s-process for [Fe/H] ≤ -2.8, and a stochastic rise to a full (solar-system) s-/r-process ratio by [Fe/H] ~ -2.

Ratios of [Ba/Eu] are subject to large uncertainties, because the four commonly employed Ba II lines are often very strong even in metal-poor stars, and because these lines suffer isotopic and hyperfine structure (hfs) splitting, so Ba abundances are very dependent on assumed microturbulent velocity and isotopic abundance fractions. Of the alternate s-process abundance indicators,

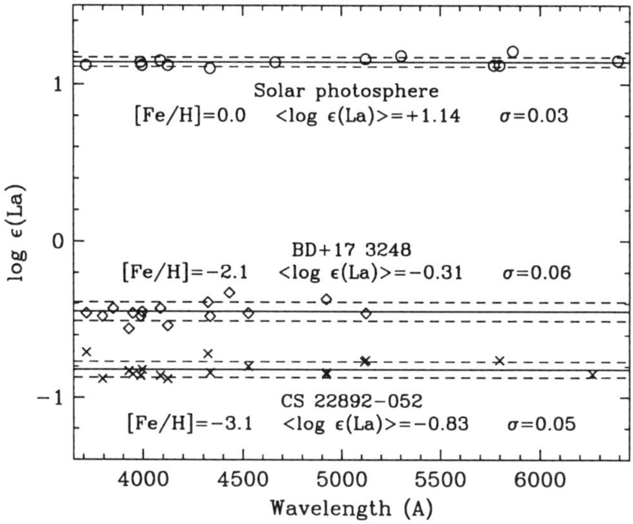

Figure 2. La abundances in the Sun and two metal-poor giants as functions of wavelength. The solid and dashed lines are drawn at the mean and mean$\pm\sigma$ abundances for each star.

Ce has only weak lines, rendering it almost useless unless [n-capture/Fe] $\gg 0$. But La has many weak and strong lines, and has only one stable isotope, ^{139}La. The atomic data for La has been not the best, but Lawler *et al.*[15] have determined new accurate gf's and hfs components for La II. In Figure 2 we show results of the application of these lab data to the spectra of the Sun and two metal-poor but n-capture-rich stars. The line-to-line scatter is satisfactorily low and a reasonable mean abundance for all these stars is obtained. In preliminary tests we have derived La/Eu abundance ratios for these and a few other stars, finding excellent accord with the predicted pure r-process ratio at the lowest [Fe/H] values, and gradual rise to solar system values at higher metallicities. We believe that La/Eu ratios can be more accurately determined than Ba/Eu ratios, and it is possible that La/Eu may become the standard s-/r-process indicator in future spectroscopic investigations of metal-poor stars.

4 The Lighter n-Capture Elements

Brief notice must be given of n-capture elements in the range $31 \leq Z \leq 55$. Until recently, the elements Rb, Sr, Y, and Zr represented the observational situation of the entire group. These elements are known to present a puzzle. Their overall abundance levels correlate poorly with heavier ($Z \geq 56$) n-capture elements (see references cited previously). Neither a pure solar system r-process, or pure s-process (or indeed any linear combination) provides a satisfactory match to their observed abundances (Cowan *et al.*[16]). New near-UV spectra for some metal-poor stars are rapidly providing abundance data for other elements in this atomic number range such as Ag (Crawford *et al.*[17]), and Sneden *et al.*[13] report first detections of five new light n-capture elements in CS 22892-052. But their derived abundances do not shed further insight on the matter, and still there is no reasonable fit to scaled solar-system abundance patterns. The study of these elements should be actively pursued in the future.

Acknowledgments

We thank all of the colleagues who have collaborated with us on various studies of n-capture elements in halo stars. This research has received support from NSF grants AST-9987162 to C.S. and AST-9986974 to J.J.C., from DOE contract B341495 to J.W.T., and from the Space Telescope Science Institute grant GO-8342.

References

1. A. G. W. Cameron, *Ap. Sp. Sci.* **82**, 123 (1982).
2. F. Käppeler *et al.*, *Rep. Prog. Phys.* **52**, 945 (1989).
3. R. E. Luck and H. E. Bond, *Ap. J.* **292**, 559 (1985).
4. K. K. Gilroy *et al.*, *Ap. J.* **327**, 298 (1988).
5. J. Westin *et al.*, *Ap. J.* **530**, 783 (2000).
6. D. L. Burris *et al.*, *Ap. J.* **544**, 302 (2000).
7. A. McWilliam *et al.*, *Astr. J.* **109**, 2757 (1995).
8. S. G. Ryan *et al.*, *Ap. J.* **471**, 254 (1996).
9. G. Wallerstein *et al.*, *Ap. J.* **137**, 280 (1963).
10. B. E. J. Pagel, *Roy. Obs. Bull.* no. 104 (1965).
11. M. Spite and F. Spite, *Astr. & Ap.* **67**, 23 (1978).
12. J. W. Truran, *Astr. & Ap.* **67**, 23 (1978).
13. C. Sneden *et al.*, *Ap. J.* **533**, L139 (2000).
14. J. Norris *et al.*, *Ap. J.* **488**, 350 (1997).
15. J. E. Lawler *et al.*, *Ap. J.* submitted (2000).
16. J. J. Cowan *et al.*, *Ap. J.* **439**, 51 (1995).
17. J. L. Crawford *et al.*, *Astron. J.* **116**, 2489 (1998).

NEW INSIGHTS INTO NEUTRON CAPTURE NUCLEOSYNTHESIS

G. J. MATHEWS
Center for Astrophsics, Univ. of Notre Dame, Notre Dame, IN 46556, USA
E-mail:gmathews@nd.edu

N. IWAMOTO, W. AOKI, T. KAJINO
National Astronomical Observatory, Mitaka, Tokyo, 181-8588, Japan

M. Y. FUJIMOTO
Department of Physics, Hokkaido University, Sapporo Hokkaido 060-0810, Japan

We explore models for s-process nucleosynthesis in the recently discovered [Fe/H]= −2.7 s-enhanced star LP625-44. We apply schematic pulsed neutron source models and show that optimum fits to the observed elemental distribution imply that either a single neutron exposure or that few multiple exposures have contributed to produce the observed abundances. These results are interpreted in the context of the Case II' third dredge-up scenario recently proposed to occur in stars with [Fe/H] $\lesssim -2.5$. We argue that this star maybe the first evidence for this new s-process paradigm in low-metallicity stars.

In this short contribution we would like to report on some new insights into neutron capture nucleosynthesis by slow neutron capture (the s-process). This insight is being gained by the study of metal-poor stars in the Galactic Halo. This work in some ways has grown out of the pioneering work of Jim Truran and his collaborators towards unraveling the sites for stellar neutron capture. Jim was the first to point out the importance of heavy-element abundances as a means to probe neutron-capture nucleosynthesis in the early Galaxy. In a remarkably intuitive paper, Jim was the first to recognize [1] that the newly emerging data from metal-poor stars exhibit an abundance pattern for the heaviest elements which looks nearly identical to the solar-system r-process abundances. This was a very important result in its time. It argued strongly for an interpretation that only a single r-process site is responsible for a large fraction of the heaviest r-process elements.

In other pioneering work, Jim and collaborators [2] were among the first to analyze what we now call the "weak" s-process component in the helium-burning cores of massive stars. Subsequently, in some seminal papers with Icko Iben [3,4] the first detailed models of the "main" s-process component in the helium-burning shells of thermally-pulsing red giants were developed.

The work reported here is complementary to Jim's earlier work in that we also are considering first observations of abundances in peculiar metal-poor stars and we are also developing detailed stellar models to explain these abundances. The present work is also in contrast to Jim's work in that we consider recent observations [5,6] of the carbon-rich metal-poor subgiants, LP625-44 and LP7006-7, which exhibit not an r-process enhancement but an s-process enhancement. The abundances on these stars reflect the nucleosynthesis from the earlier red-giant phase of a binary companion that deposited its outer envelopes onto the surfaces of these stars during an episode of Roch-lobe overflow. Contrary to the r-process enhanced stars, these stars show an abundance pattern which looks nothing like the solar s-process abundances. Although one could adjust [7] parameters of the conventional radiative ^{13}C burning s-process in asymptotic giant branch stars to reproduce these observed stellar abundances, here we argue that these data may be evidence for a new s-process paradigm.

Figure 1 shows measured abundances for LP625-44. For comparison, a "best fit" parameterized s-process model for this star is also shown. The s-process model consists of repeating thermal pulses of constant neutron density ($n_n = 10^7$ cm^{-3}) and temperature ($T_9 = 0.15$). The most sensitive parameter for these fits is the mean neutron exposure τ_0. This quantity relates to the fraction of material surviving from one pulse to the next r, and to the neutron exposure per pulse $\tau = n_n v_T \Delta t$, where Δt is the pulse duration. The mean exposure per pulse is then $\tau_0 = \tau/(-\ln r)$. The large required mean exposure implies a small surviving fraction $r \approx 0.1$ and large exposure per pulse, $\tau = 0.71$. This is significant. It implies that very few pulses could have contributed to make the s-process abundances for this star. In fact, an almost equivalent fit to these abundances can be obtained for only a single neutron exposure instead of multiple thermal pulses.

This result is interesting because this is exactly the behavior recently anticipated by Fujimoto et al. [8] in what was referred to as their Case II' CN-rich third dredge-up. In Fujimoto et al. it was proposed that metal-poor stars with [Fe/H]$\lesssim -2.5$ should not experience multiple neutron exposures but only one or two thermal pulses before the convective helium shell falls too far below the outer convective shell to allow further dredge up. The metallicity for LP625-44 is [Fe/H] $= -2.7$. Hence, in this star we may be seeing observational confirmation for a new s-process paradigm, i.e. an s-process occurring in a single neutron exposure for metal-poor stars. If this interpretation is correct it is testable. Stars with a metallicity greater than [Fe/H] > -2.5, should begin to exhibit an abundance distribution which is consistent with a more solar-like multiple distribution of exposures. Hence, there is motivation to search for and study other similar s-process enhanced halo stars, albeit those with slightly higher metallicity.

One of the authors (GJM) wishes to thank the hospitality of the National

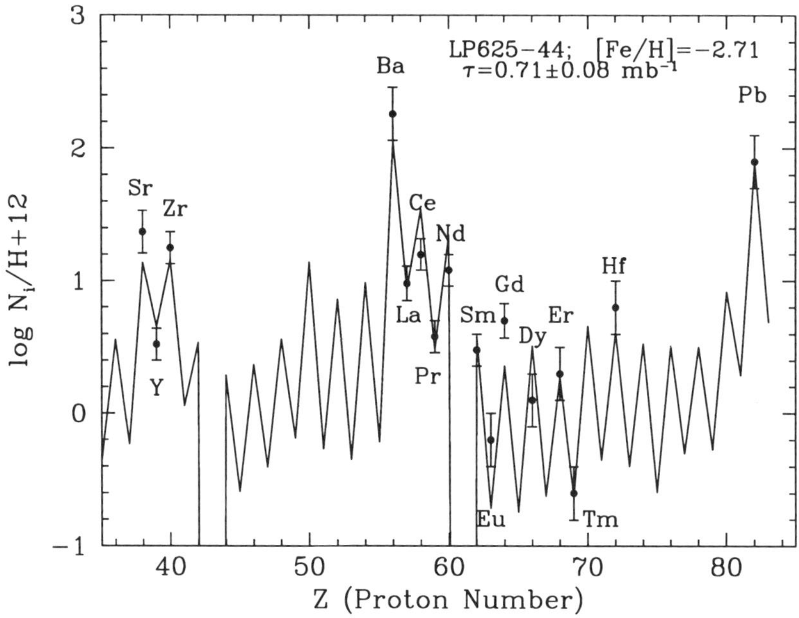

Figure 1: Best fit to observed abundances for a model with $n_n = 10^7$ cm^{-3}, $\tau = 0.71 \pm 0.08$ mb^{-1}, and $r = 0.1$.

Astronomical Observatory of Japan. Work at UND supported in part by DoE Nuclear Theory Grant DE-FG02-95ER40934.A2.

References

1. J. Truran, Astron. Astrophys., 97, 392 (1981).
2. S.A. Lamb *et al*, Astrophys. J., 217, 213 (1977).
3. I. Iben and J. Truran, Astrophys. J., 220, 980 (1978).
4. K. Cosner, I. Iben and J. Truran, Astrophys. J. Lett., 238, L91 (198).
5. W. Aoki *et al*, Astrophys. J. Lett., 529, L25 (2000).
6. W. Aoki *et al*, Astrophys. J., submitted (2001).
7. M. Busso, R. Gallino and G.J. Wasserburg, Ann. Rev. Astron. Astrophys., 37, 239 (1999).
8. M.Y. Fujimoto, Y. Ikeda and I. Iben, Astrophys. J. Lett., 529 L25 (2001).

NUCLEAR CHRONOMETERS

J. J. COWAN
Department of Physics & Astronomy, University of Oklahoma, Norman, OK 73019, USA
E-mail: cowan@mail.nhn.ou.edu

C. SNEDEN
Department of Astronomy, University of Texas, Austin, TX 78712, USA
E-mail: chris@verdi.as.utexas.edu

J. W. TRURAN
Department of Astronomy, University of Chicago, Chicago, IL 60637, USA
E-mail: truran@nova.uchicago.edu

Observations of metal-poor Galactic halo stars indicate that the abundance pattern of the (heaviest) neutron-capture elements is consistent with the scaled solar system r-process abundances. Utilizing the radioactive (r-process) element thorium, age determinations have been made for several of these same stars, placing constraints on both Galactic and cosmological age estimates.

1 Introduction

Stellar observations of neutron-capture elements (*i.e.*, those produced in the slow- or rapid-neutron capture processes) provide a number of important clues about the formation history, chemical evolution and age of these elements in the Galaxy. In particular, the abundances of the n-capture elements in metal-poor (*i.e.*, low iron abundance) Galactic halo stars can be employed to determine the nature of the progenitors and the nucleosynthesis history in the early Galaxy. Further, the changes in the stellar abundance trends with respect to metallicities (and times) can provide clues about the nature of Galactic chemical evolution and the sites for neutron-capture synthesis. Just as importantly, certain long-lived radioactive elements, such as thorium and uranium, are produced entirely in the r-process. The abundance levels of these nuclear chronometers in the most metal-poor halo stars provide direct age determinations and hence set lower limits on Galactic and cosmological age estimates.

2 Stellar n-Capture Abundance Distributions

There have been a number of recent ground-based (see *e.g.* Sneden *et al.*[1], McWilliam[2], Burris *et al.*[3], Sneden *et al.*[4]) and space-based (Cowan *et al.*[5], Sneden *et al.*[6]) observations of metal-poor Galactic halo stars. Several of these studies have concentrated on the star CS 22892–052 (Cowan *et al.*[7], Sneden *et al.*[1], Sneden *et al.*[4]). This star, while ultra-metal-poor ([Fe/H] = –3.1), is neutron-capture rich with an average abundance level of 30-40 times solar (with respect to iron) for the elements heavier than Ba. These large over-abundances have allowed the detection of a number of elements for the very first time in very metal-poor halo stars. In Figure 1 (at the top) the neutron-capture abundances in CS 22892–052 for elements with atomic number, Z, $\geq$ 56 are shown. Also shown in comparison is the solar system elemental r-process abundance distribution, indicated by a solid line. This distribution has been determined by summing the individual isotopic contributions from the s- and the r-process in solar system material, as determined by Käppeler *et al.*[8] from neutron-capture cross section measurements. (See Burris *et al.*[3] for more details.) It is clear in Figure 1 that there is a remarkable agreement between the elemental abundances in this star and the scaled solar system r-process abundances – the relative abundances of the heavy n-capture elements are similar in CS 22892–052 and the solar system. While this argument has been made previously, the detection of additional elements in CS 22892–052 has strengthened that argument.

While the agreement between scaled solar system r-process abundances and CS 22892–052 is very suggestive, it might be considered coincidental if it was a unique situation. This is not in fact the case. Recent observations of other neutron-capture rich halo stars have shown a similar pattern. The n-capture abundances in one such star, HD 115444, with a metallicity similar to CS 22892–052, is shown in the lower part of Figure 1 (Westin *et al.*[9]). The abundances of the elements in this star have been arbitrarily (uniformly) reduced (i.e., moved downward) for illustration purposes - the actual abundances are approximately a factor of 5 lower than in CS 22892–052. The agreement between the abundances in HD 115444 and the scaled solar system elemental r-process abundance curve is again excellent and extends through the heaviest stable elements including Os and Pt in the so-called 3rd r-process peak. Still further support for this agreement is provided by recent observations of stars in the globular cluster M15. The abundances of the heaviest neutron-capture elements in three giants in that cluster are again consistent with a scaled solar system r-process abundance distribution (Sneden *et al.*[10]).

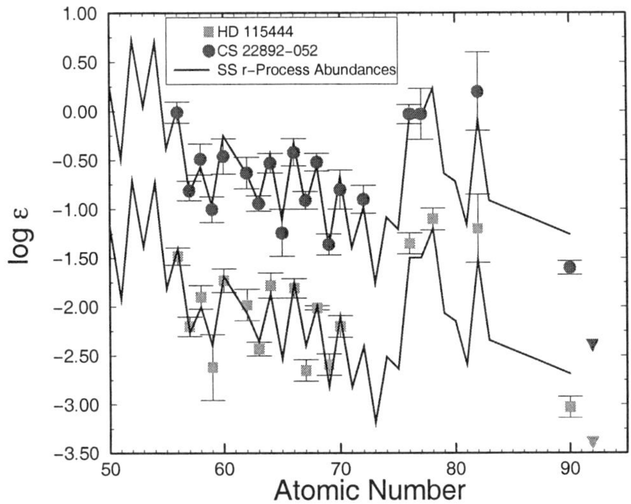

Figure 1. The heavy n-capture abundance patterns for the two stars CS 22892–052 and HD 115444 are compared with the solar system r-process abundance distribution (solid line) .

It is also apparent from Figure 1 that elements, such as barium, in CS 22892–052 and HD 115444 that are typically thought of as s-process elements (and predominantly produced by that process in solar system material – see Burris *et al.*[3]) were almost exclusively produced by the r-process early in the history of the Galaxy. This result affirms earlier predictions (Truran[11]) concerning the dominance of r-process nucleosynthesis early in the history of the Galaxy. The detection of r-process material in the metal-poor and old halo stars, including the heaviest r-process elements, indicates an early presence of these elements in the Galaxy. That in turn also places constraints on the type of progenitors of the r-process elements and, in particular, suggests rapidly evolving progenitors of the halo stars. This follows since the early Galaxy appears to be chemical inhomogeneous in n-capture elements (see Sneden *et al.* in this volume) implying a relatively short (with respect to stellar evolutionary) timescale between the death of the progenitors and the formation of the halo stars.

Until now the element domain between Zr and the heavier n-capture elements, *i.e.* $40 < Z < 56$, has been largely unexplored. For the first time Sneden *et al.*[4] have detected six elements in this region in CS 22892–052, as

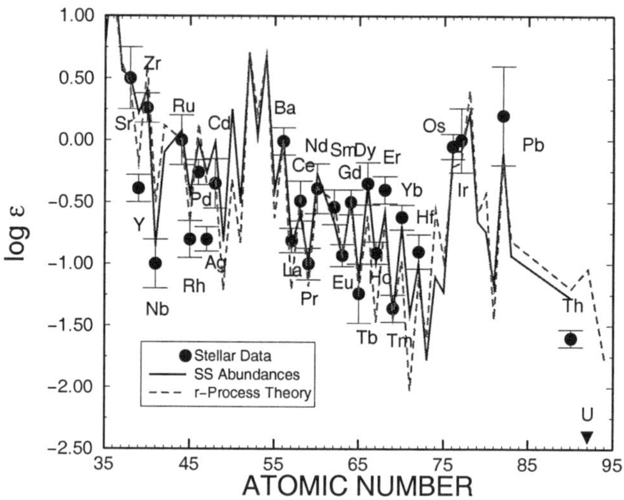

Figure 2. The total neutron-capture abundance pattern in CS 22892–052 compared with the solar system *r*-process abundances (solid line) and a theoretical *r*-process abundance curve (dashed line).

shown in Figure 2. It is clear from the abundance comparisons in Figure 2, however, that the lighter neutron-capture elements, specifically the six newly detected elements in CS 22892–052, do not fall on the same curve as the heavier neutron-capture elements. These recent observations lend support to earlier suggestions - based upon analyses of solar system meteoritic material - of a second *r*-process with the heavier elements (Ba and above) produced in more rapidly occurring events and the lighter elements in less commonly occurring syntheses (Wasserburg *et al.*[12], Wasserburg and Qian[13]; see Sneden *et al.* in this volume for discussion.)

3 Ages from Cosmochronometry

The long-lived radioactive nuclei (known as chronometers) in the thorium-uranium region are formed entirely in the *r*-process and can be used to determine the ages of stars and the Galaxy. The technique of using the observed stellar thorium, particularly as a ratio to a stable element, to obtain ages was first introduced by Butcher[14] and developed by François *et al.*[15]. The detection of thorium in CS 22892–052 and HD 115444 (see Figure 1), therefore, allows for chronometric stellar age estimates. The age is derived by comparing the observed stellar abundance ratio Th/Eu with predictions of the initial (zero-decay) value of Th/Eu at the time of the formation of these elements. (Eu is employed since it is also a predominantly an *r*-process element and has spectral features that are easily observed in most metal-poor stars.) Cowan *et al.*[16] and Westin *et al.*[9] compared the observed values of Th/Eu in CS 22892-052 and HD 115444 with that in solar system material to obtain an average age of 13.8 Gyr. This estimate is in fact a lower limit on the age of these very low metallicity stars. This follows since the solar Th/Eu ratio (at the formation of the solar nebular 4.5 Gyr ago) represents a lower limit to the zero decay-age *r*-process abundances – Eu is stable and Th (although constantly produced and ejected into the interstellar medium) is partially decayed.

Further refinements in the age estimates can be obtained by determining the zero-decay abundance of thorium when produced in an *r*-process site. This, however, depends critically upon theoretical predictions for very neutron-rich nuclei, whose properties are in general not obtainable by experimental determination. We can, however, employ the stable abundances of the neutron-capture elements and the solar system *r*-process abundances to constrain the theoretical calculations and, hence, assess the reliability of any nuclear mass formulae utilized in those calculations. One such calculation, based upon the work of Pfeifer *et al.*[17] (see also Cowan *et al.*[16] for details), is shown in Figure 2 by a dashed line. It is seen in the figure that this calculation gives very good agreement for the abundance predictions of the heavy *n*-capture elements in CS 22892–052 and the solar system abundances. In this same calculation, constrained by the stable abundances, abundance predictions for the radioactive heavy elements are simultaneously derived. There is still an uncertainty associated with the abundance predictions for the radioactive chronometers thorium and uranium. Unlike the stable heavy solar system elements, such as Pt, which can be compared directly with theoretical predictions, no such comparison is possible with Th or U. However, the α-decay of heavier nuclei ($209 < A < 255$), including those from Th and U, is responsible for the production of the stable lead and bismuth isotopes, whose abundances are known and thus can be compared with theoretical estimates. Employing this important solar system abundance constraint on the theoretical *r*-process calculations to determine the zero-decay Th/Eu abundances, Cowan *et al.*[16] and Westin *et al.*[9] found an average age for CS 22892–052 and HD 115444 of 15.6 Gyr, with an estimated uncertainty of $\simeq$ 4 Gyr. Sneden *et al.*[10] report on thorium detections in several stars in the globular cluster M15 and obtain ages, $\simeq$ 14 Gyr, that are also consistent with those found for the two halo stars.

While there are still uncertainties associated with the nuclear physics predictions, this technique offers great promise for stellar dating. One large advantage is that such radioactive age determinations of very low metallicity stars are independent of, and thus avoid the inherently large uncertainties in, Galactic chemical evolution models. With the addition of more observations and better nuclear physics predictions, the current radioactive age uncertainties should be reduced. Finally, we note that while uranium has not been observed so far in either CS 22892–052 or HD 115444 (upper limits on U are indicated by inverted triangles in Figures 1 & 2), in the near future it may be possible to observe this element in some metal-poor halo stars. The addition of this other nuclear chronometer would help to constrain further age estimates for the Galaxy and the Universe.

Acknowledgments

We thank all of our colleagues who have collaborated with us on various studies of n-capture elements in halo stars. This research has received support from NSF grants AST-9986974 to J.J.C., AST-9987162 to C.S. and from DOE contract B341495 to J.W.T., and from the Space Telescope Science Institute grant GO-8342.

References

1. C. Sneden *et al.*, *ApJ* **467**, 819 (1996).
2. A. McWilliam, *AJ* **115**, 1640 (1998).
3. D. L. Burris *et al.*, *ApJ* **544**, 302 (2000).
4. C. Sneden *et al.*, *ApJ* **533**, L139 (2000).
5. J. J. Cowan *et al.*, *ApJ* **460**, L115 (1996).
6. C. Sneden *et al.*, *ApJ* **496**, 235 (1998).
7. J. J. Cowan *et al.*, *ApJ* **439**, L51 (1995).
8. F. Käppeler *et al.*, *Rep. Prog. Phys.* **52**, 945 (1989).
9. J. Westin *et al.*, *ApJ* **530**, 783 (2000).
10. C. Sneden *et al.*, *ApJ* **536**, L85 (2000).
11. J. W. Truran, *A&A* **97**, 391 (1981).
12. G. J. Wasserburg *et al.*, *ApJ* **466**, L109 (1996).
13. G. J. Wasserburg and Y.-Z. Qian, *ApJ* **529**, L21 (2000).
14. H. R. Butcher, *Nature* **328**, 127 (1987).
15. P. François *et al.*, *A&A* **274**, 821 (1993).
16. J. J. Cowan *et al.*, *ApJ* **521**, 194 (1999).
17. B. Pfeiffer *et al.*, *Z. Phys. A* **357**, 235 (1997).

UNCERTAINTIES IN THE TH COSMOCHRONOMETRY

S. GORIELY

IAA - ULB - Campus de la Plaine CP226, B-1050 Brussels

Recent observations of Th in ultra-metal poor stars revived the old idea that the Th cosmochronometry could provide an age estimate of the oldest stars in the Galaxy. Numerous astrophysics and nuclear physics uncertainties still keep the Th cosmochronometry on unreliable grounds, unless Th and U lines could be observed accurately and simultaneously in ultra-metal-poor stars.

One of the methods called for estimating the age of the Galaxy is based on the analysis of observed abundance of some long-lived radionuclides. The recent observation of r-process elements, including Th, in ultra-metal-poor halo stars, such as CS 22892-052 or HD 115444[1], has brought some renewed excitement in the estimate of the age of the Galaxy on grounds of the Th cosmochronometry. With a metallicity as low as [Fe/H] = –3 and a composition enriched in some pure r-elements, these stars provide strong evidence that the production of heavy elements by the r-process already took place early in the history of the Galaxy. Moreover, the abundance pattern of the r-elements heavier than Ba shows a striking similarity with the solar system r-abundance distribution, leading to the tempting (though hazardous) conclusion that the r-process mechanism is "unique". This conclusion has been critically analyzed by Goriely & Arnould[2] who showed that this assumption may be valid indeed, but is by far *not the only possible one*, as the observations in the limited $56 \leq Z \leq 76$ range are equally compatible with an abundance distribution that does not fit the solar one outside the observed domain. This ambiguity is assigned to the fact that the observed CS 22892-052 pattern of abundances reflects primarily nuclear physics properties, and not one or another specificity of a blend of r-process events. This universality assumption is a fundamental prerequisite to build a Th cosmochronometry upon the abundance analysis of metal-poor stars at the present time.

If we disregard the difficulty associated with the fundamental assumption relative to the unicity of the r-process site in the Galaxy, the Th cosmochronometry reduces to the evaluation of the ratio of the observed to the original (i.e produced by the universal r-process) Th abundances. A high accuracy has already been achieved observationally with errors reduced to $\log \epsilon = 0.08$[1] affecting the stellar age by about 3.7 Gyr. However, the estimate of the Th production is obviously model dependent. The Th abundance has to be determined within less than 16% if we hope to predict the age of the star within less than 3 Gyr. The r-process remains the most complicated nucleosynthetic process to model from the astrophysics as well as nuclear physics point of view[3]. On the nuclear physics side, the nuclear structure properties (such as the nuclear masses, deformation, ...) of thousands of exotic nuclei located between the valley of β-stability and the neutron drip line have to be known, as well as interaction properties such as the (n,γ) and (γ,n) rates, α- and β-decay half-lives and the fission probabilities. Despite much recent experimental effort, none of those quantities are known for the nuclei involved in the r-process, so that they have to be extracted on theoretical grounds and are subject to the associated uncertainties. In particular, the nuclear masses play a crucial role in the Th abundance prediction. The extrapolation to the Th abundance obtained with different mass models lead to a star age that can differ by more than 20 Gyr[4].

On top of these nuclear difficulties, the question of the astrophysical conditions under which the r-process can develop is far from being settled. The site(s) of the r-process is (are) not identified yet, all the proposed scenarios facing serious problems. For this reason, only simplistic parametric approaches, such as the so-called canonical model, can be used to estimate the Th production. This approach is not free from uncertainties. Among the different thermodynamic parameters entering the canonical model, the most critical one affecting the Th synthesis is obviously the maximum number of neutrons captured by the initial seed nuclei. This quantity remains so far unknown and can affect the stellar age by about 16 Gyr[4]. Finally, another difficulty lies in the very s- or r-origin of the Pb and Bi isotopes. The predicted Th abundance is directly correlated to the estimated r-contribution to the solar Pb and Bi abundances. The uncertainties in the solar r-abundances affect the age determination by about 20 Gyr[4].

For all these astrophysics and nuclear physics difficulties, it is impossible at the present stage to determine reliably the age of a ultra-metal-poor star from parametric models of the r-process. The age of CS 22892-052 is found to lie in the $7 \lesssim T^*[\mathrm{Gyr}] \lesssim 39$ range. The Th cosmochronometry could be put on safer grounds, if Th and U lines could be observed accurately and simultaneously in ultra-metal-poor stars[4].

References

1. C. Sneden et al., *ApJ* **496**, 235 (1998).
2. S. Goriely and M. Arnould, *A&A* **322**, L29 (1997).
3. M. Arnould and K. Takahashi, *Rep. Prog. Phys.* **62**, 393 (1999).
4. S. Goriely and B. Clerbaux, *A&A* **342**, 881 (1999).

THE P- AND R-PROCESSES: REVIEWS AND OTHER VIEWS

M. ARNOULD, S. GORIELY AND M. RAYET
IAA - ULB - Campus de la Plaine CP226, B-1050 Brussels, Belgium

A review is presented of the p-process in Type II supernovae, one of its goals being to enlighten the changes in views on this nucleosynthesis mechanism since the work of Jean and Jim on the subject in 1975. Specific discussions are also devoted to cases of particular interest, like the light Mo and Ru stable isotopes, the rare nuclide ^{138}La or the radionuclide ^{146}Sm. Some comments of diverse natures are also made on the r-process. These considerations do not aim at really providing an exhaustive review of the many nuclear physics and astrophysics intricacies of this process. In contrast, they are hoped to complement or to put in perspective other views that are often expressed in relation with this nucleosynthesis mechanism

1 AN INTRODUCTION TO THE P-PROCESS IN TYPE II SUPERNOVAE

The p-nuclides are those neutron-deficient stable isotopes of the elements heavier than iron that cannot be produced by the s- or r-processes of neutron captures. In view of the fact that the number of research papers devoted to the p-process called for their production remains inferior to the number of p-nuclides after more than about 50 years of nuclear astrophysics research, we like to refer to the 'nuclear astrophysics p-nuts' when talking about the p-nuclides.[a] Either independently or in collaboration, Jean and Jim have had the good taste to contribute to the increase in the number of papers devoted to the p-process. In ref.[1], Jean put forth the possible spallogenic origin of selected p-nuclides, while in ref.[2], Jean and Jim pushed to its limits the original proposal[3] to produce the p-nuclides at temperatures in excess of 10^9 K in the H-rich envelopes of Type II supernovae (SNII). Already at the end of the sixties, it was realized that such high temperatures were unlikely to be reached in the envelopes of massive star explosions. This is why one of the authors of this review (M.A.) proposed in his 1971 thesis work (see also ref.[4]) to locate the p-process in the deep O-Ne rich layers of massive stars either in their pre-supernova or supernova phases.

[a] The fact that the p-nuclides are much less abundant (by factors of the order of 100 to 1000) than the corresponding more neutron-rich isotopes in the solar system, which is the only astrophysical location to-date where they are observed, is certainly not a viable explanation for the scarcity of the efforts devoted to the understanding of their origin. The validity of this statement appears in a crystal-clear way when referring to the light elements Li, Be and B, to which myriads of astrophysics papers have been devoted (a scan of the contributions to this conference is quite convincing in this respect)

The SNII p-nuclide synthesis has been the most actively studied since then, and is certainly to-date the most developed model and the most successful one in reproducing the p-nuclide content of the bulk solar system material (e.g. ref.[5], hereafter RAHPN, and references therein). The extensive RAHPN calculations rely on SNII explosion models for stars with initial masses in the $13 \lesssim M \lesssim 25$ $M_\odot$ range[6]. The p-nuclides are produced in layers with explosion temperatures peaking in the $(1.8\text{-}3.3)\times 10^9$ K range. The nuclear flow in these conditions is dominated by photodisintegrations of the (γ,n), (γ,p) or (γ,α) types, complemented mainly with some neutron captures. Figure 1 shows the p-process overproduction factors obtained for the 25 $M_\odot$ model star selected by RAHPN, and for two sets of reaction rates. It appears that about 60% of the overproductions displayed in Fig. 1 fit the solar system composition within a factor 3. This conclusion is in general not drastically affected either by the selected set of reaction rates, or by the stellar mass, at least in the range explored by RAHPN. However, the predictions are not free from shortcomings, some of them being discussed below.

2 THE PUZZLE OF THE LIGHT Mo AND Ru ISOTOPES

One of the most embarassing problems which is evidenced by Fig. 1 concerns the severe underproduction of the light Mo ($^{92-94}$Mo) and Ru ($^{96-98}$Ru) isotopes, this conclusion being robust to various changes in the input physics (see RAHPN). Exotic solutions have been proposed to remedy this puzzle, calling in particular for accreting neutron stars or black holes (e.g. ref. [9]). The level of the contribution of such sites to the solar system content of the p-nuclides is clearly impossible to assess in any reliable way. In contrast, we have emphasized many times in the past that the problem might just be due to some misrepresentation of the production in the He-burning cores of massive stars of the s-nuclides which are the seeds for the p-process. This 'non-exotic' idea has been put on quantitative grounds recently[10]. More specifically, the seed abundances from ref.[11] used to construct Fig. 1 are replaced by results obtained with the 'adopted' NACRE reaction rates, except for the rate r_{22} of ^{22}Ne$(\alpha, n)^{25}$Mg which is the main neutron source in the considered stars. This rate is in fact varied in the range of uncertainties defined in the NACRE compilation.

Use of the NACRE 'adopted' value for r_{22} leads to the classical 'weak' s-process component pattern exhibiting a decrease of the overproduction (with respect to solar) of the s-nuclides by a factor ranging from about 100 to unity when the mass number A increases from about 70 to 100. This 'canonical' picture changes gradually with an increase in the neutron production rate, the

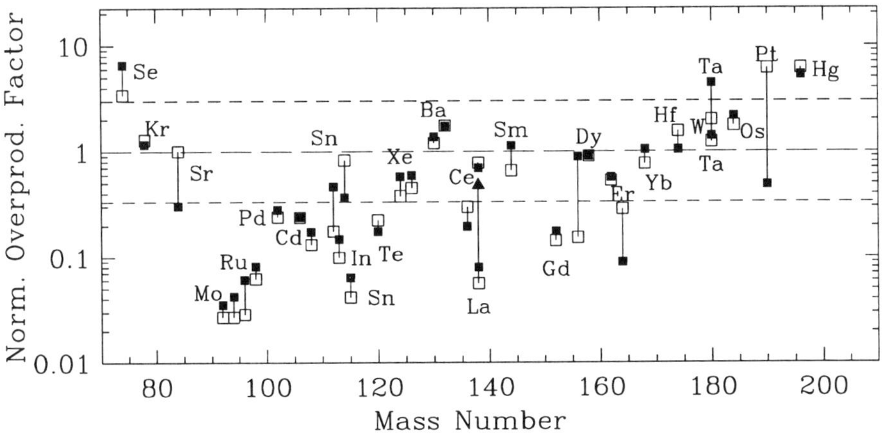

Figure 1. Normalized p-nuclide overproduction factors (with respect to solar) in the 25 $M_\odot$ star model adopted by RAHPN. The open squares are obtained with the reaction rates considered by RAHPN, while the black squares relate to the use of the 'adopted' rates from the NACRE compilation[7] and of the Hauser-Feshbach predictions from the code MOST[8]. In both cases the s-process seed abundances are those of ref.[11]. The NACRE and MOST rates are available in the Brussels Nuclear Astrophysics Library (http://www-astro.ulb.ac.be). The black triangle at ^{138}La indicates the yield increase by a factor of 6 obtained by adopting an evaluated upper (lower) limit of the rate of ^{138}La production (destruction) (see Sect. 3)

overproduction of heavier and heavier s-nuclides increasing steadily. More specifically, adoption of the NACRE upper limit for r_{22} leads to s-nuclide overproduction factors from about 10^3 to 10^4 in the $70 \lesssim A \lesssim 90$ mass range, decreasing to values around unity for mass numbers as high as about 150 (ref.[10]). These results obtained with increased r_{22} values may shake some aspects of a long tradition in our views concerning the s-process. In particular, massive stars appear to be able to produce quite substantial amounts of s-nuclides around Ba. At the same time, Asymptotic Giant Branch stars synthesize s-elements in the Zr region. It is our opinion that none of these predictions can really act as a deterrent to r_{22} substantially in excess of the NACRE adopted value. Quite on the contrary, some specific problems might even find a solution, like the (admitedly quite uncertain) Ba abundance in SN1987A[10].

Use of the s-process seed abundances obtained with the NACRE adopted and upper values for r_{22} leads to the 25 $M_\odot$ p-nuclide overproduction factors

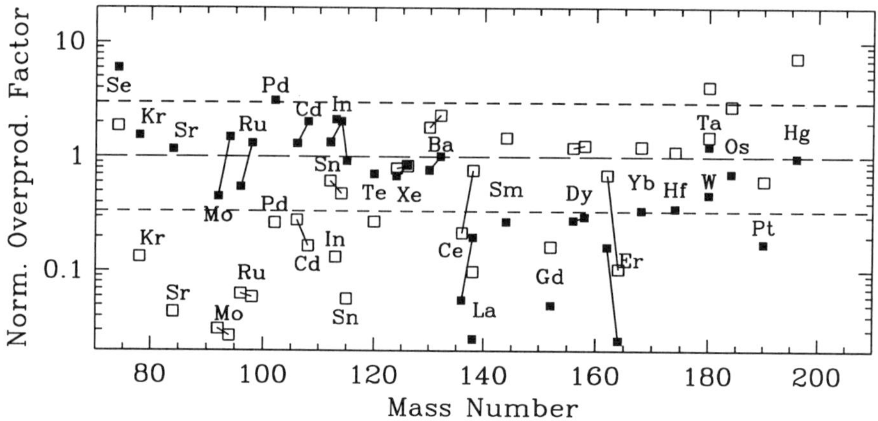

Figure 2. Same as Fig. 1 (black squares), but for s-process seeds obtained with the NACRE rates. Open and black squares correspond to the NACRE 'adopted' and 'upper' $^{22}Ne(\alpha, n)^{25}Mg$ rates, respectively. Solid lines join different p-isotopes of the same element

shown in Fig. 2. Changes in the shape of the p-nuclide abundance distribution are clearly noticeable. In particular, the Kr-Sr-Mo-Ru trough observed for the NACRE adopted r_{22} disappears when the upper limit for that rate is considered. Details about this effect as well as other interesting by-products of the increase of r_{22} are discussed in ref.[10].

3 SOME OTHER SPECIFIC QUESTIONS OF p-PROCESS NATURE

Other important questions directly connected with the p-process have been the subject of specific research. One of them concerns the experimental study of the $^{144}Sm(\alpha, \gamma)^{148}Gd$ rate and the analysis of its implication in the development of a possible nucleo-cosmochronology of the p-process. The measurements have proved to be of substantial nuclear physics interest, and have been used to estimate the p-process production ratio of ^{144}Sm to the radionuclide ^{146}Sm ($t_{1/2} \approx 10^8$ y). From the confrontation of these evaluations with the corresponding solar system abundance ratio inferred from the analysis of primitive meteorites, one might hope to estimate the age of the p-nuclides. Large uncertainties still remain in this chronology in spite of the nuclear physics advances reported above. The interested reader can find a detailed account of this study in ref.[12].

On the other hand, some study has been devoted to ^{138}La which is, along with ^{180}Ta, an odd-odd p-nuclide with very low solar system abundances. In spite of their scarcity, the very origin of these two species has long been a nucleosynthesis puzzle. As discussed by RAHPN and as shown in Figs. 1 and 2, ^{180}Ta is found to be a natural p-process product. In contrast, ^{138}La is systematically underproduced in all p-process calculations performed so far. This situation has triggered an examination of the sensitivity to nuclear uncertainties of its predicted yields, which result from a subtle balance between its production by $^{139}\mathrm{La}(\gamma,n)^{138}\mathrm{La}$ and its photodisintegration at temperatures around $T \simeq 2.4 \pm 0.1\ 10^9$K. No experimental cross sections are available, so that the ^{138}La yields of Fig. 1 entirely rely on the predictions of the code MOST. Related reasonable uncertainties in the neutron capture rates on ^{138}La and on ^{137}La (from which the ^{138}La destruction rate is evaluated) at typical p-process temperatures are found to amount to factors of about 10 and 3, respectively. If the corresponding lower limit of the ^{138}La destruction rate and upper limit of its production rate are adopted, the ^{138}La yield from the 25 $M_\odot$ supernova shown if Fig. 1 is increased by a factor of about 6. This puts the ^{138}La overproduction at the same level as the one of the neighboring p-nuclides. Nuclear physics measurements are needed in order to shed light on a possible nuclear solution to the ^{138}La mystery.

Other studies relating to the p-process are currently pursued. They concern in particular the computation of this synthesis mechanism in C-deflagrating white dwarfs (in collaboration with J. Jose, Barcelona), or in the framework of 2-D simulations of the O-burning zone of a 20 $M_\odot$ presupernova star (in collaboration with D. Arnett and co-workers, Tucson).

4 SOME VIEWS ABOUT THE r-PROCESS AND RELATED QUESTIONS

Since the seminal work of ref.[13], as well as of some early research by Jim and his collaborators, much nuclear physics and astrophysics efforts have been devoted to the r-process aimed at accounting for the production of the neutron-rich stable isotopes of the elements heavier than iron. In spite of that, one is clearly left with a variety of difficult questions concerning the very site(s) of the r-process, and the properties of the very exotic neutron-rich nuclides it involves. In the following, we raise some questions and make some short sketchy comments, a fraction of which might be considered by certain researchers in the field as highly unorthodox:

Question 1. *What is (are) the true astrophysical site(s) of the r-process?*
Comment 1. The lack of answer to this essential question prevents any firm (sometimes far-reaching) conclusion to be drawn in relation with the r-process.

Question 2. *What are the true properties of the highly exotic neutron-rich nuclides involved in the r-process?*
Comment 2. One cannot derive the properties of exotic nuclei and identify the 'best' nuclear models from a confrontation between r-nuclide observations and predictions from a (toy) r-process model. This exercise (which is performed by some) is a kind of highly complex inverse problem the solution of which is unreliable, and even possibly non-unique.

Question 3. *Can one build reliable r-process cosmic clocks (which go beyond age limits, or are more reliable than other dating methods) from long-lived actinides observed in the solar system?*
Comment 3. In several papers, Jim and his collaborators have attempted to answer this important question. Our views may be summarized as follows (e.g. ref.[14] for references): the classical r-process chronometers (^{232}Th-^{238}U, ^{235}U-^{238}U) are most likely quite poorly reliable if all possible sources of uncertainties (arising from the r-process production ratios, the models for the chemical evolution of the Galaxy, ...) are taken into account in a fair way. In fact, the best hope in the field may well come from the ^{187}Re-^{187}Os pair, even if it is not free from uncertainties. This is due to the s-process nature of ^{187}Os (apart from the radiogenic contribution from ^{187}Re). There is also reasonable hope for the nuclear problems raised by that pair to find answers in the nuclear physics laboratory in a foreseeable future.

Question 4. *Can one build a reliable cosmic clock from Th and other r-nuclides observed in some very metal-poor stars?*
Comment 4. The recent observation of r-nuclides, including Th, in some ultra-metal-poor halo stars (see the review by C. Sneden in these Proceedings) has been the subject of some excitement and of several studies by Jim and some of his collaborators. Quite generally, these Th observations are considered to provide a r-process clock of unmatched quality. This conclusion relies heavily on some critical assumptions. One of them concerns the 'uniqueness' of the r-process in view of the striking similarity of the r-nuclide abundances in the analyzed stars and the corresponding solar system abundances. This view is challenged in ref.[15]; see also S. Goriely, these proceedings). On the other hand, and even if the r-process uniqueness is taken for granted, it is close to impossible to evaluate the level of Th production in the r-process with the required accuracy (i.e. within about 16%; ref.[16] and S. Goriely, these pro-

ceedings). The uncertainties in the predictions of these yields related to the difficulties in the involved nuclear physics and their impact on age determinations are examined in detail in ref.[16] (see also S. Goriely, these Proceedings). Other possible sources of embarassment for a Th-based chronology are discussed and analyzed in the same references. A note of hope might come from a reliable determination of the Th/U ratio at the surface of ultra-metal-poor stars.

5 A FINAL COMMENT

There is no doubt that Jean and Jim have contributed in an important way to the understanding of the p- and r-processes. They have quite fortunately left unanswered questions, so that there has been room for others to tackle some interesting problems. More generally, they have been actively involved in the shaping of many chapters of nuclear astrophysics in a very professional manner, but also with a precious sense of humanism.

References

1. J. Audouze, *A&A* **8**, 436 (1970).
2. J. Audouze and J.W. Truran, *ApJ* **202**, 204 (1975).
3. E.M. Burbidge et al., *Rev. Mod. Phys.* **29**, 547 (1957).
4. M. Arnould, *A&A* **8**, 436 (1976).
5. M. Rayet et al., *A&A* **298**, 517 (1995).
6. M. Hashimoto, *Progr. Theor. Phys.* **94**, 663 (1995).
7. C. Angulo and the NACRE collaboration, *Nucl. Phys.* **A656**, 3 (1999).
8. S. Goriely, in *Capture Gamma-Ray Spectroscopy and Related Topics*, AIP Conf. Proc. **529**, ed. S. Wender, 287 (2000).
9. H. Schatz et al., *Phys. Rep.* **294**, 167 (1998).
10. V. Costa et al., *A&A* **358**, L67 (2000).
11. N. Prantzos et al., *A&A* **234**, 211 (1990).
12. E. Somorjai et al., *A&A* **333**, 1112 (1998).
13. P.A. Seeger et al., *ApJS* **11**, 121 (1965).
14. M. Arnould and K. Takahashi, *Rep. Prog. Phys.* **62**, 393 (1999).
15. S. Goriely and M. Arnould, *A&A* **322**, L29 (1997).
16. S. Goriely and B. Clerbaux, *A&A* **342**, 881 (1999).

ABUNDANCES IN TURN-OFF AND SUBGIANT STARS IN GLOBULAR CLUSTERS

R.G. GRATTON et al.
Osservatorio Astronomico di Padova, Vicolo dell'Osservatorio, 5
35122 Padova, Italy
E-mail: gratton@pd.astro.it

UVES on Kueyen (VLT UT2) was used to gather high dispersion spectra ($R > 40,000$) for a quite large number of stars at the main sequence turn-off and at the base of the RGB in NGC6397 and NGC6752. The [Fe/H] values we found are $-2.03\pm0.02\pm0.04$ and $-1.42\pm0.02\pm0.04$ respectively. In both clusters the [Fe/H] obtained for TO stars agrees perfectly with that obtained for stars at the base of the RGB. The star-to-star scatter both in O and Fe is very small in NGC6397. Our results show clearly and for the first time that the O-Na anticorrelation is present among unevolved stars in NGC 6752.

1 INTRODUCTION

Observations of turn-off (TO) and lower RGB stars in globular clusters (GC) are basic in a number of astrophysical issues: (i) Most critical features in the determination of distances to GC (and hence ages) using the main sequence fitting method are related to possible systematic differences in the ways reddening and metallicity are derived for field subdwarfs and clusters [1]. Error bars can be reduced by using cluster stars that are intrinsically very similar to nearby stars with good parallaxes. (ii) Metal-poor stars near the TO offer a basic way to derive the Lithium abundance produced by Big Bang [2]. A critical feature is here related to the impact of diffusion, that may reduce the surface Li abundance. Most critical observations are provided by a comparison between Li abundances in TO stars and stars at the base of the RGB, where Li survived in the outer layers is diluted by the growth of the convective envelope. (iii) Abundances of the lighter elements (from C to Al) in GC stars show a complex pattern. A Na-O anticorrelation among GC red giants was discovered by the Lick-Texas group [3]. This is an indication of the presence of elements processed through the complete CNO-cycle; at temperatures where complete CNO cycle occurs, Na may be produced by proton capture on ^{22}Ne nuclei [4]. Proposed explanations include an "in situ" mechanism (very deep mixing in RGB stars, possibly activated by core rotation) and an external source of material (either primordial proto-cluster gas or processed material from a polluting companion). Observation of a Na-O anti-correlation among TO stars in GC would clearly rule out in-situ mechanisms.

Earlier explorative studies of stars close to the TO and at the base of the RGB of a few clusters have been based on HIRES at Keck (M92) and EMMI at NTT (exploiting the presence of closer clusters in the southern emisphere). However, only a handful of stars could be observed, and at very low S/N. Only UVES at the Kueyen (VLT UT2) Telescope is able to provide extensive data, crucial to clarify most of these issues. Here we present first results of the analysis of high dispersion spectra ($R > 40,000$) of a rather large sample of stars at the TO and at the base of the RGB in NGC6752 and NGC6397 (ESO Large Program 165.L-0263).

2 OBSERVATIONS AND ANALYSIS

Observations were carried out in June and September 2000. In the same runs we also acquired data for 25 field stars with good parallaxes from the Hipparcos catalogue and metallicity similar to that of the GCs. For each cluster, we selected two groups of stars: (i) dwarfs about 0.2 mag brighter than the TO; and (ii) stars at the base of the RGB. Stars with likely strong and weak CN bands were selected from Strömgren photometry [5]. Observed spectral ranges are from about 3,500 Å to about 4,700 Å, and from 5,700 Å to about 8,700 Å. This very broad spectral range allows observations of a large number of spectral features. Equivalent widths for a large number of lines were measured using an authomatic procedure; typical errors are $\pm 3 - 5$ mÅ.

One of the aims of the present analysis was to derive new reddening estimates for GCs by comparing the observed colours with effective temperatures $T_{\rm eff}$'s obtained from the spectra; our temperatures are based on fitting of the wings of the H_α profiles. Internal errors of individual temperatures are estimated to be 150-200 K for individual stars. The adopted temperature scale is confirmed (with larger error bars) by analysis of line excitation. For field stars, we averaged temperatures from colours (after correcting them to the scale given by H_α) and those derived from H_α. We find that these $T_{\rm eff}$'s are on the same scale of Alonso et al. [6]. In the following analysis, we have adopted for all cluster stars the same average temperatures.

3 RESULTS

(i) The [Fe/H] value for NGC6397 is [Fe/H]$=-2.03 \pm 0.02 \pm 0.04$ (internal and systematic errors), less than derived from giants by Carretta & Gratton [7] and Zinn & West [8], but in agreement with the recent analysis of giants and subgiants by Castilho et al. [9]. The [Fe/H] value for NGC6752 ([Fe/H]$=-1.42\pm0.02\pm0.04$) coincides with the value derived from giants by Carretta & Gratton [7], and is slightly higher than the value quoted by Zinn & West [8]

(ii) In both clusters, [Fe/H] obtained for TO stars agrees perfectly (within a few percents) with that obtained for stars at the base of the giant branch; this is a constraint on the impact of diffusion in stellar models. The star-to-star scatter is extremely small: the r.m.s. scatter is only 0.04 dex (i.e. 10%) for NGC 6397. This seems a very homogeneous cluster as far as Fe abundances are considered.

(iii) For NGC6397 we obtain [O/Fe]=+0.21 ± 0.05, a value lower than usually found in metal-poor stars, but in agreement with previous determinations for red giants. The scatter of individual values is small, and none of the program stars seem to be oxygen-poor.

(iv) For NGC6752, there is a clear anticorrelation between Na and O among TO stars and subgiants (similar to that seen among giants in this and other clusters). Also, an anticorrelation is observed between Mg and Al. Na-poor stars (i.e. stars with [Na/Fe]< 0.2) in NGC6752 have [O/Fe]=+0.30 ± 0.04 (7 stars, r.m.s.=0.10 dex). Extremes in Al abundances differ by over 1 dex, while for Mg the star-to-star scatter is smaller. Note that the most Al-rich stars are also the most Na-rich ones. Given the adopted sample selection criteria, extreme cases of Na-poor and Na-rich stars may be overrepresented among observed stars.

(v) The abundance of Li in NGC6397 TO stars and subgiants is similar to that of metal-poor field stars. The same occurs for O-rich, Na-poor stars in NGC6752. However, O-poor, Na-rich stars in NGC6752 have lower Li abundances, although Li lines could be detected in all TO stars.

References

1. Gratton, R.G., Carretta, E., Clementini, G. 1998, in Post-Hipparcos Cosmic Candles, Heck A. & Caputo, F. eds., Kluwer, Dordrecht, p. 89
2. Spite, F., Spite M. 1982, A&A, 115, 357
3. Kraft, R.P, Sneden, C., Smith, G.H., Shetrone, M.D., Langer, G.E., & Pilachowski, C.A. 1997, AJ, 113, 279
4. Denissenkov, P.A., & Denissenkova, S.N. 1990, SVAL, 16, 275
5. Grundahl, F., Vandenberg, D.A., Stetson, P.B., Anderson, M.I., & Briley, M. 1999, in The Galactic Halo: from Globular Clusters to Field Stars (astro-ph/9909447)
6. Alonso, A., Arribas, S., & Martinez-Roger, C. 1996, A&A, 313, 873
7. Carretta, E., & Gratton, R.G. 1997, A&AS, 121, 95
8. Zinn, R., & West, M.J. 1984, ApJS, 55, 45
9. Castilho, B.V., Pasquini, L., Allen, D.M., Barbuy, B., & Molaro, P., 2000, A&A 361, 92

METALLICITY OF OLD OPEN CLUSTERS

A. BRAGAGLIA, M. TOSI
Osservatorio Astronomico di Bologna, via Ranzani 1, I-41027 Bologna, Italy
E-mail: angela@bo.astro.it, tosi@bo.astro.it

E. CARRETTA, R.G. GRATTON
Osservatorio Astronomico di Padova, vicolo Osservatorio 5, I-35122 Padova, Italy
E-mail: carretta@pd.astro.it, gratton@pd.astro.it

The metallicity of four old open clusters has been measured from high resolution, high S/N FEROS spectra, as part of a program to study the Galactic disk properties through open clusters.

1 Motivation, collection of data and preliminary analysis

Elemental abundances of Galactic old open clusters (OCs) are important to study the chemical evolution of the disk, and in particular to determine the time behaviour of chemical abundance gradients. From the theoretical point of view, a gradient steepening or flattening with time implies quite different scenarios for the disk formation, but Galactic chemical evolution models are not sufficiently constrained yet to unequivocally discriminate between these two possibilities (see e.g. Tosi 1996, 2000). OCs represent a fine tool to understand whether and how the abundance gradients change with time, but they do not provide yet a reliable answer, since 1) lack of homogeneity in the derivation of cluster parameters (like age, distance, and reddening) confuses the picture, and 2) analysis of high resolution, high S/N spectra, the best way to determine elemental abundances, has been performed only on less than twenty old open clusters.

To overcome these problems, we have started a program to homogeneously analyse with great accuracy a significant sample of open clusters at various galactic locations and covering a wide range in age and metallicity. Deep photometry and the synthetic Color-Magnitude diagrams method (e.g. Sandrelli et al. 1999) provide age, distance, reddening and approximate metallicity, while high resolution spectroscopy is used to determine accurate element abundances (e.g. Carretta et al. 2000, Bragaglia et al. 2001).

We present here preliminary results on four old OCs, namely NGC 2506, Cr 261, IC 4651, and NGC 6134, as part of our project. The spectra were obtained with the ESO 1.52m telescope mounting FEROS on April 28 to May 1, 2000. The resolving power is R=48,000, and the wavelength coverage is wide

Table 1. Average values of iron abundance for the four observed OCs. Also given are the Galactocentric distances in kpc and ages in Gyr.

Cluster	< [Fe/H] >	Stars	Clump stars	$R_{GC}(kpc)$	age (Gyr)
Cr 261	$+0.37 \pm 0.02$	6	2	6.00	6–7
NGC 6134	$+0.47 \pm 0.07$	3	3	7.10	0.7
IC 4651	$+0.16 \pm 0.01$	5	3	7.65	2.5
NGC 2506	-0.02 ± 0.01	4	2	10.50	1.5–2

(3700-8600 Å). The spectra were reduced using the dedicated Data Reduction Software. All objects were chosen among probable cluster members, and membership is confirmed by the radial velocities determined from our spectra. We have examined first the red clump stars (leaving red giant branch stars for later) since they have rather uniform parameters within each cluster, and are also warmer than RGB stars, making abundance analysis of their spectra easier.

Equivalent widths were measured on the unidimensional spectra, after removing the blaze function and defining a local continuum. EWs were used to derive the best set of atmospheric parameters (effective temperature from line excitation, gravity from the equilibrium of ionization, microturbulence by zeroeing trends with expected line strength) and elemental abundances.

We have determined only [Fe/H] values up to now, leaving α-elements, Li, etc. for future investigation, and preliminary results are presented in Table 1. Extensive comparisons with synthetic spectra will later be used to confirm the adopted abundances. From these preliminary results the four OCs seem to conform to the observed radial metallicity gradient (see e.g. Friel 1995), but further analysis is required, and more clusters have to be observed with high resolution spectroscopy.

References

1. A. Bragaglia et al., *AJ* **121**, 327 (2001).
2. E. Carretta, A. Bragaglia, M. Tosi, G. Marconi, in *ASP Conf. Ser.*, eds. R. Pallavicini, G. Micela, S. Sciortino, **198**, 273 (2000).
3. E.D. Friel, *ARAA* **33**, 381 (1995).
4. S. Sandrelli, A. Bragaglia, M. Tosi, G. Marconi, *MNRAS* **309**, 739 (1999).
5. M. Tosi, *ASP Conf. Ser.*, eds. C. Leitherer, U. Fritze-von-Alvensleben, J. Huchra, **98**, 299 (1996).
6. M. Tosi, in *The Evolution of the Milky Way. Stars vs Clusters*, eds. F. Giovannelli, F. Matteucci (Kluwer, Dordrecht, 2000).

OBSERVATIONAL TESTS OF INTERMEDIATE MASS STAR YIELDS USING PLANETARY NEBULAE

K.B. KWITTER

Department of Astronomy, Williams College, Williamstown, MA 01267, USA
E-mail: kkwitter@williams.edu

R.B.C. HENRY

Department of Physics & Astronomy, University of Oklahoma, Norman, OK 73019, USA
E-mail: henry@mail.nhn.ou.edu

This paper summarizes a project designed to study abundances in a sample of planetary nebulae representing a broad range in progenitor mass and metallicity. We collect abundances of C, N, and O determined for the entire sample and compare them with theoretical predictions of planetary nebula abundances from a grid of intermediate-mass star models. We find very good agreement between observations and theory, lending strong support to our current understanding of nucleosynthesis in stars with progenitor masses below 8 $M_\odot$. This agreement between observation and theory also supports the validity of published stellar yields of C and N in the study of the abundance evolution of these two elements.

1 Introduction

The cores and shell-fusing regions of intermediate-mass stars ($1<M<8\ M_\odot$) become hot enough to synthesize C and N. Planetary nebulae contain the synthesis products of intermediate-mass stars. We determined C and N abundances for 20 PNe and compared derived abundances with stellar model predictions for PNe with progenitors over this same mass range. Since these same models predict chemical yields for C and N, our analysis constitutes an observational check on the validity of these theoretical yields.

2 Results & Conclusions

Abundances are determined using emission-line data from ground-based (KPNO) and orbiting (IUE) telescopes. Calculations using appropriate atomic data for reaction rates provide abundances of ions associated with observed emission lines. The sum of observed ionic abundances is multiplied by an ionization correction factor to account for the contribution of unobserved ions to the total element abundance. A second correction is made using photoionization models to produce the final abundances.

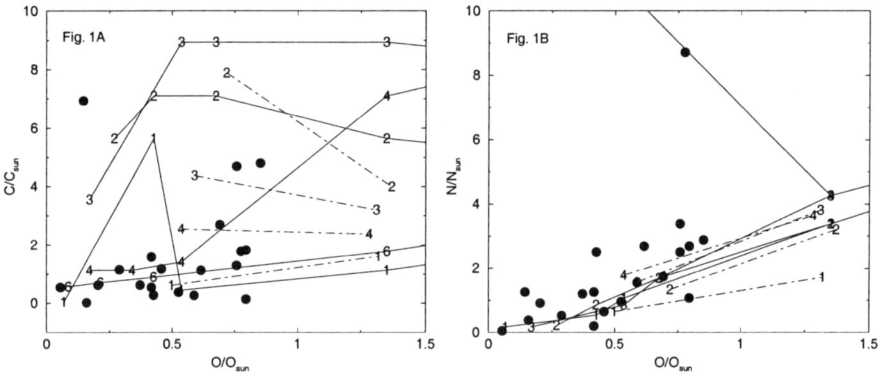

Figs. 1A,B compare our abundance results (filled circles) for C and N versus O with model predictions of van den Hoek & Groenewegen[1] (solid lines) and Marigo et al.[2] (dashed lines). Lines connect models of indicated progenitor mass over a range of metallicity. Modest agreement is shown.

Our overall *conclusions* are: (1) C and N abundances in PNe show much more scatter than they do in H II regions, implying that PN progenitors produce and expel these two elements; (2) the abundances of C and N in our sample of PNe are consistent with model predictions of intermediate-mass stars; and (3) our abundance patterns are consistent with the production of N via hot-bottom burning in intermediate-mass stars with progenitors above 3.5-4 $M_\odot$.

Our complete study is available in Henry, Kwitter, & Bates[3] and earlier references cited therein.

Acknowledgments

We are grateful for support from NSF grant AST-9819123 and NASA grant NAG 5-2389. In addition we appreciate the support from the staff of KPNO while carrying out the observing portion of this program.

References

1. van den Hoek, L.B., & Groenewegen, M.A.T. 1997, A&ApS, 123, 305
2. Marigo, P., Bressan, A., & Chiosi, C. 1996, A&Ap, 313, 545
3. Henry, R.B.C., Kwitter, K.B., & Bates, J.A. 2000, ApJ, 531, 928

THE CALIBRATION OF MASS LOSS BASED ON OBSERVATIONS OF AGB STARS IN THE MAGELLANIC CLOUDS

P.VENTURA AND F.D'ANTONA
Osservatorio Astronomico di Roma, Via Frascati 33, 00040 Monteporzio Catone, Italy

I.MAZZITELLI
Istituto di Astrofisica Spaziale del CNR, Via del Fosso del Cavaliere 100, 00133 Rome, Italy

We present the evolution of intermediate mass stars up to the end of the AGB phase at various metallicities, focusing our attention on the Lithium production during the phase of thermal pulses. We find that the Lithium yield provided by these sources is extremely dependent on the assumed mass loss rate, which we calibrate by looking for an agreement between the prediction of our models and the observation of Lithium rich sources in the Magellanic Clouds (MCs). The role which these stars can play for the lithium enrichment of the interstellar medium is discussed.

1 The way to manifacture lithium and the calibration of mass loss

Lithium production is achieved in AGBs via the Cameron & Fowler (1971) mechanism: once the temperature at the base of the external envelope exceeds $\sim 40{\cdot}10^6$K, Beryllium is produced via the reaction $^3He{+}^4He \longrightarrow^7 Be$. Under these conditions the process of Beryllium decay is much slower than mixing, therefore it is carried outwards to more external layers before it decays into Lithium. The latter is produced in the coolest regions of the star, therefore it can at least partly survive to proton fusion.

The range of masses involved is slightly dependent on some input macrophysics parameters assumed to build the models; yet, even considering all the uncertainties connected with the overshooting distance and mass loss, we expect, within a level of uncertainty of $\sim 0.5M_{\odot}$, Lithium production to be achieved within the interiors of stars with initial masses $3.5M_{\odot} \leq M \leq 6M_{\odot}$. The above limits hold for chemical compositions typical of the MCs, and are slightly shifted downwards for stars of very low metallicity.

Mass loss has been calibrated by requiring the most luminous AGBs to be lithium rich and s-process enriched, a well known results of the MCs optical surveys led by Smith & Lambert (1989, 1990). A low $\dot{M}$ would lead to a situation in which all the stars could be observed up to very late phases, when their luminosity is still large, but their surface Lithium has dropped to extremely low values, dut to the exhaustion of ^3He whithin their convective envelope. On the other hand, an extremely large mass loss rate would make only the most massive models to be detected at large luminosities, but just during phases before the beginning of TPs, thus without any possibility of being s-process enriched.

By relying on Blöcker (1995) prescription for mass loss, we find that an appropriate value for the free parameter η_R entering the quoted prescription is $\eta_R = 0.01$, though larger values up to $\eta_R = 0.02$ cannot be completely ruled out.

2 Conclusions

The calibration of $\dot{M}$ allows to predict that the large luminosity, Lithium rich sources currently observed in the MCs are the descendants of massive progenitors with masses $M = 4, 4.5 M_\odot$, in a phase when they are still Lithium rich, and before they become completely obscured in the optical due to formation of dust surrounding their outermost regions. Once we calibrated mass loss, our next step was to compute the Lithium yields for all the masses involved in the process of Lithium production at different metallicities, ranging from $Z = 2 \cdot 10^{-4}$ up to $Z = 0.02$. The interested reader can find elsewhere the complete list of the yields (Ventura et al. 2000). What can be seen is that these results seem to indicate that the contribution provided by these sources to the lithium enrichment of the interstellar medium is very poor.

Acknowledgments

One of us (P.V.) is very grateful to M.Groenewegen for providing very useful informations concerning the observability of these stars in the optical during their AGB evolution.

References

1. T.Blöcker, A&A, 297, 727, 1995.
2. A.G. Cameron and W.A. Fowler, ApJ, 164, 111, 1971.
3. V. Smith and D.Lambert, ApJ, 345, L75, 1989.
4. V. Smith and D.Lambert, ApJ, 361, L69, 1990.
5. P.Ventura, F.D'Antona and I.Mazzitelli, A&A, 363, 605.

NON-COSMIC ABUNDANCES IN THE POLAR BY CAM?

M. MOUCHET[1], J.-M. BONNET-BIDAUD[2], K. BEUERMANN[3], D. DE MARTINO[4], R. FERLET[5], R. FRIED[6], B. GÄNSICKE[3], S. HOWELL[7], A. LECAVELIER DES ETANGS[5], K. MUKAI[8], D. PORQUET[2], E. ROUEFF[1] AND P. SZKODY[9]

[1]*DAEC, Observatoire de Paris, Meudon, F,* [2]*Service d'Astrophysique, CEN, Saclay, F,* [3]*Universtäts-Sternwarte, Göttingen, D,* [4]*Astronomical Observatory Capodimonte, Naples, I,* [5]*Institut d'Astrophysique de Paris, F,* [6]*Braeside Observatory, Flagstaff, USA,* [7]*Planetary Science Institute, Tucson, USA,* [8]*LHEA, NASA/GSFC, Greenbelt, USA,* [9]*University of Washington, Seattle, USA*

The Polar BY Cam is characterized by a UV emission resonance line ratio of NV/CIV much greater than observed in other Polars. The far-UV FUSE observations reveal a broad resonance OVI line weaker than in the prototype of Polars, AM Her, which, together with the presence of a strong NIII line and weak CIII lines would indicate a mix of CNO redistribution following thermonuclear runaways at the surface of the magnetized white dwarf, and of the chemical composition of the companion atmosphere.

1 Properties of BY Cam and FUSE Observations

Polars are semi-detached synchronized systems composed of a magnetic white dwarf (WD) accreting matter from a red dwarf filling its Roche lobe. Accretion occurs along the magnetic field lines, resulting in strong emission lines superimposed on a blue continuum.

Though identified to a Polar on the basis of a strong circular polarisation, BY Cam differs from this class, being a hard X-ray source, slightly desynchronized (1% of 3.3h) (Mouchet et al. 1997 and ref. herein) and exhibiting pathological UV emission resonance lines, with a ratio NV/CIV $\sim$ 30 times larger than observed in Polars (Bonnet-Bidaud & Mouchet 1987).

High resolution far UV spectra in the range 910-1180Å have been obtained with the FUSE satellite (Moos et al. 2000) for a total of 20 ks exposure time. These observations were simultaneous with RXTE and EUVE observations as well as optical spectroscopy and photometry (APO, HET). Pronounced H_2 absorption lines of the Lyman and Werner series are clearly present in these spectra, indicating a column density of about 10^{19} cm^{-2}. The spectra also reveal broad emission lines of the H Lyman series, OVI, HeII, NIII and SIV. Contrary to the prototype of Polars, AM Her, observed with the ORFEUS telescope (Mauche & Raymond 1998), no narrow components are detected in BY Cam and the OVI line is weaker.

2 Emission line modelisation and possible CNO redistribution

The broad components of the resonance lines are believed to be formed in the accretion column which is photoionized by the X-ray source. The predicted line intensities of CIV, NV and OVI have been computed using the code Cloudy (Ferland et al. 1998) for typical X-ray spectra emitted by Polars. The 3 ion species are all present for an ionization parameter U $\sim$ 0.01, and the corresponding line ratios are NV/CIV = 0.4 ; OVI/CIV = 0.4 ; OVI/NV = 1.0. These values are in agreement, within a factor 3, with those observed in AM Her but fail to reproduce those of BY Cam.

The presence of strong lines of NIII (990Å) and NIV (1718Å) together with the weaknesses of the CIII lines (977Å, 1175Å) favors an overabundance of N and a depletion of C. As observed in many nova ejecta, such non-solar abundances can result from the CNO redistribution following the dredge-up of matter from the underlying CO core during a nova thermonuclear runaway (Livio & Truran 1994). 2D calculations of turbulent nuclear burning indeed predict self-enrichement of the envelope composition (Kercek et al. 1998). A large quantity of C is reprocessed into N while O might be non affected or almost completely destroyed (Kovetz & Prialnik 1997). Enhancement of N and depletion of C can also be obtained in the context of evolution models of semi-detached systems in which the secondary surface is polluted by re-accretion of material ejected during nova outburts (Marks & Sarna 1998). However this increase of metallicity is less sensitive for Population I CVs (Stehle & Ritter 1999). Models suitable for orbital parameters of BY Cam and other Polars, aimed to reproduce the weak fraction of Polars (3/$\sim$50) with peculiar UV emission lines, have still to be computed. The influence of the WD magnetic field onto the nova explosions has also to be elucidated.

References

1. J.M. Bonnet-Bidaud and M. Mouchet, *A&A* **188**, 89 (1987).
2. G.J. Ferland *et al*, *PASP* **110**, 761 (1998).
3. A. Kercek *et al*, *A&A* **337**, 379 (1998).
4. A. Kovetz and D. Prialnik, *ApJ* **477**, 356 (1997).
5. M. Livio and J.W. Truran, *ApJ* **425**, 797 (1994).
6. P.B. Marks and M.J. Sarna, *MNRAS* **301**, 699 (1998).
7. C.W. Mauche and J.C. Raymond, *ApJ* **505**, 869 (1998).
8. H.W. Moos *et al*, *ApJ* **538**, L1 (2000).
9. M. Mouchet *et al*, *A&A* **324**, 109 (1997).
10. R. Stehle and H. Ritter, *MNRAS* **309**, 245 (1999).

ABUNDANCE ANOMALIES IN RGB STARS AS PROBES OF STELLAR NUCLEOSYNTHESIS AND GALACTIC CHEMICAL EVOLUTION

A. PALACIOS, C. CHARBONNEL
Laboratoire d'Astrophysique de Toulouse, 14 av. Edouard Belin, 31400 Toulouse, France
E-mail: apalacio@ast.obs-mip.fr or corinne@ast.obs-mip.fr

RGB stars with low masses and metallicities present chemical features that remain unexplained by classical models of stellar evolution. These "abundance anomalies" can be primordial and/or evolutionary. In the last case, the H-burning products are brought up to the surface during the RGB phase, so that these stars may be important contributors to the chemical enrichment of the interstellar medium. We present evolutionary models including a preliminary treatment of rotational-induced transport on the RGB. This extra mixing process leads to important changes of the surface chemical composition of the stars. It allows thus to sketch a new picture of these stars as contributors to the galactic chemical evolution (Palacios *et al.*) [3].

1 Setting the stage

Numerous abundance anomalies of the isotopes involved in proton capture nucleosynthesis appear in globular cluster RGB stars (and to a lower extent in low metallicity field RGB stars) and are unexplained by canonical stellar evolution. While some data require a primordial origin, part of the observations can be explained by a purely evolutionary picture. This implies that a non-standard mixing process connects the convective envelope of the low mass red giants to the deeper external wing of the hydrogen burning shell where a rich nucleosynthesis (CNO cycle, NeNa chain) occurs.

2 Rotation-induced mixing : Presentation and results

2.1 Physical inputs of the models

All the relevant data indicate that the non-standard mixing process starts modifying the surface abundances at the RGB bump, and that it occurs in $\sim$ 95% of low-mass (M$\leq$2 - 2.2 M$_\odot$) stars. Various observations put contraints on its nature, and indicate that it could be due to rotation.
We present preliminary results taking into account recent progress in the description of the transport of chemicals and angular momentum in stellar interiors by Zahn and collaborators [1]. We use a simplified two-slope global diffusion

coefficient acting between the base of the convective envelope and the outer parts of the H-burning shell ($\log_{10}(D) \sim 11$ to 6).

2.2 Predictions vs observations

The models show that Na and O remain constant during the first dredge-up which lowers the carbon isotopic ratio. After the RGB bump, the rotation-induced mixing starts acting. While it leads to a further decrease of the carbon isotopic ratio, it also modifies the Na and O abundances at the stellar surface.

The values predicted by this simplified model reproduce the O-Na anticorrelation observed in globular cluster stars and the Na anomalies in field stars. The differences in terms of abundance anomalies between both populations can be reproduced when assuming that the mixing acts in a different way according the environment (the duration of effective mixing may be shorter in field stars). These differences remain to be studied in details.

Our simulations show for the first time the existence of a short-lived phase of lithium enrichment when the mixing starts acting. This gives some clues to answer the "Li-rich giants problem" (see Charbonnel & Balachandran 2000) [4], and confirms that these stars should not reject important quantities of lithium in the ISM (Palacios & Charbonnel 2001) [2]. Meanwhile, ^{3}He is strongly depleted in these stars.

Even when the extra-mixing leads to the most extreme O and Na anomalies observed, the change in the surface abundance of ^{4}He is modest. This affects only slightly the position of the star on the Horizontal Branch and excludes the extra-mixing on the RGB as a "second parameter" candidate.

Acknowledgments

We would like to sincerely thank the organizing committee for having been so kind to let us present our work in this conference in spite of our absence.

References

1. A. Maeder and J.P. Zahn, *A.&A.* **334**, 1000 (1998).
2. A. Palacios, C. Charbonnel, A.&A. in preparation, 2001.
3. A. Palacios, C. Charbonnel, M. Forestini, S. Talon,A.&A. in preparation, 2001.
4. C. Charbonnel and S. Balachandran, *A.&A.* **359**, 563 (2000)

THEORETICAL INTERPRETATION OF ABUNDANCE TREND SEEN IN EXTREMELY METAL-POOR STARS

T. Tsujimoto
National Astronomical Observatory, Mitaka-shi, Tokyo 181-8588, Japan
E-mail: taku.tsujimoto@nao.ac.jp

T. Shigeyama, Y. Yoshii
Graduate School of Science, University of Tokyo, Tokyo, Japan

Abundance patterns seen in extremely metal-poor stars with [Mg/H]< -2.0 follow those seen in the individual supernova remnants. This suggests that stars are born from individual supernova events. This hypothesis sheds light on the entire process of early-epoch star formation. Galactic chemical evolution is envisioned as a successive sequence of supernova explosions, shell formation, and resulting star formation. In this picture, the abundance pattern of each star is set not only by the heavy elements ejected from the supernova explosion which directly preceeds local star formation, but also incorporates those elements already present in the interstellar gas which are swept up by the supernova remnants. From this viewpoint, we present how we can interpret abundance trends seen in n-capture elements and odd elements.

1 Barium

Figure 1a shows a unique feature in the observed distribution of [Ba/Mg] *vs.* [Mg/H] for extremely metal-poor stars. We associate this feature, a bifurcation of the observed elemental ratios into two branches in the Mg abundance interval $-3.7 \leq$ [Mg/H] ≤ -2.3, with two distinct processes. The first branch, which we call the "y"-branch, is associated with the production of Ba and Mg from individual massive supernovae. This y-branch reflects the different nucleosynthesis yields of [Ba/Mg] for each SN II with $20M_\odot \lesssim M_{\rm ms} \lesssim 25M_\odot$. The second branch, which we call the "i"-branch, is associated with the elemental abundance ratios of stars which were formed in the dense shells of the interstellar medium swept up by SNe II with $M_{\rm ms} < 20M_\odot$ that do *not* synthesize r-process elements. The Ba abundances in these stars reflect those of the interstellar gas at the time of their formation. The existence of a [Ba/Mg] i-branch strongly suggests that SNe II which are associated with stars of progenitor mass $M_{\rm ms} \leq 20M_\odot$ are infertile sources for the production of r-process elements. We predict the existence of this i-branch for other r-process elements, such as europium (Eu), to the extent that their production site is in common with Ba.

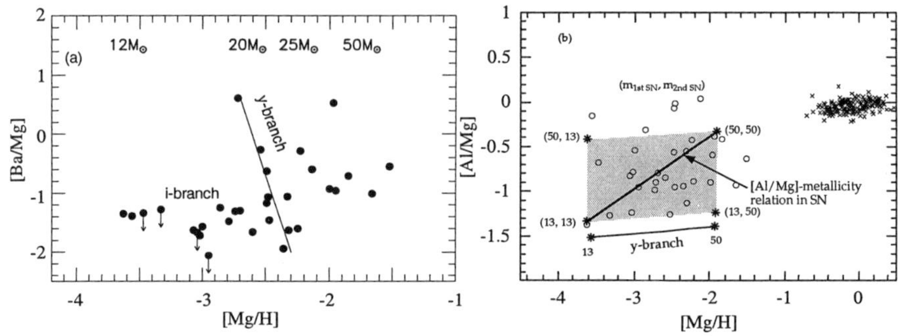

Figure 1: (a) Correlation of [Ba/Mg] with [Mg/H] for metal-poor stars. (b) The observed correlation of [Al/Mg] with [Mg/H] for metal-poor halo stars, together with that for disk stars. The shaded area cornered with asterisk symbols is the predicted location of the stars born in the first few generations.

2 Aluminum

Figure 1b shows the values of [Al/Mg] plotted against [Mg/H]. Contrary to the small scatter in [Al/Mg] observed for disk stars, the values of [Al/Mg] for [Mg/H] $\lesssim$ -2 scatter over 1 dex at any given [Mg/H]. The y-branch is indicated by the thin line. The thick line indicates the relation between synthesized [Al/Mg] and *initial* metallicity [Mg/H], which SNe should satisfy in order to reproduce the observed scatter in [Al/Mg]. Given this metallicity-dependent Al yield, we can account for the resultant distribution of long-lived stars in the [Al/Mg] versus [Mg/H] diagram. Second-generation stars located along the y-branch have [Al/Mg]~ -1.4 by assumption, but have various metallicities, indicated by [Mg/H], over a range of almost 2 dex. For example, because SNe with larger progenitor mass eject a larger amount of metals, second-generation stars born in a SNR of a Pop III progenitor star having a mass of 50 $M_\odot$ have [Mg/H]~ -1.8 on the right end of the y-branch, and those stars from a 13 $M_\odot$ SN have [Mg/H]~ -3.6 on the left end. Third-generation stars born in a SNR from a progenitor star with a higher metallicity, of order [Mg/H]~ -1.8, must have a higher ratio of [Al/Mg]~ -0.4 due to the metallicity-dependent Al yield. On the other hand, stars of the same generation from a progenitor star with lower metallicity, of order [Mg/H]~ -3.6, must have a lower ratio of [Al/Mg]~ -1.4. In either instance, apart from [Al/Mg], stars that originated from SN explosions must have very different metallicities of [Mg/H] if the progenitor stars have different masses.

Part 4. Dynamical and Chemical Evolution

CHEMICAL EVOLUTION OF THE GALAXY: THE G-DWARF PROBLEM AND RADIOACTIVE CHRONOLOGY REVISITED TAKING ACCOUNT OF THE THICK DISK

B.E.J. PAGEL

Astronomy Centre, CPES, University of Sussex, Brighton BN1 9QJ, UK
E-mail: bejp@star.cpes.susx.ac.uk

Thin and Thick disk metallicities overlap, but α(Fe) relations differ; there is also a hiatus in time before the first thin-disk stars, with a minimum metallicity [Fe/H] $\simeq -0.6$. Thus there are two 'G-dwarf' problems. We fit metallicity distribution functions (MDFs) for the two disks derived by Wyse & Gilmore (1995) with analytical models (for O and Mg). The Thick disk fits a simple inflow model with effective yield and final [Mg/H] about the same as for the Thin disk, excluding continuous mass loss from the former as a source of 'prompt initial enrichment' (PIE) for the latter. The Thin disk has a narrow MDF fitted (rather poorly) with a closed model having PIE from some other source, or more elegantly with a slow infall model similar to the 'two-inflow' scenario of Matteucci et al., but with a PIE and a severe break in the age-metallicity relation. Implications for radio-active cosmochronology are investigated.

1 Introduction

The 'G-dwarf' problem has been with us for nearly 40 years and extended to other galaxies. There is still no definitive universally accepted solution; rather too many, the most popular being 'prompt initial enrichment' (PIE; Truran & Cameron 1971) and inflow (Larson 1972). Truran & Cameron postulated a first generation of massive stars, sometimes called Population III, collapsing to black holes or 'collapsars' in the halo, thereby accounting for its dark mass. Later the idea gained ground that the halo consists mostly of dark matter, wholly or partly non-baryonic, but PIE came in by another route. After discovery of the Thick Disk, Gilmore & Wyse (1986) suggested that (a) the Thick disk was initially enriched by gas shed from the halo as in the model by Hartwick (1976); and (b) the Thin disk was similarly enriched by gas shed from the Thick one. I never took much notice of this paper, because the relative numbers are not right unless there is inflow as well, and Wyse & Gilmore (1992) themselves abandoned proposition (a) on angular momentum grounds. This still leaves open a connection between the two disks, and Wyse & Gilmore (1995) have published separate MDFs for them which, with new data on O/Fe and Mg/Fe, have implications that I discuss in this talk.

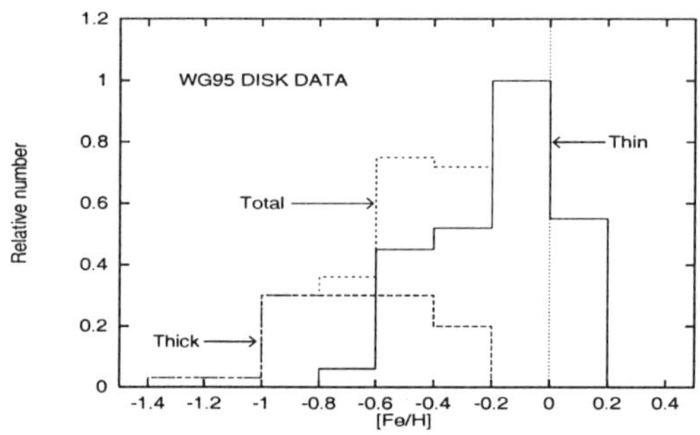

Figure 1. Iron MDFs for Thick and Thin disks, after Gilmore & Wyse (1995).

2 Iron and magnesium MDFs and AMRs

Numerous models based on inflow give a fair fit to the MDF of nearby disk stars as a whole (Matteucci & François 1989; Pagel 1989; Pagel & Tautvaišienė 1995). PIE models based on outflow of metals from a rapidly evolving bulge have also been quite successful (Köppen & Arimoto 1990; Samland et al. 1997). Strobel (1991), Pardi et al. (1995), Chiappini et al. (1997), Fuhrmann (1998) and Gratton et al. (2000), among others, have investigated chemical evolution implications of the different time-scales of evolution of the two disks, and Matteucci and co-authors have developed 'two-inflow' models which I consider a good idea, but I shall argue that they do not go far enough.

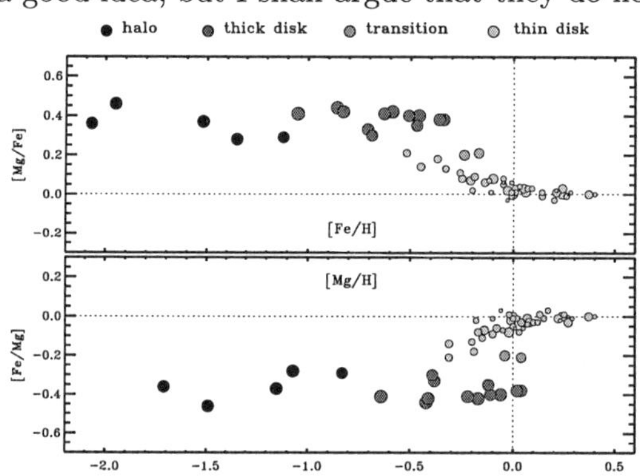

Figure 2. [Mg/Fe] vs [Fe/H] and [Fe/Mg] vs [Mg/H] for stars of the Galactic halo, thick disk and thin disk, after Fuhrmann (1998). Courtesy Klaus Fuhrmann.

Wyse & Gilmore's iron MDFs are shown in Figure 1, where it is seen that the Thick disk has a mean [Fe/H] just at the sharp lower limit of the Thin one, which made it seem natural for Gilmore & Wyse (1986) to suggest a Hartwick-like model for the Thick disk where gas expelled from it over time led to a low effective yield and had just the right metallicity to start off the Thin disk, the mass of which can be made up as desired by subsequent inflow. This, however, assumes a single α(Fe) relation, which now seems incorrect (Fuhrmann 1998; Gratton et al. 2000), as is illustrated in Figure 2. The Thick-disk stars have enhanced α/Fe (and O/Fe) even when their α-metallicity is as high as solar.

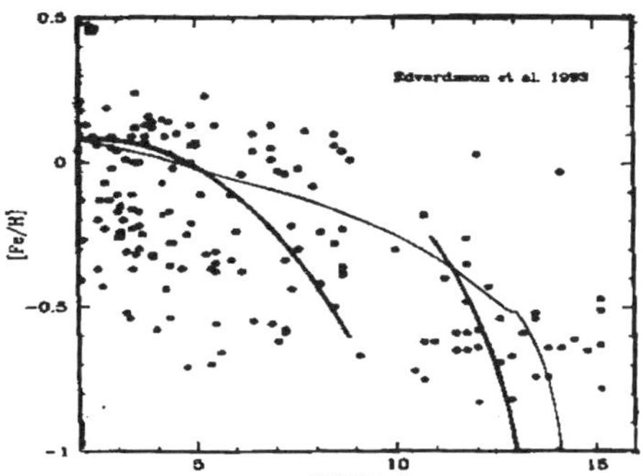

Figure 3. Thin curves: age-metallicity relation from the 'two-inflow' model of Chiappini, Matteucci & Gratton (1997). Thick curves: sketch of suggested separate AMRs for the Thick and Thin disks. Adapted from Chiappini et al. (1997).

This gives further support to a short time-scale for Thick-disk star formation, confirmed by HIPPARCOS parallaxes (Ng & Bertelli 1998), which show a hiatus between that and the first stars of the Thin disk, also visible in the data from Edvardsson et al. (1993) shown in Fig 3. The need for two time-scales and a hiatus in star formation, as postulated by Chiappini et al., is clearly evident, but their model does not go far enough, and it seems more realistic to draw in two entirely separate age-metallicity relations as shown schematically by the thick curves.

We also face the complication that the iron-to-α-element conversion is different in the two disks. Making use of the relationships embodied in Fig 2, we can convert Fig 1 into the Mg MDF shown in Fig 4.

Fig 4 shows some interesting features not previously apparent. For one thing, the upper abundance limits are identical and the means nearly so, so there is no longer any basis for assuming Hartwick-style mass outflow from the Thick disk with a resulting reduction in effective yield. It does have a low-metallicity tail, also noticed by Beers & Sommer-Larsen (1995), whereas the Thin one has a sharp lower limit suggestive of PIE; but if the previous

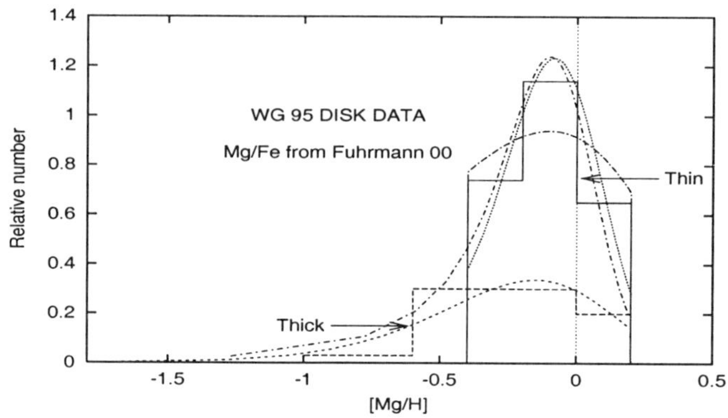

Figure 4. Magnesium MDFs. Histograms: data after Wyse & Gilmore (1995). Curves: dashed, simple inflow model for Thick disk; dotted, PIE plus inflow model for Thin disk; dash-dotted, inflow model without PIE; long-dash dotted, PIE without inflow.

metallicity came from the Thick disk I would rather call it Prompt Initial Impoverishment!

The curves in Fig 4 show some analytical fits to the two MDFs. The Thick disk is readily fitted by a simple decaying inflow model as in Pagel (1997) with a yield 0.15 dex below solar, a final gas fraction of 0.1 and any time scale less than that of SNIa, say around 1 Gyr.

For the Thin disk there are more choices. A pure PIE model without inflow gives a barely acceptable fit, while a pure inflow model (Clayton 1985 standard model with $k = 4$) produces too much of a low-metallicity tail. A combined PIE-inflow model fits very well, possibly representing terminal Thick-disk gas diluted by inflow of unprocessed material before the Clayton episode — thus a 'three-inflow' model.

3 Implications for radio-active cosmochronology

Solar-system actinide abundances have been used in the past to guess the age of the Galaxy, although there is also a clear dependence on chemical evolution models. More recently, with the discovery of r-process rich halo stars, it has become possible to use stellar thorium abundances in a manner that is less model-dependent, but the age of the halo does depend explicitly or implicitly on the Th/Eu production ratio. This has been estimated on the basis of nuclear physics, but with a fair degree of uncertainty (see Westin et al. 2000), while Truran (1999) has appealed to the proto-solar ratio, but this is clearly only a lower limit, thereby giving just a minimum possible age for the halo stars.

Table 1. K-ratios of actinide abundances at Solar System formation

	235, 238	232, 238	Eu, Th
'Observed' after Fowler (1987)	$0.25 \pm .04$	$1.35 \pm .10$	
Modified Fowler model	0.24	1.45	1.23
Inflow model (Clayton $k = 4$)	0.49	1.13	1.06

Now we know more about stellar ages and the gap between the two disks, it may be of interest to turn the argument round and check for consistency. We use our Galactic models to estimate the K-ratios defined by Fowler (1987):

$$K_{ij} \equiv \frac{N_i}{N_j}(\text{protosolar})/\frac{\mathrm{p_i}}{\mathrm{p_j}} \tag{1}$$

and use the resulting $K_{\mathrm{Eu\,Th}}$ to predict the production ratio from the known proto-solar abundance ratio.

We recall Fowler's own model, a closed PIE model from which he deduced an age for the disk of just 10 Gyr. If we think only of the Thin disk, then this was remarkably prescient, but now we also have to think of the Thick disk or bulge where any PIE would probably have originated and which is 2 or 3 Gyr older. We assume the Thick disk to have formed instantaneously 12 Gyr ago and the Thin disk to have started forming after an interval $\epsilon = 3$ Gyr. The Solar System was formed after a further interval $\Delta = 4.4$ Gyr. Fowler's equations are then modified to:

$$N_i(\text{radioactive}) \propto p_i \left[\frac{S}{1-S}e^{-\lambda_i(\Delta+\epsilon)} + \frac{1-e^{-\lambda_i\Delta}}{\lambda_i\Delta}\right] \tag{2}$$

$$N_k(\text{stable}) \propto p_k/(1-S) \tag{3}$$

where in our case the PIE contribution $S = 0.4$ rather than 0.17. Corresponding formulae for the inflow model have been given by Clayton (1985). Resulting K-ratios in the Solar System are given in Table 1, where it is seen that the modified Fowler model still gives excellent agreement with 'observed' K-ratios, whereas the inflow model gives numbers that are barely compatible if at all and presumably represents a generous upper limit to the modifications from inflow.

Implications for halo ages are given in Table 2. Fowler's model implies a large correction from protosolar to production ratio and hence a large (and perhaps unacceptable) figure for the halo age, while the inflow model gives only a minimal correction and a more acceptable age range. It is intriguing that the range of production ratios permitted by Galactic chemical evolution models matches so closely the range deduced from nuclear physics.

Table 2. Th/Eu ratios and halo age

	Th/Eu ratio	halo age (Gyr)
Halo stars	0.22 to 0.25	
Proto-solar	0.46	
Nuclear theory prediction	0.48 to 0.55	13.2 to 18.6
Modified Fowler model prediction	0.57	16.7 to 19.3
Inflow model prediction	0.49	13.7 to 16.3

References

1. Beers, T.M. & Sommer-Larsen, J. 1995, ApJS, 96, 175
2. Chiappini, C., Matteucci, F. & Gratton, R. 1997, ApJ, 477, 765
3. Clayton, D.D. 1985,, in W.D. Arnett & J. Truran (eds), *Nucleosynthesis: Challenges and New Developments*, University of Chicago Press, p. 65
4. Edvardsson, B. et al. 1993, A & A, 275, 101
5. Fowler, W.A. 1987, QJRAS, 28, 87
6. Fuhrmann, K. 1998, A & A, 338, 161
7. Gilmore, G. & Wyse, R.F.G. 1986, Nature, 322, 806
8. Gratton, R. et al. 2000, A & A, in press, astro-ph/0004157
9. Hartwick, F.D.A. 1976, ApJ, 209, 418
10. Köppen & Arimoto, N. 1990, A & A, 240, 22
11. Larson, R.B. 1972, Nature Phys. Sci., 236, 7
12. Matteucci, F. & François, 1989, MNRAS, 239, 885
13. Ng, Y.K. & Bertelli,G. 1998, A & A, 329, 943
14. Pagel, B.E.J. 1989, Rev. Mex. Astr. Astrofis., 18, 153
15. Pagel, B.E.J. 1997, Nucleosynthesis and Chemical Evolution of Galaxies, Cambridge University Press, p. 243
16. Pagel, B.E.J. & Tautvaišienė, G. 1995, MNRAS, 276, 505
17. Pardi, M.C., Ferrini, F. & Matteucci, F. 1995, ApJ, 444, 207
18. Samland, M., Hensler, G. & Theis, Ch. 1997, ApJ, 476, 544
19. Strobel, A. 1991, A & A, 247, 35
20. Truran, J.W. 1999, in R. Diehl & D. Hartmann (eds), *Astronomy with Radioactivities*, Ringberg Workshop, MPE Rep. 274, p. 261
21. Truran, J.W. & Cameron, A.G.W. 1971, Ap. Sp. Sci., 14, 179
22. Westin, J., Sneden, C., Gustafsson, B. & Cowan, J.J. 2000, ApJ, 530, 783
23. Wyse, R.F.G. & Gilmore, G. 1992, AJ, 104, 144
24. Wyse, R.F.G. & Gilmore, G. 1995, AJ, 110, 2771

ON THE COSMIC ORIGINS OF CARBON AND NITROGEN

R.B.C. HENRY
Department of Physics & Astronomy, University of Oklahoma, Norman, OK 73019, USA
E-mail: henry@mail.nhn.ou.edu

M.G. EDMUNDS
Department of Physics & Astronomy, Cardiff University, P.O. Box 913, Cardiff, CF23YB Wales, UK
E-mail: mge@astro.cf.ac.uk

J. KÖPPEN
Observatoire Astronomique, UMR. 7550, 11 rue de l'Université, F-67000 Strasbourg, France
E-mail: koppen@astro.u-strasbg.fr

We employ analytical and numerical chemical evolution models to study observed trends in abundance ratios involving carbon, nitrogen, and oxygen. Several sets of published stellar yields for both intermediate-mass and massive stars are considered, and the most appropriate sets are selected through the use of analytical models. These yields are then used in the numerical models to match observed data trends in C/O, N/O, and O/H. We conclude that the principal production site for carbon is massive stars, while that for nitrogen is intermediate-mass stars.

1 Introduction

Carbon and nitrogen are among the most abundant chemical elements, and of obvious importance for life. The main nuclear processes that generate these two elements are reasonably well understood: the carbon must come from the triple-alpha reaction of helium while nitrogen originates in CNO processing by the conversion of carbon and oxygen during hydrogen burning. A lingering problem, though, has been the lack of knowledge of which *sites* are most important for their generation: in particular, do they come mainly from short-lived massive stars (MS) or from longer lived progenitors of asymptotic giant branch stars, i.e intermediate-mass stars (IMS)? Since necessary threshold temperatures for the production of both carbon and nitrogen are reached in both stellar types, identification of C and N production sites is difficult.

We report on the results of a study in which numerical chemical evolution models are used to match trends in nebular emission line and stellar absorption line measurements of C, N, and O, where oxygen in this case is employed as the tracer of metallicity. The basis of our study is the abundance data compiled by Henry & Worthey[1]. Figure 1A is a plot of log(C/O) versus 12+log(O/H) by number, while Fig. 1B is the same but for log(N/O). Typical uncertainties are shown in the upper left of both figures; heavy lines are model results discussed below.

2 Results & Conclusions

Our goal is to calculate numerical chemical evolution models to match trends visible in the data, at the same time employing analytical models of the data to select appropriate stellar yields for C, N, and O. Despite the scatter, we move forward by assuming a positive trend between C/O and O/H in Fig. 1A, indicating a metallicity dependence for carbon production. In Fig. 1B, we assume that the zero-slope behavior at low metallicity is commensurate with primary nitrogen production (independent of metallicity), while the clear ascending threshold for O/H above 12+log(O/H)=8.3 signifies the dominance of secondary nitrogen production (metallicity-sensitive).

The numerical models assume a one-zone open box and include infall but not outflow in the calculations. The code comprises standard differential equations, a quadratic Schmidt law with adjustable efficiency for star formation, a Salpeter IMF, and an infall rate which was taken to be a decreasing exponential function of time with a characteristic time of 4 Gyr.

Stellar yields are the most crucial bit of model input. Their selection was based on the construction of analytical fits to the data trends by assuming simple functional forms of stellar yield dependence on metallicity. Coefficients of successful fits were then compared with coefficients describing published yields. In this way, we chose to adopt Maeder's[2] yields for massive stars, in which carbon production is greatly enhanced by stellar mass loss whose rate is metallicity-sensitive. This important feature explains the secondary behavior of carbon production seen in Fig. 1A. For IMS, we adopted the yields of van den Hoek & Groenewegen[3], whose primary and secondary nitrogen production rates predict results which are very consistent with observed behavior of N/O.

Our model results are shown with the heavy lines in Figs. 1A,B. Models A, B, and C differ in star formation efficiency, which increases by five times each between A/B and B/C. Behavior of C/O among these models is unchanged because all C and O are produced by the same massive stars, so the abundances of these elements are unaffected by variation in star formation rate. In contrast, since N is synthesized by IMS, the evolutionary time delay of these stars compared with MS render the behavior of N/O sensitive to the

FIG 1A: CARBON

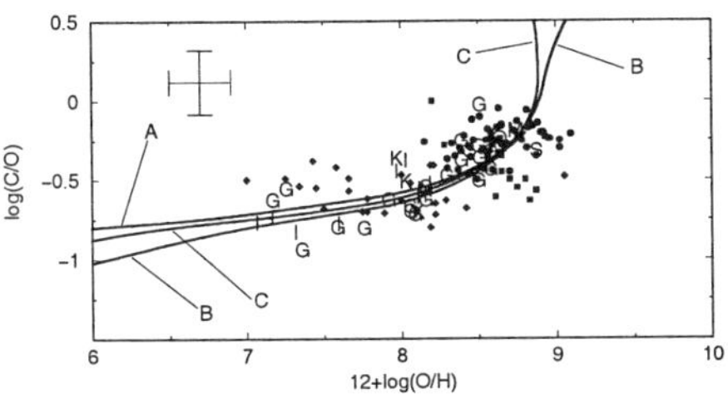

FIG. 1B: NITROGEN

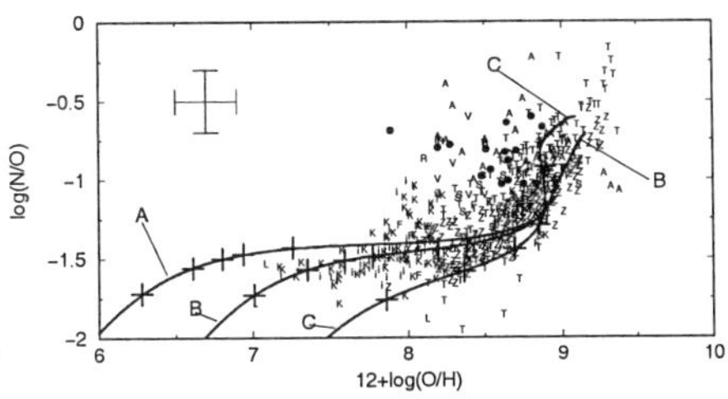

temporal changes in the star formation rate. So, the greater the SFR at early times (e.g. models C vs. A), the higher O/H rises (greater rightward shift) before N/O can catch up.

Our *conclusions* are: (1) Carbon is produced mostly (>90%) by massive stars, and thus this element experiences no production delay relative to the synthesis of oxygen. The secondary character of its production stems from the metal-sensitive mass loss rate assumed in Maeder's massive star yields; (2) Nitrogen is produced mostly (>90%) by intermediate-mass stars. At 12+log(O/H)<8.3, N synthesis is independent of metallicity, while above this value its production is sensitive to it. (3) The characteristic delay in nitrogen production relative to that of carbon and oxygen is 250 Myr, and the relevant stellar masses for its production are 4-8$M_\odot$. (4) The low N/O and O/H values observed in low surface brightness galaxies do not necessarily imply young ages for these systems, but may instead be explained by invoking low efficiency star formation rates. A full discussion of our analysis is available in Henry, Edmunds, & Köppen[4].

Acknowledgments

RBCH gratefully acknowledges support from NSF grant AST-9819123.

References

1. Henry, R.B.C., Worthey, G. 1999, PASP, 111, 919.
2. Maeder, A. 1992, A&Ap, 264, 105
3. van den Hoek, L.B., & Groenewegen, M.A.T. 1997, A&ApS, 123, 305
4. Henry, R.B.C., Edmunds, M.G., & Köppen, J. 2000, ApJ, 541, 660

CHEMICAL EVOLUTION OF THE MILKY WAY

N. PRANTZOS
Institut d'Astrophysique de Paris, 98bis Bd Arago, 75014 Paris
E-mail: prantzos@iap.fr

We present a comprehensive study of the evolution of the abundances of intermediate mass elements, from C to Zn, in the Milky Way halo and in the local disk, using metallicity dependent stellar yields for SNII and SNIa. We compare our results to an extensive body of observational data and make a critical analysis of the successes and shortcomings of current stellar yields.

1 The model for halo + local disk evolution

We use a consistent model to describe the evolution of the halo and the disk. Contrary to most previous studies, we assume that those two galactic subsystems evolve independently, both starting with gas of primordial composition, and in different ways: strong outflow is assumed to take place during the ~1 Gyr of the halo formation, while the disk is built by slowly infalling gas, on a time-scale of 7 Gyr. This description of the halo+disk evolution can correctly account for the main observational constraints (at least in the framework of simple models of galactic chemical evolution), i.e:

(a) The metallicity distributions of low-mass stars in the halo and the disk; (b) The current disk surface densities of gas (Σ_G), stars (Σ_*), the total amount of matter (Σ_T) and the current star formation rate (Ψ_0); (c) The disk age-metallicity relationship, traced by the Fe abundance of long-lived, F-type stars; (d) The abundances of various elements and isotopes at solar birth ($X_i, \odot$) and today ($X_i, 0$); (e) The present day mass function (PDMF), resulting from the stellar IMF (initial mass function) and SFR (star formation rate) history and providing an important consistency check for the adopted SFR and IMF.

2 Abundance ratios

The evolution of the abundance ratios X/Fe of intermediate mass elements (C to Zn) in our models is presented in Fig. 1, where it is compared to a large body of observational data (see Goswami and Prantzos 2000 - GP2000 -for references). The main conclusions of the comparison of our results to those data are the following:

- C and N require other sources than those studied here. For C, it could be WR or low mass stars, contributing to C production in the disk. For N, the

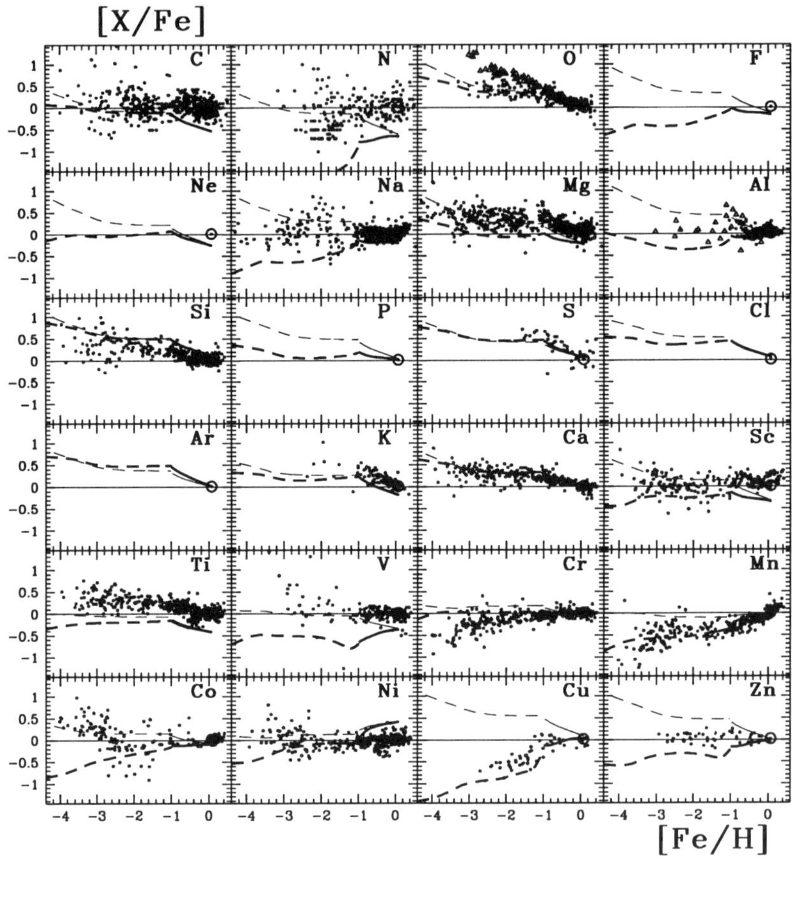

Figure 1: Abundance ratios [X/Fe] of stars in the halo and the local disk, as a function of [Fe/H]. Theoretical results are obtained with models that treat properly the halo (*dashed curve* assuming *outflow*) and the disk (*solid curve* assuming *slow infall*). Two sets of massive star yields are used, both from WW1995: at constant (=solar) metallicity (*thin curves*, Case A, only for illustration purposes) and at variable metallicity (*thick curves*, the reference Case B). Yields of the W7 and W70 models of Iwamoto et al. (1999) for SNIa are used in both cases (properly interpolated as a function of metallicity); intermediate mass stars are not considered. Data references are given in GP2000, while the "new" data on O vs Fe (Israelian et al. 1998; Boesgaard et al. 1999) are represented by *triangles*.

source of primary N required in the halo could be either IMS with hot-bottom burning or rotationally induced mixing in massive stars.

- The evolution of the α-elements O, Si, S and Ca is well understood (baring the discrepant "new data" for O, see below) with the assumption that SNIa contribute most of Fe in the disk; however, the WW1995 yields underproduce Mg/Fe and Ti/Fe.

- Similarly, the odd-Z elements Sc and V are underproduced at all metallicities by WW1995 and other available yields; this discrepancy points to some important revision required in current models of nucleosynthesis in massive stars, at least for those elements. It is significant that observationally, neither Sc nor V show the theoretically expected behaviour of odd-Z elements, suggesting that the "odd-even" effect may be overestimated in current nucleosynthesis models.

- Among the Fe-peak elements, several important discrepancies between theory and observations are found when results are plotted w.r.t. Fe. The theoretical trends of Cr, Co, Ni and Zn deviate from the observed ones to various extents; in the case of Ni, the adopted W7 model for SNIa largely overproduces the main isotope ^{58}Ni in the disk, as well as^{54}Cr, a minor Cr isotope. We notice that, when results are plotted w.r.t. Ca (GP2000), the observed behaviour of Cr is well reproduced by the model; this might imply that it is the Fe yields that are problematic at low metallicities. We notice that Cr is produced at layers lying at larger distance from the core than Fe, and are thus less subject to the uncertainties of the mass-cut.

- There is a remarkably good agreement between the theoretical and the observed behaviour of the odd-Z Fe-peak elements Mn and Cu, when their evolution is plotted w.r.t. Fe (or w.r.t. Ca, in the case of Mn).

References

1. Boesgaard A. M., King J. R., Deliyannis C. P., Vogt S. S., AJ, 117 (1999) 492
2. Goswami A. & Prantzos N. A&A, 359 (2000) 151 (GP2000)
3. Israelian G., Garcia-Lopez R. & Rebolo R., ApJ, 507 (1998) 805
4. Iwamoto K., Brachwitz F., Nomoto K., Kishimoto N., Hix R. & Thielemann K.-F., ApJS, 125 (1999) 439
5. Kroupa P., Tout C. & Gilmore G., MNRAS, 262 (1993) 545
6. Woosley S. E. & Weaver T. A., ApJS, 101 (1995) 181

TWO-COMPONENT MODEL FOR CHEMICAL EVOLUTION OF THE GALACTIC DISK

R.X. CHANG, J. L. HOU, C.G. SHU, C.Q. FU

Shanghai Astronomical Observatory, 80 Nandan Road, Shanghai, 200030, China

E-mail: crx, hjlyx, cgshu@center.shao.ac.cn

We introduce a two-component model of the Galactic disk to investigate its chemical evolution. It is assumed that the formation of the thick and thin disks occurs in two main accretion episodes with both infall rates being Gaussian. Both the pre-thin and post-thin scenarios for the formation of the thick disk are considered. The best fit is obtained through χ^2-test between the models and the observed metallicity distribution function of G dwarfs in the solar neighbourhood. Our results show that, the predictions of both the pre-thin and post-thin model agree well with the new observed G-dwarf metallicity distribution. However, the present-day local density of the thick disk, required for the pre-thin model to get the best fit, is larger than most of results from star counts, yet this disagreement does not exist for the post-thin model. From this point of view, it would seem that the post-thin model is more viable. Finally, other comparisons between model predictions and observations are given.

1 The model

Models for the formation of thick disk fall into either "top-down" scenarios (the pre-thin model), where the formation of the thick disk precedes that of the thin disk, or the "bottom-up" scenarios (the post-thin model), where the thick disk is the result of some action on or by the thin disk. In the pre-thin disk model, the formation of the thick disk is a transitional phase during the general contraction of the Galaxy. The post-thin model resorts to formation of the thick disk after the gas has collapsed into a thin disk. It should be emphasized that the required events to form the thick disk must not occur too late in the disk's life time so that the gas can cool again and form stars in the thin disk.

In this study, both the pre-thin and post-thin model for the formation of thick disk are considered. Following Chiappini, Matteucci & Gratton (1997) , we assume that there are two main infall episodes in both cases. In the pre-thin model, the first infall episodes forms the thick disk, the second one forms the thin disk. However, the post-thin model assumes that the thin disk begins to form earlier than the thick disk. But, due to the fact that the formation timescale of the thin disk is longer than that of the thick disk, most of thin disk stars are older and metal-richer than thick disk stars.

The rate of mass accretion in each ring could be expressed as an Gaussian form (Prantzos & Silk 1998).

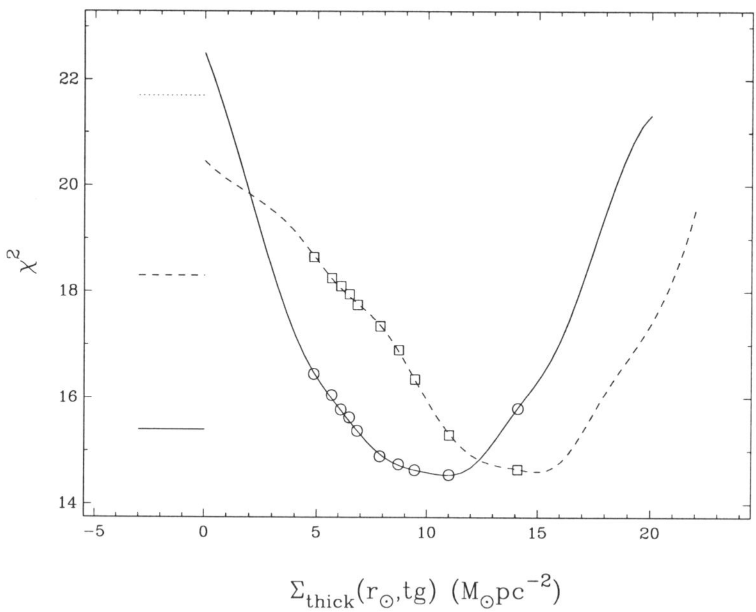

Figure 1. The χ^2 as a function of the $\Sigma_{thick}(r_\odot, t_g)$ within a reasonable range for both the pre-thin model (long dashed curve) and post-thin model (full curve). Three short horizontal lines indicate the model confidential levels of 30% (dotted line), 50% (long dashed line) and 70% (full line), respectively. The points indicate the different results of χ^2 if the value of $\Sigma_{thick}(r_\odot, t_g)$ from ten literature values (Table 3) are adopted for both the pre-thin model (open squares) and post-thin model (open circles), respectively.

2 Results

We perform model calculations for a broad range of free parameter combinations. The best fit is obtained through χ^2-test between model results and the observed metallicity distribution function of G dwarfs in the solar neighbourhood. Figure 1 shows the χ^2 as a function of the $\Sigma_{thick}(r_{\odot}, t_g)$ within a reasonable range for both the pre-thin model (long dashed curve) and post-thin model (full curve). The results suggest that the post-thin model is more viable.

References

1. C. Chiappini, F. Matteucci & R. Gratton, *ApJ* **477**, 765 (1997).
2. N. Prantzos & J. Silk, *ApJ* **507**, 229 (1998).

A GRID OF CHEMICAL EVOLUTION MODELS ALONG THE HUBBLE SEQUENCE

M. MOLLÁ AND A. I. DíAZ
Universidad Autónoma de Madrid, 28049 Cantoblanco, Spain
E-mail: mercedes@pollux.ft.uam, angeles@pollux.ft.uam.es

AND F. FERRINI
Università di Pisa, Piazza Torricelli 2, 56100 Pisa, Italy
E-mail: federico@astro2.difi.unipi.it

We have computed a grid of multiphase chemical evolution models whose results are valid for any spiral galaxy, using as input the maximum rotation velocity and the morphological type or index T

1 Introduction

In previous works, we have calculated the chemical evolution of the Solar Neighborhood (Ferrini et al [1]), the Galactic Disc and Bulge (Ferrini et al [2],[4]), and a sample of spiral galaxies (Mollá et al [5,6]) with the multiphase model. All of them reproduce with success the observational data, in particular those related with the disk radial distributions. They also predict the correlations found along the Hubble Sequence (Mollá & Roy [7]), although the number of computed models is small to ensure it. Now, we have computed an extended grid of multiphase models, which allow the comparison of observations with theoretical models for a large range of galaxy masses and morphological types.

2 The Multiphase Model Description

We assume a spherical protogalaxy with a gas mass which collapses to fall onto the equatorial plane by forming the disk as a secondary structure. The sphere is divided up into concentric cylindrical regions 1 kpc wide, with a *halo* and a *disk* region. The Universal Rotation Curve from Persic, Salucci & Steel [8], an analytical expression for V(R), is used to calculate the radial distributions of the mass included in each of our cylinders ΔM (R).

The characteristic gas infall rate from the halo to the disk is proportional to a parameter f, which, in turn, is inversely proportional to the collapse timescale τ_0, dependent on the total mass (Gallagher et al [3]). It is easily computed from its value for the MWG model, $\tau\odot$, and the ratio between total masses: $\tau_0 = \tau_{\odot}(M_9, gal/M_{9,MWG})^{-1/2}$.

Stars form out in the halo, by a Schmidt law with a proportionality factor K. In the disk, stars form in two steps: molecular clouds form from the diffuse gas with a proportionality factor μ; then cloud-cloud collisions produce stars by a *spontaneous* process, at a rate proportional to a parameter H. Moreover a *stimulated* star formation process, by the interaction of massive stars with surrounding molecular cloud, is assumed,s proportional to a parameter a,

These parameters are variable along the radius, by the volume effect, through the use of the process efficiencies (ϵ's), valid for the whole galaxy. We assume that the two efficiencies ϵ_K and ϵ_a are constant for all halos and galaxies, while ϵ_μ, and ϵ_H change according the Hubble type T by a function $\epsilon = \exp(-T^2/a)$. We have computed 10 types of galaxies for each mass.

3 Preliminary Results and Conclusions

A biparametric grid of models has been obtained for types 1 to 10 and different luminosities. We obtain radial distributions for the star formation rate, the diffuse and molecular gas densities and the elemental abundances.

The most important result concerns the oxygen abundances: the radial gradient only appears for the intermediate types ($7 < \mathrm{T} < 4$) at all galaxy masses, being larger for the less massive galaxies. However, the latest ones ($\mathrm{T} \geq 8$) have not developed a gradient in a Hubble time, their abundances being $12 + \log(O/H) \sim 7.5$-8. The early types T≤ 5 reach a saturation level, flattening the gradient once again, after having a steep one early times.

This important result reproduces the observations and it solves the apparent inconsistency of the largest gradients appearing in late type spirals, while some irregulars shows no gradient at all. The explanation resides in the stimulated star formation which, being a local process, maintains a minimum level of star formation constant for all regions.

References

1. Ferrini, F., Matteucci, F., Pardi, C., & Penco, U., ApJ **387**, 138 (1992).
2. Ferrini, F., Mollá, M., Pardi, C., & Díaz, A. I, ApJ **427**, 745 (1994).
3. Gallagher, J. S., Hunter, D. A., & Tutukov, A. V., ApJ **284**, 544 (1984)..
4. Mollá, M. & Ferrini, F. ApJ **454**, 726 (1995)
5. Mollá, M., Ferrini, F., & Díaz, A. I,ApJ **466**, 668 (1996).
6. Mollá, M., Hardy, E., & Beauchamp, D., ApJ **513**, 695 (1999).
7. Mollá, M. & Roy, J.-R., in *Chemical Evolution from zero to high redshift*, eds. J.R.Walsh & M.R. Rosa (Springer-Verlag, Berlin, 1999)
8. Persic, M., Salucci, P. & Steel, F., MNRAS **281**, 27 (1996).

A NEW PROJECT FOR THEORETICAL COLORS OF GLOBULAR CLUSTER STARS

J. MONTALBAN AND F. D'ANTONA
Osservatorio Astronomico di Roma, I-00040 Monteporzio, Italy
E-mail: dantona@coma.mporzio.astro.it

F. KUPKA AND W. SCHMIDT
Institute for Astronomy, University of Vienna, A-1180 Wien, Austria
E-mail: Kupka@galileo.astro.univie.ac.at

1 Summary

At $M \geq 0.7$ M$_\odot$ convection in the atmosphere and envelope is over-adiabatic, and the shape of the isochrones depends on the efficiency of the convective transport of the energy. Therefore, the $T_{\rm eff}$ and colors of the upper main sequence (MS), turnoff (TO) and Red Giant Branch (RGB) stars in globular clusters depends on the treatment of over-adiabatic convection, both in the interior and in the atmosphere.

Here we present some results of the study of this dependence (for details, Montalban et al. 2001). We have computed several sequences of models for masses $M \geq 0.5$ M$_\odot$ with different treatment of convection in the interior: full spectrum of turbulence (FST), and mixing length theory (MLT) with $\alpha = 1.25$ (as in the model atmospheres) and $\alpha = 1.6$ (as given by the solar calibration). As boundary conditions at the surface we use two new sets of model atmospheres (for metallicity $Z = 2.10^{-4}$) in which convection is treated either in the MLT or in the FST formulations (Heiter et al. 2001). We also check the effect of changing the optical depth at which we fit the interior with the atmosphere ($\tau_{\rm ph}$=1,10,100). If the atmospheric integration is consistent with the interior physics, the model should not depend on the choice of $\tau_{\rm ph}$. In Fig. 1a we show that using a FST treatment of convection both for the interior and atmosphere we obtain consistent results; while a match with a MLT atmosphere is not adequate: the TO $T_{\rm eff}$ of the $\tau_{\rm ph} = 1$ track is 270 K cooler than the TO of $\tau_{\rm ph} = 100$ track.

The theoretical isochrones have been translated in the Johnson system (B,V magnitudes) using $T_{\rm eff}$–color relations from the same model atmospheres, and comparisons with observational data shows that full FST models provide

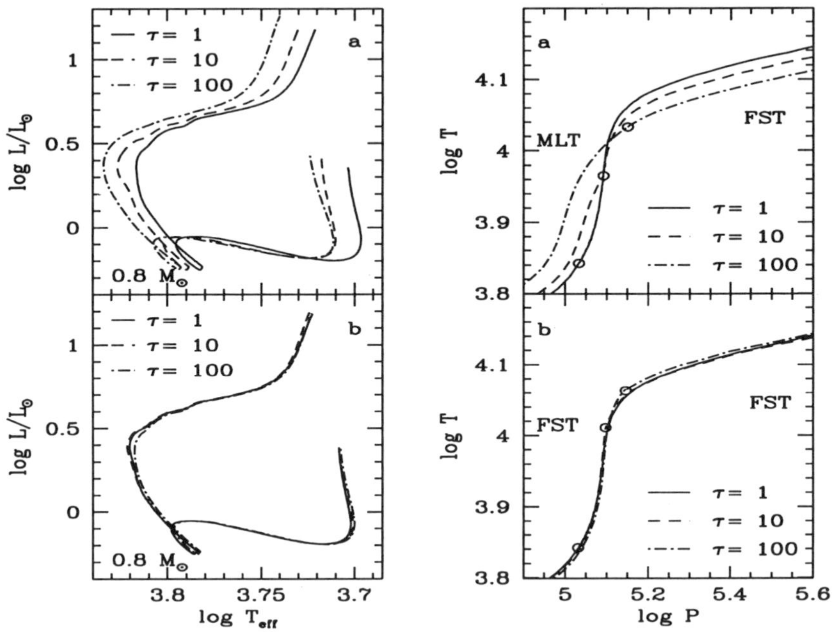

Figure 1. a)Evolutionary track for $0.8M_\odot$ using MLT atmospheres and FST interiors (left panel); temperature stratification in atmosphere and interior for models at 10 Gyr. Big empty dots represent the fitting points (right pannel). b) Same than a) for full FST models.

relative locations of TO and RGB in better agreement with the HR diagram morphology of the Globular Cluster M92 than the full MLT models.

We conclude that a unified model of stellar evolution including stellar atmospheres as an outer boundary conditions requires a consistent formulation of convective energy transfer. And therefore, new sets of model atmospheres for higher metallicities should be developed in order to allow us to do a complete study of the HR diagram morphology of Globular Clusters Stars.

References

1. U. Heiter *et al.*, A&A **in preparation**, (2001).
2. J. Montalbán *et al.*, A&A **submitted**, (2001).

THE FORMATION OF THE MILKY WAY IN THE COSMOLOGICAL CONTEXT

A. BURKERT
Max-Planck-Institut für Astronomie, Königstuhl 17, D-69117 Heidelberg, Germany
E-mail: burkert@mpia-hd.mpg.de

The formation of the Milky Way is discussed within the context of the cold dark matter scenario. Several problems arise which can be solved if the Galaxy experienced an early phase of gas heating and decoupling from the dark matter substructure. This model combines the Eggen, Lynden-Bell and Sandage picture of a monolithic protogalactic collapse with the Searle and Zinn picture of an early merging phase of substructures into one consistent scenario of Galactic formation.

1 ELS versus Searle & Zinn

Because of its proximity, the Milky Way Galaxy is an ideal candidate to investigate the formation and evolution of galaxies. In 1962 one of the most influential papers on galaxy formation was published by Eggen, Lynden-Bell and Sandage who studied the motions of a sample of high-velocity stars. They found a correlation between metallicity and orbital eccentricities and angular momentum and inferred that the Galaxy formed quickly, through the collapse of a uniform, isolated protogalactic cloud. In 1978, this scenario was challenged by Searle and Zinn who analysed Galactic globular clusters and concluded from the lack of a metallicity gradient and from a large age spread that the halo was built on a timescale of several Gyrs from independent low-mass fragments.

Subsequent surveys, coupled with more precise determinations of stellar ages and stellar element abundances have revealed in greater details the chemodynamical formation history of the Galaxy. No significant metallicity gradient has been found in the Galactic halo, confirming the predictions by Searle and Zinn. In addition, strong evidence for a merging history of the Galaxy came from coherent moving groups of halo stars, that might represent remnants of the original substructures (Majewski et al. 1994, Helmi et al. 1999).

Element abundances have also provided constraints on the timescale of early Galactic nucleosynthesis and by this on the evolution of the Milky Way (Wheeler, Sneden & Truran 1989). In the Galactic halo, oxygen and intermediate mass nuclei are enriched by a factor of order 3, relative to iron. This is characteristic of pure high-mass star contamination by supernovae Type II with lifetimes of less than 10^8 yrs (Tinsley 79, Truran 91). Deviations from this trend seem first to appear at the point where the first disk stars appear, indicating the onset of substantial enrichment from Type Ia supernovae (Matteucci & Greggio 86). Unfortunately, the lifetime of Type Ia supernova progenitors is still poorly understood and might even depend on metallicity (Kobayashi et al. 98). These objects therefore cannot be used as clocks to measure the timescale of disk formation. One interesting coincidence is however the similarity of the epoch where SN Ia begin to contribute to the metal enrichment in the Milky Way and the formation epoch of the galactic disk. It is not clear why these two very different processes should occur on the same timescale.

Truran (1981) was the first to suggest that low $[Ba/Eu]$-ratios of very metal-deficient halo stars with $[Fe/H] < -2.5$ reflect a pure r-process contamination. At a metallicity of $[Fe/H] \approx -2.5$, the ratio of s-process/r-process, as reflected by [Ba/Eu] increases which indicates the entry of products from AGB stars, corresponding to lifetimes of order 10^9 yrs. If one assumes a typical disk lifetime of order 10-11 Gyrs, as inferred from the white dwarf luminosity function and a galactic age of 13 Gyrs as predicted by cosmological models, the galactic halo formed on timescales of 2×10^9 yrs with a roughly constant SN II rate of order 0.01 yr^{-1} which is consistent with the Searle & Zinn model.

2 Cosmological predictions

Up to now, the formation of the Milky Way has been studied neglecting its cosmological formation history. Recently however, cosmological simulations have been able to resolve structure formation on galactic scales. For the first time we are now able to construct a self-consistent model of the formation of the Milky Way, starting from realistic, cosmologically motivated initial conditions. The standard cold dark matter (CDM) scenario predicts that the Galaxy formed as suggested by Searle & Zinn from many small substructures on timescales of order 10^9 yrs, in agreement with the observational evidence.

One of the fundamental assumptions of cold dark matter is its collisionless nature and the fact that the CDM particles formed with negligible random velocities and subsequently interacted only through gravitational forces. Numerous simulations have by now shown that this assumption leads to dark matter halos which have universal density distributions that can be approximated well by the simple function (Navarro, Frenk & White 1997; Moore et al. 1999) $\rho(r) = \rho_0 / \left(r^{1.5}(r + r_s)^{1.5}\right)$ where r_s is the scale radius which for the Galaxy is of order 10 kpc. The Galactic disk therefore lies within the central power-law cusp of the Milky Way's dark halo. This universal density structure is not sensitive to the adopted cosmological model and is a universal property of CDM.

Interestingly, this strong prediction of CDM models leads to several unsolved problems. (i) Observations of dark matter dominated dwarf galaxies indicate that their dark halos have isothermal, constant density cores instead of compact power-law cusps (Moore 1994, Burkert 1995). (ii) Due to their compact cores, the substructures from which halos form are not completely disrupted, leading to the cosmological prediction that galaxies like the Milky Way should have of order a thousand satellites (Klypin et al. 99, Moore et al. 99). Only a dozen satellites have been found so far. (iii) Cosmological simulations, including gas dynamics predict a disk scale length for the Milky Way of order 300 pc, whereas its observed scale length is a factor of 10 larger (Steinmetz & Navarro 1999). Again, this problem is related to dense substructures from which the Galaxy formed. As their cores are not disrupted during the merging epoch they experience dynamical friction by this loosing 90% of their angular momentum. The gas cools and is trapped in these cores. Dynamical friction will therefore also remove 90% of the angular momentum of the gaseous component, leading eventually to galactic disks with very small scale lenghts.

In summary, although cosmological models can by now resolve structure formation on galactic scales, the dark matter core problem makes it difficult to construct a consistent model of the formation of the Galaxy.

3 Evidence for an Early Heating Phase of the Milky Way

One solution of the cosmological satellite and disk problem is an early phase of gas heating and decoupling from the dark matter substructure (Weil et al. 1998; Sommer-Larsen et al. 1999). Suppose that gas was settling into the low-mass progenitors of the Milky Way. A first stellar population II formed at their center. Subsequent energetic feedback processes ejected the gas in a violent galactic wind. If the gas could be kept from falling back, the gas-free, dark substructures could merge, forming the dark halo of the Milky Way. The population II stars in their cores would now represent the halo component. As star formation in the substructures was very inefficient, most of the satellites would not easily be detectable. Their stars could represent some of the moving groups that have been found by recent surveys. The hot gas component, on the other hand, cooled and settled into the equatorial plane through a monolithic collapse phase, conserving its angular momentum and forming a disk-like substructure with the observed scale length. During the first phase of disk formation the dark halo was still violently relaxing with massive substructures frequently interacting with the proto-disk. During this early phase which might have lasted a Gyr the thick disk formed. Subsequently, after the major merging epoch, gas could settle into a thin disk.

This scenario can in fact explain an interesting puzzle that has been discussed by Wyse & Gilmore (92). They noted that the specific angular momentum of the Galactic halo stars is a factor of 10 smaller than required to form the Galactic disks. As a result, the halo could not have been the progenitor of the disk. Its metal-enriched gas should instead have settled into the Galactic bulge. This seems to be in contradiction with the bulge mass and with the chemical continuity between the halo stars and the thick and thin disk component. If on the other hand the gas was ejected by a first population of stars before the major merging event, the gas could keep its initial angular momentum whereas the halo stars, residing inside the cores of the dark matter substructures would experience angular momentum loss by dynamical friction as predicted by the cosmological models, leading to a low-angular momentum halo and a disk with large angular momentum.

The ELS and Searle & Zinn picture have been considered as contradicting theories in the past. An early phase of galactic heating would however combine both models in a new, consistent model of Galaxy formation. Whereas the galactic halo formed by hierarchical merging of substructures, the Galactic thick and thin disk components resulted from the monolithic infall of a diffuse gaseous component that was ejected from the substructures prior to the merging event.

Acknowledgments

I would like to thank the LOC for the excellent organization of an exciting conference.

References

1. Burkert, A., 1995, ApJ, 447, L25
2. Eggen, O.J., Lynden-Bell, D. & Sandage, A., 1962, ApJ, 136, 748
3. Helmi, A. et al., 1999, Nature, 402, 53
4. Klypin, A. et al., 1999, ApJ, 522,8
5. Kobayashi, C. et al., 1998, ApJ, 503,155
6. Majewski, S.R., Munn, J.A. & Hawley, S.L., 1994, ApJ, 427, L37
7. Matteucci, F. & Greggio , L. 1986, AA, 154, 279
8. Moore, B. 1994, Nature, 370, 620

9. Moore, B. et al., 1999, MNRAS, 310, 1147
10. Navarro, J.F., Frenk, C.S. & White, S.D.M. 1996, ApJ, 462, 563
11. Searle, L. & Zinn, R., 1978, ApJ, 225, 357
12. Sommer-Larsen, J., Gelato, S. & Vedel, H., 1999, ApJ, 519, 501
13. Steinmetz, M. & Navarro, J.F., 1999, ApJ, 513, 555
14. Tinsley, B.M., 1979, ApJ, 229, 1046
15. Truran, J.W., 1981, AA, 97, 391
16. Truran, J.W., 1991, IAU Symp. 145, eds. G. Michaud & A.V. Tutukov (Kluwer, Dordrecht), p. 13
17. Weil, M.L., Eke, V.R. & Efstathiou, G. 1998, MNRAS, 300, 773
18. Wheeler, J.C., Sneden, C. & Truran, J.W., 1989, ARAA, 27, 279

A NEW SURVEY OF GALACTIC HIGH-VELOCITY HI CLOUDS

FELIX J. LOCKMAN
National Radio Astronomy Observatory, P.O. Box 2, Green Bank, WV 24944, USA
E-mail: jlockman@nrao.edu

EDWARD M. MURPHY
Astronomy Department, University of Virginia, P.O. Box 3818 University Station, Charlottesville, VA 22903-0818, USA
E-mail: emm8x@virginia.edu

Very sensitive 21cm observations for high-velocity HI clouds too faint to have been seen in other surveys have been made in more than 850 directions around the sky. Preliminary results from the survey indicate that high-velocity 21cm HI emission is found in about one-third of all directions. Clouds with the lowest column density have kinematics similar to that of the brighter clouds, and a location on the sky which suggests that much of the faint emission comes from the outskirts of known high-velocity complexes.

1 Introduction

A significant fraction of the sky shows 21cm HI emission which does not participate in Galactic rotation, yet cannot be classified unambiguously as extragalactic. These "high-velocity" HI clouds may arise from a number of phenomena including nearby supernova remnants, a Galactic fountain, clouds in the Local Group, and infall of primordial matter [6,1]. Some high-velocity gas comes from a tidal tail of the Magellanic Clouds [3]; there may be other remnants of tidal events as well. High-velocity clouds provide the most convincing evidence that there is continuing accretion of gas onto the Milky Way, which will alter the chemical mix of the interstellar medium with time [7].

Most studies of high-velocity clouds have mapped large areas. We chose instead to make very sensitive observations in selected directions to search for clouds at low column density. The primary goals were to establish the existence (or absence) of low-N_H clouds, to determine their number, and to examine their relationship to the brighter, previously-known objects. A report on the first part of the survey appeared several years ago [2]. Here we summarize results from a preliminary analysis of the complete survey, which contains three times more data than were available earlier.

The survey was made on the 140 Foot radio telescope of the NRAO in

Green Bank, WV, USA, at an angular resolution of 21′ and a velocity resolution of 4 km s^{-1}. The survey is complete to a beam-averaged hydrogen column density of 8×10^{17} cm^{-2}. Depending on the structure of the clouds, these data are between three and eight times more sensitive than previous observations. More than 850 directions were studied, about one-third of which were centered on an extragalactic object (QSO, Seyfert, or other AGN).

2 Nature of the Faint High-Velocity Clouds

We have measured more than two hundred high-velocity HI features, half of which could not have been detected in previous, less sensitive, radio surveys. We find that at least one-third of the sky is covered with high-velocity HI, a significantly larger fraction than the 18% implied by previous radio measurements [5] and approaching the frequency of detections in the UV [4]. Most of the faint clouds appear to be associated with the brighter clouds and are not a new population. Most likely they are faint because they have little neutral hydrogen, not because they are at a great distance.

This result implies that the contribution of the faint clouds to the total mass of high-velocity gas is small – most of the mass (in neutral gas) must lie in the brighter clouds. The clouds we observe are usually extended implying that their diameter is $D(pc) \geq 30\ d$ where d is the distance in kpc. Complete analysis of these data is underway with the hope that they will allow us to distinguish between competing models for the origin of the clouds, and give good estimates of mass and mass flux onto the Milky Way, and, in conjunction with observations at other wavelengths, elemental abundances in this gas.

The National Radio Astronomy Observatory is operated by Associated Universities, Inc., under a cooperative agreement with the National Science Foundation.

References

1. Blitz, L., Spergel, D.N., Teuben, P.J., Hartmann, D., & Burton, W.B. 1999, *ApJ*, 514, 818.
2. Murphy, E.M., Lockman, F.J. & Savage, B.D. 1995, *ApJ*, 447, 642.
3. Putman, M.E., 2000, PhD Thesis, The Austrilian National University.
4. Savage, B.D., *et al.*, 1993, *ApJ*, 413, 116.
5. Wakker, B.P. 1991, *Astr.Ap.*, **250**, 499.
6. Wakker, B.P. & van Woerden, H., 1997, *Ann. Rev. Astr. Ap.* **35**, 217.
7. Wakker, B.P. *et al.* 1999, in *Stromlo Workshop on High-Velocity Clouds*, ed. B.K. Gibson and M.E. Putman, ASP Vol. 166, p. 26.

GALAXY EVOLUTION FROM MULTIWAVELENGTH OBSERVATIONS

LAURENT VIGROUX
CEA/DSM/DAPNIA/ Service d'Astrophysique,
CEA Saclay, F91191 Gif sur Yvette cedex, France

Observations of nearby and distant galaxies span now the whole range of the electromagnetic spectrum, from X Rays to radio waves. Combined observations between infrared and X rays have been used to study the star formation rate in galaxies up to redshift of 1. A strong evolution of L15 μm/LB luminosity ratio is observed in large distant galaxies. This is attributed to strong starburst events triggered by a larger number of galaxy-galaxy interaction at redshift > 0.5. This evolution is at variance from what have been inferred from HST results, where the evolution was dominated an increase of the number of small blue galaxies.

1 Introduction

The observations of galaxies have been confined for a very long period of time to the visible and radio domains. Stellar content was probed by the galaxy emission in the visible and near infrared, while the interstellar matter , HI or molecules, was observed by its emission in the radio domain. New observatories, in space or on ground provide new capabilities, which can be used for detailed studies of nearby galaxies, as well as for studies of integrated properties of galaxies at cosmological distances. Deep imaging with HST, followed by spectroscopic surveys with the Keck telescopes or the VLT has expanded over previous classical works on galaxy evolution. Infrared observatories, IRAS, ISO and more recently SCUBA on the JCMT have demonstrated the importance of the infrared to study the galaxy evolution. X ray observatories, ROSAT, CHANDRA and XMM provide a unique opportunity to disentangle the origin of activity in distant galaxies. All these tools have produced a new view on the evolution between galaxies at redshift about 1 and local galaxies. The next generation of observatories, NGST, FIRST and ALMA will provide the capabilities to study galaxies at higher redshift, and hopefully to detect the primeval galaxies.

2 Mid infrared and star formation in nearby galaxies

IRAS has demonstrated the importance of the infrared for understanding the energetic budget of galaxies. The infrared flux of galaxies is dominated by the radiation from dust heated by the interstellar radiation field. The shape of the galaxy spectral energy distribution is dominated by the fraction of the interstellar radiation field which is absorbed by the dust and reradiated in the infrared. IRAS observations have shown that the range of the fraction of the interstellar radiation field which is absorbed by the dust component is very large, from 10% to almost 100%. Observations in the visible alone cannot be used to determine the bolometric luminosity of a galaxy since they do not provide information about absorption. ISO has improved our observing capabilities in several ways: spatial resolution improved by a factor of 20 at mid infrared wavelength with ISOCAM, measurements at wavelength longer than 100 μm with ISOPHOT, and high spectral resolution capabilities with the two spectrographs.

In normal galaxies, the infrared emission is thought to be due to two main components. The diffuse interstellar medium associated with cold dust (T = 15 - 30 K) which can be detected by the dust thermal emission in the far infrared, and the mid infrared emission of carboneous materials which are present in the cirrus clouds. Star forming complexes which can be detected by the emission of the central HII regions or the surrounding photodissociation regions. HII regions spectra are dominated by ionic lines, hot dust (T is about 100 K) thermal continuum, and thermal emission of very small grains which dominates the emission between 10 and 40 μm.

The mid infrared emission of the interstellar matter is characterized by the presence of large emission bands at 3.3, 6.2, 7.7, 8.6, 11.3 and 12.7 μm, superimposed on a continuum. These bands and the continuum can be explained by the emission of low thermal capacitance species heated by an impulsive mechanism. It was the base of the Polycyclic Aromatic Hydrogeneous molecules (PAH) model [13], in which a single molecule is excited by a single UV photons, and then go back to fundamental state emitting IR photons in specific emission bands. Beyond 14 microns, the PAH cannot explain the shape of the spectrum. An additional component is needed; most probably Very Small Grains (VSG) in a transition state between thermal equilibrium and impulsive heating [5]. ISO observations have confirmed the overall picture, but have put much more constraints on the exact nature of PAH. The impulsive heating have been strongly confirmed by the presence of the emission bands in a very large range of astrophysical environments , see Tielens 98 [20] for a review). However, the observed PAH spectra cannot be explained by small molecules like the one which were originally advocated, but rather by aggregates of several hundreds of atoms [4].

While PAH are ubiquitous in neutral medium, they disappear in presence of a strong radiation field (> 10000 the intensity of the interstellar radiation field in the Sun vicinity) like in the core of HII regions [21]. In these conditions, the VSG emission dominates the mid IR spectrum. The most likely origin for the VSG are carboneous material, graphite like. The transition between the PAH band dominated regions toward the VSG dominated band is smooth. Transition regions still exhibit bands in the 6 to 9 microns regions, but individual bands disappear and are replaced by a very wide bump, which could be due to a graphitization of the PAH [18].

Before ISO, it was supposed that the 7 μm map would trace the diffuse component, while the 15 microns maps will outlined the star formation complexes. Instead, both emissions are dominated by star forming complexes. A detailed analysis of a sample of spiral galaxies [15] have shown that 7 μm maps are dominated by PDRs with only a minor contribution from the neutral ISM which never represents more than 30% of the total emission at 7 microns.

The close correlation between star formation complexes and mid infrared emission in the disk of spiral galaxies indicates that the mid IR emission should be a good star formation efficiency indicator. Roussel et al (2001) [16] have shown that in the disks of spiral galaxies, a good correlation exists between mid IR fluxes, at 7 and 15 microns and Hα. This correlation holds when the total emission remains dominated by PDRs, which is the as long as the star formation rate integrated over the disk remains $< 0.1\ Mo.yr^{-1}kpc^{-2}$. At larger star formation rate, this relation does not hold since the mid IR emissions are dominated by the central HI regions and since Hα is strongly absorbed. This is the case in central regions of barred galaxies where a high star formation rate is induced by gas mass transfer along the bar [10], or in interacting galaxies which experienced a strong starburst due to shocks in the colliding gaseous disks [19]. In the starburst regime, the mid infrared emission exhibits a larger contribution of the thermal continuum of VSG, compared to PAH bands, and Hα cannot be used as a reliable star formation indicator. From detailled studies of starburts galaxies with the Short Wavelength Spectrograph (SWS) of ISO, or the Circular Variable Filter (CVF) of ISOCAM, a strong correlation have been found between the number of ionizing photons deduced from mid infrared Ar lines and the flux density at 14 μm. Mid infrared emission remains a good star formation indicator, but 15 μm flux density alone should be used, as variance from the normal galaxies where the 7 and 15 μm flux densities might be used almost in an equivalent manner. In most extreme cases, close to active galactic nuclei, the radiation field is so strong that all PAH are destroyed and the mid infrared spectra is totally dominated by the thermal continuum of the VSG. PAH band intensities versus VSG have been used to disentangle AGN and starburts contribution in Ultra Luminous Infrared Galaxies [22,23].

The mid infrared, from 5 to 25 μm, provides reliable star formation indicators, which are less prone to extinction that normal star formation index in the visible or the near infrared. The mid infrared can be used from quiet normal galaxies up to the most active galaxies. The mid infrared star formation indicators are based on the comparaison between the strength of the PAH bands near 7 μm, and the VSG continuum flux density at 15 μm. In normal galaxies, the 15 μm to 7 μm flux density ratio is close to 1; in starburst galaxies this ratio is close to 2, due to the increase of the VSG emission. Close to AGN, this ratio get even larger, due to the disapearance of PAH bands.

The Antennae galaxies (NGC 4038/4039) is an ideal template to compare the different star formation indicator. It is closed enough to be observed with a spatial resolution suitable to identify all the star formation complexes, which span a wide range of activity, from normal HII regions to strong starburst regions. The Antennae system is due to the interaction of two large spiral galaxies. It has been extensively studied in the past. In the visible, the most dramatic images have been obtained with the refurbished Wide Field Camera on the HST [24]. The northen galaxy exhibits a bright ring of HII regions and a strong nucleus. The ring of HII regions extends south towards southern galaxy. This galaxy is much more distorded and absorbed. In the transition region, a dark area is supposed to be the region of overlapp between the two galaxy disks [26]. As often in starburst galaxies, massive star clusters have been found in the northen part of the system and in the overlap region [24]. They are supposed to be the relics of previous starburst. In the overlap, a few clusters are very red, indicating a strong absorption. In the UV, the brightest regions are the northen HII regions. Even the north nucleus is dimmed by dust. The overlapp region is totally dark. In the mid infrared, the overlapp region radiates 50% of the total 15 μm flux density of the whole system [19]. The structure of the overlapp regions in the mid infrared is very similar to the complex patches observed in radio continuum [27]. The brightest of these spots radiates almost 20% of the 15 μm flux density of the two galaxies. The 15 to 7 μm flux density ratio shows that this spot is the most active of the whole system, even more active than the two galaxy nuclei. CO distribution [28] is very similar to the mid infrared emission. Most of the CO is concentrated in the overlapp region and in the two nuclei. However, the CO intensity is not directly related to the mid infrared emission. The CO in the overlapp region is only 25% of the whole CO. Difffuse 2 keV emission follows the mid infrared distribution and some of the brightest IR spots have X ray counterpart. This is the case for the brightest IR spot which is also a strong X ray source, probably due to supernova remnants associated with the starburst. In conclusion, in the Antennae, the UV and the visible reveal old starbursts and normal HII regions; Mid infrared and X rays are associated with the most active star forming regions, while CO is present in each of the star forming knots, but its intensity is not directly related to the present star formation rate. These results should be kept in mind, to analyse the observations of distant galaxies, where only integrated properties can be measured.

3 Distant galaxies and Cosmology

The last few years have brought the first direct evidence for strong galaxy evolution in the last billions years. The first hint for such an evolution was provided by the Canada- France-Redshift-Survey [8]. These authors have found a large excess of small blue galaxies at redshift larger than 0.7. Translated in age, this corresponds to an age about half of the present age. This result was confirmed from observations with the Hubble Space Telescope and the Keck telescopes [9]. These results have shown that the star formation rate was higher by a factor 3 to 5 at redshift = 1, compared to what is observed now [9]. However, there is a strong possibility that this result underestimates the exact star formation rate due to the large extinction, which is present in the most active star formation regions. The observations done in the visible correspond to emission in the UV in the galaxy rest frames, which is where the extinction is the largest. Several estimates of this effect where proposed [12], but without being very convincing. ISO was expected to provide a more reliable estimate of the star formation rate, since the extinction is much smaller in the infrared. Several deep surveys were done with ISOCAM (see Elbaz 2000[29] for a review), and ISOPHOT [7], to look for dust enshrouded star formation in distant galaxies. ISOCAM deepest surveys have been able to detect sources as faint as 50 μJy, a factor 10^4 fainter than the faintest source observed with IRAS !

The source counts obtained in these surveys are in good agreement with a straight extrapolation from the source counts obtained in the infrared with IRAS down to 1 mJy, that is 100 times fainter than the faint IRAS sources. However, for fainter sources, a strong increase of source counts is observed. This increase cannot be explained without invoking a strong evolution of galaxies. A very detailed analysis of the sources, which are responsible for this increase was done by [1] on the Hubble Deep Field North (HDF). About 40 sources have been detected by ISOCAM in the HDF and the flanking fields. This represents only a small fraction of all the HDF galaxies. All the galaxies detected by ISOCAM have their counterpart in the HST images. The identification is generally unambiguous, with the better than 3" position accuracy of ISOCAM images. All the galaxies detected are large spiral galaxies, most of them with distorted morphologies, which are the sign of recent interaction with another galaxy. At variance from the visible, where the evolution is due to a large number of small blue galaxies, in the mid infrared, the evolution shows an increase of the mid IR luminosity of some galaxies. However, in the visible and in the UV, the colors and the spectra of these bright IR galaxies do not depart from those of the other galaxies observed in this field. The redshift distribution of ISOCAM detected galaxies is very similar to that of the HDF galaxies [30]. The peculiarities of these galaxies are visible only in the infrared.

Dust enshrouded active nuclei or strong starbursts could be at the origin of this infrared luminosity excess in some galaxies. Six sources have been detected in the deepest Chandra observation of the Hubble Deep Field [32], with a detection limit of $4.10^{-16} erg.cm^{-2}s^{-1}$. One is an AGN, four are elliptical galaxies, and the last one is a nearby spiral. Only this nearby object is in common with the ISOCAM sources. On a larger field, 20 x 20 $arcmin^2$ in the Lockman hole, XMM have detected more than 100 sources at a sensitivity limits of $2.10^{-16} erg.cm^{-2}s^{-1}$ [25]. Several of these sources have very hard spectra, which can be explained by absorbed sources. However, in the area in common with ISOCAM observations, only 15% of the 15 μm sources have their counterpart in the X ray domain [31]. The small level of coincidence between X ray and 15 μm sources demonstrates that the IR sources are not dusty AGN. On the other hand, 7 of the HDF ISOCAM sources have radio continuum counterparts. There is a very good correlation between the 15 μm flux and the radio continuum intensity [2]. This is a very strong evidence for a starburst origin for the IR excess of galaxies at z=1.

Additional constraints on the nature of these galaxies can be obtained from the extragalactic background, which have been derived from COBE [7]. Spectral energy distribution (SED) templates from the UV to the far infrared can be calculated for typical galaxies. A wide range of galaxies can be represented by SEDs, which are bracketed by the normal spiral M51 and the most extreme local Ultra Luminous Infrared Galaxy (ULIRG), Arp 220 [1]. A global SED can be calculated with the assumption that all ISOCAM detected galaxies have the same SED weighted by their luminosity function. With a SED similar to that of Arp 220, the extragalactic background calculated is far above the COBE far infrared observations. With an M51 SED, the FIR extragalactic background is a factor 10 too low. With a mildly active starburst, M82 type, SED, the prediction agrees with the observations within a factor of 2. It means that the galaxies detected by ISOCAM are not ULIRGS, nor the population of small blue galaxies detected in the visible. They are normal large galaxies, which are suffering a strong star formation event, probably related with a collision. They looks more to the M82-M81 or to the Antennae

(NGC4038-4039) interacting systems, that to extreme object like Arp 220 or Arp 299. Presumably, the IR emission is due to a starburst hidden in a very dusty region which are barely detectable in the visible and the UV, similar to local counterparts [11]. However, the ISOCAM observations imply a very strong evolution of the IR luminosity, related with a strong increase of the star formation rate at distant redshifts. While in the local Universe, only 3% of large galaxies are strong IR sources, they are 30% at a redshifts of 0.7 - 1. This is confirmed by the analysis of the far infrared surveys, where a large number of objects have also been detected [14]. The amount of evolution, in terms of energy production by galaxy mass is three times larger in the IR than in the visible. When transformed in star formation rate, the total star formation rate was 50 times larger at z=1 than now [29].

4 Acknowledgements

This work would not have been possible without the collaboration of the people of Saclay involved in the ISOCAM data analysis. I am specially redevable to Hervé Aussel, David Elbaz, Dario Fadda, Hector Flores, Hélène Roussel and Marc Sauvage.

References

1. H. Aussel, D. Elbaz, C.J. Cesarsky, J.L. Starck, The Universe as seen by ISO, Edts P. Cox and M. Kessler, ESA SP.427, 1031, (1999)
2. H. Aussel, L. Vigroux, A. Franceschini, D. Elbaz, M. Dennefeld, C.J. Cesarsky, *AAS* **195**, 917 (1999)
3. F. Boulanger, W.T. Reach, V. Abergel, et al, *A&A* **315**, L325 (1996)
4. F. Boulanger, P. Boisssel, D. Cesarsky, C. Ryter, *A&A* **339**, 194 (1998)
5. F.X. Désert, F. Boulanger, J.L. Puget, *A&A* **237**, 215 (1990)
6. D. Elbaz, H. Aussel, C.J. Cesarsky, F.X. Désert, D. Fadda, A. Franceschini, M. Harwit, J.L. Puget, J.L. Starck, The Universe as seen by ISO, Edts P. Cox and M. Kessler, ESA SP.427, 999, (1999)
7. G. Lagache, J.L. Puget, *A&A* **355**, 17 (2000)
8. S.J. Lilly, O. Le Fevre, F. Hammer, D. Crampton, *ApJ* **460**, L1 (1996)
9. P. Madau, H.C. Ferguson, M.E. Dickinson, M. Giavalisco, C.C. Steidel, A. Fruchter, *MNRAS* **283**, 1388 (1996)
10. L. Martinet, D. Friedli, *A&A* **323**, 363 (1997)
11. I.F. Mirabel, L. Vigroux, V. Charmandaris, M. Sauvage, P. Gallais, D. Tran, C.J. Cesarsky, S.C. Madden, P.A. Duc, *A&A* **333**, L1 (1998)

12. M. Pettini, C.C. Steidel, M.E. Dickinson, M. Kellogg, M. Giavalisco, K.L. Adelberger, uulh.conf, 279, (1997)
13. J.L. Puget, A. Leger, *ARA&A* **27**, 161 (1989)
14. J.L. Puget, G. Lagache, D.L. Clements, W.T. Reach, H. Aussel, F.R. Bouchet, C.J. Cesarsky, F.X Désert, H. Dole, D. Elbaz, A. Franceschini, B. Guiderdoni, A.F.M. Moorwood, *A&A* **345**, 29 (1999)
15. H. Roussel, L. Vigroux, M. Sauvage, C. Bonoli, A. Bosma, P. Gallais, T. de Hawarden, S.C. Madden, P. Mazzei, The Universe as seen by ISO, Edts P. Cox and M. Kessler, ESA SP.427, 957, (1999)
16. H. Roussel, M. Sauvage, L. Vigroux, *A&A* **in press**, (2001)
17. M. Sauvage, Trinh X Thuan, *ApJ* **396**, L69 (1992)
18. L. Verstraete, J.L. Puget, E. Falgarone, S. Drapatz, C.M. Wright, R. Timmermann, *A&A* **315**, L337 (1996)
19. L. Vigroux, F. Mirabel, B. Altieri, F. Boulanger, C.J. Cesarsky, D. Cesarsky, A. Claret, C. Fransson, P. Gallais, D. Levine, S. Madden, K. Okumura, D. Tran, D., *A&A* **315**, L93 (1996)
20. A.G.G. M. Tielens, *ApJ* **499**, 267 (1998)
21. C.J. Cesarsky, A. Abergel, P. Agnese, B. Altieri, J.L. Augueres, H. Aussel, A. Biviano, J. Blommaert, J.F. Bonnal, F. Bortoletto, O. Boulade, F. Boulanger, S. Cazes, D.A. Cesarsky, A. Chedin, A. Claret, M. Combes, J. Cretolle, J.K. Davies, F.X. Desert, D. Elbaz, J.J. Engelmann, G. Epstein, A. Franceschini, P. Gallais, R. Gastaud, M. Gorisse, S. Guest, T. Hawarden, D. Imbault, M. Kleczewski, F. Lacombe, D. Landriu, J. Lapegue, P. Lena, M.S. Longair, R. Mandolesi, L. Metcalfe, N. Mosquet, L. Nordh, K. Okumura, S. Ott, M. Perault, F. Perrier, P. Persi, P. Puget, T. Purkins, Y. Rio, T. Robert, D. Rouan, A. Roy, O. Saint-Pe, J. SamLone, A. Sargent, M. Sauvage, F. Sibille, R. Siebenmorgen, F. Sirou, A. Soufflot, J.L. Starck, D. Tiphene, D. Tran, G. Ventura, L. Vigroux, F. Vivares, R. Wade, *A&A* **315**, L32 (1996)
22. R. Genzel, D. Lutz, E. Sturm, E. Egami, D. Kunze, A.F.M. Moorwood, D. Rigopoulou, H.W.W. Spoon, A. Sternberg, L.E. Tacconi-Garman, L. Tacconi, N. Thatte, *ApJ* **498**, 579 (1998)
23. O. Laurent, I.F. Mirabel, V. Charmandaris, P. Gallais, S.C. Madden, M. Sauvage, L. Vigroux, C.J. Cesarsky, *A&A* **359**, 887 (2000)
24. B.C. Whitmore, Q. Zhang, C. Leitherer, S.M. Fall, F. Schweizer, B.W. Miller, *Astron. Journ.* **118**, 1551 (1999)
25. G. Hasinger, B. Altieri, M. Arnaud, X. Barcons, J. Bergeron, H. Brunner, M. Dadina, K. Dennerl, P. Ferrando, A. Finoguenov, R.E. Griffiths, Y. Hashimoto, F.A. Jansen, D.H. Lumb, K.O. Mason, S. Mateos, R.G. McMahon, T. Miyaji, F. Paerels, M.J. Page, A.F. Ptak, T.P. Sasseen,

N. Schartel, G.P. Szokoly, J. Trumper, M. Turner, R.S. Warwick, M.G. Watson, *A&A* **365**, L45 (2001)

26. J.E. Barnes, *ApJ* **331**, 699 (1998)
27. E. Hummel, J.M. van der Hulst, *A&A* **155**, 151 (1986)
28. C.D. Wilson, N. Scoville, S.C. Madden, V. Charmandaris, *ApJ* **542**, 120 (2000)
29. D. Elbaz, American Astronomical Society Meeting 197, 77.07, (2000)
30. A.J. Barger, L.L. Cowie, D.B. Sanders, *ApJ* **518**, L5 (1999)
31. D. Fadda, H. Flores et al, in preparation, (2001)
32. A.E. Hornschemeier, W.N. Brandt, G.P. Garmire, D.P.Schneider, A.J. Barger, P.S. Broos, L.L. Cowie, L.K. Townsley, M.W. Bautz, D.N. Burrows, G. Chartas, E.D. Feigelson, R. Griffiths, D. Lumb, J.A. Nousek, W.L.W. Sargent, American Astronomical Society, HEAD meeting 32, 26.13, (2000)

DETAILED CHEMICAL ANALYSIS OF TWO GIANTS IN THE SGR DSPH

P. BONIFACIO, P. MOLARO
Osservatorio Astronomico di Trieste, Via G.B. Tiepolo 11,
34131 - Trieste, Italy
E-mail: bonifaci@ts.astro.it, molaro@ts.astro.it

The 8m class telescopes allow for the first time to study stars of external galaxies with the same resolution and S/N ratio which has been so far used for Galactic stars. It is quite likely that this study will shake some of our current beliefs. In this poster we highlight some of the results which have been obtained for two giants in the Sgr dSph thanks to the UVES spectrograph on the ESO 8.2m Kueyen telescope. Further details on the observations and data analysis may be found in Bonifacio et al [1].

1 Results

The two stars turned out to be rather metal-rich with [Fe/H] ≈ -0.3. This was somewhat a surprise since the stars have roughly the same $V-I$ colour but differ in V by 0.18 mag and were thought to be representative of the spread in Sgr RGB which common–wisdom has attributed to a spread in metallicity[2,3,4,5,6,7] . Given the steepness of the RGB in this colour range it may well be that the two stars lie in fact on the same isochrone and their $V - I$ colours are different but photometric errors of $\approx$ 0.04 mag make them equal. However there is another possible explanation whose consequences are rather intriguing and cannot be easily discarded: the two stars have a different distance. A difference of about 2 kpc would be enough to explain the difference in V mag. This value is not unreasonable since Ibata et al[5] estimate a line–of–sight depth of about 1.2 kpc. ¿From our spectra we may find evidence which further supports this possibility: the Na I D lines of the brighter star, # 143, appear perfectly symmetric with no hint of the presence of interstellar components at the radial velocity of Sgr, while the fainter star, # 139, shows a weak but definite asymmetry in both Na I D lines, which may be interpreted as due to interstellar material associated with Sgr.

The high metallicity found for both stars implies that Sgr has undergone a high level of chemical processing; it also suggests that a sizeable population of Sgr is this metal–rich. The abundance pattern is interesting, although most of the elements with $A \leq 39$ are consistent within error with the solar ratio, taken at face value they are all slightly sub-solar. On the other hand Ba and more massive elements show over-solar ratios. The pattern may be summarized as follows: Na and Sc are over-deficient with respect to iron by about 0.4 dex; heavy neutron–capture elements Ba, La, Ce, Nd, Eu are over–abundant with respect to iron by 0.3 to 0.5 dex; Y is over-deficient with respect to iron by about 0.4 dex.

It is interesting to compare the abundances of these two Sgr giants with those obtained by Shetrone et al[8] for three other Local Group Galaxies (Draco, Ursa Minor and Sextans). One should also keep in mind that Sagittarius is considerably more massive than the other three galaxies and this is, perhaps, the key to understand its high–metallicity population. The [Mg/Fe] ratio seems to drop, with increasing [Fe/H], faster in Local Group Galaxies than it does in the Galactic Halo or disk and this could be the signature of a very low or episodic star formation. It shall be very interesting to see what is the [Mg/Fe] ratio of the metal–poor population of Sagittarius, when its existence is confirmed. Damped Lyman α systems, observed in absorption in the line–of–sight of QSO's have metallicities in the range -2,-1 and show $[\alpha/\mathrm{Fe}]$ ratios which are solar [9,10]. Could Damped Lyman α systems be linked to dwarf galaxies ? This possibility is appealing, although these galaxies are now devoid of gas, they must have been relatively rich in gas when the stars we observe now were formed.

The over-deficiency of Na seems a common occurrence in Local Group galaxies, although there is no definite trend and a large scatter. On the other hand Y is always over-deficient with very few exceptions and this over-deficiency is mirrored by an overabundance of heavy elements.

References

1. Bonifacio P., et al *A&A* **359**, 66 (2000)
2. Ibata R.A., Gilmore G., Irwin M.J. *MNRAS* **277**, 781 (1995)
3. Sarajedini A., Layden A.C. *AJ* **109**, 1086 (1995)
4. Mateo M., et al *AJ* **109**, 588 (1995)
5. Ibata R.A., et al, *AJ* **113**, 634 (1997)
6. Marconi G., et al *A&A* **330**, 453 (1998)
7. Bellazzini, M., Ferraro, F.R., Buonanno R., *MNRAS* **307**, 619 (1999)
8. Shetrone M.D., Côté P., Sargent W.L.W. *ApJ* in press astro-ph/0009505 (2000)
9. Centuriòn M, Bonifacio P., Molaro P., Vladilo G. *ApJ* **536**, 540 (2000)
10. Molaro P., et al *ApJ* **541**, 54 (2000)

N-CAPTURE ELEMENTS IN LMC GLOBULAR CLUSTERS

V. HILL
ESO, Karl Schwarschild str. 2, D-85748 Garching bei München
E-mail: vhill@eso.org

M. SPITE
Observatoire de Meudon, 2 pl. J. Jansen. F-92195 Meudon CEDEX
E-mail: Monique.Spite@obspm.fr

Since the work of Spite et al. (1989), it is known that the Magellanic Clouds young population (cool supergiants) is enriched in heavy neutron-capture elements compared to similar objects in the solar neighborhood, by factors of respectively [n-capture/Fe]=+0.3 and +0.5dex in the SMC and LMC (Russell & Bessell 1989, Hill 1997). However, the nature and source of this enhancement is not understood. s- process elements are produced in AGB stars, whereas r- process elements are thought to be produced by massive supernovae (SN II). Whether the excess of neutron-capture elements is a feature of the young generations only or also of older generations of stars would tell us which of the two process is responsible for the enhancement, and hence the importance of the respective production sites in the LMC.

The LMC possess a large population of stellar clusters of a whole range of ages and metallicities. A sample of 9 giants in the four LMC globular clusters NGC 1866 NGC 1978 ESO 121-SCO3 and NGC 2210 of respective ages 0.2, 2, 9 and 14 Gyrs were observed at a resolution of R=45,000 with UVES, the newly installed high resolution spectrograph of ESO at VLT. Details concerning the abundance analysis are given in Hill et al. (2000).

In Fig1-a are displayed the mean [Y/Fe] [Ba/Fe], [La/Fe], [Ce/Fe], [Nd/Fe] and [Eu/Fe] values for each of the four clusters, as a function of the cluster age. Note that while all the abundance of elements which likely receive an s- process contribution (Y, Ba, La, Ce, Nd) increase along time (relative to iron), the abundance of the pure r-process products Eu shows a flat trend.

In Fig.1-b are displayed the individual values of the [Ba/Eu] ratio for each giant in the four clusters, as a function of [Fe/H]. The [Ba/Eu] ratio indicates the relative contribution of the s- and r- process to the neutron-capture elements enrichment (Ba is mainly synthesised by s- and Eu by r- process). The clear positive slope of the [Ba/Eu] ratio versus metallicity (and age) is the signature of an increasing s- process contribution along time. The increase is already visible in the cluster ESO-121, ie 9 Gyrs ago. It therefore seems that

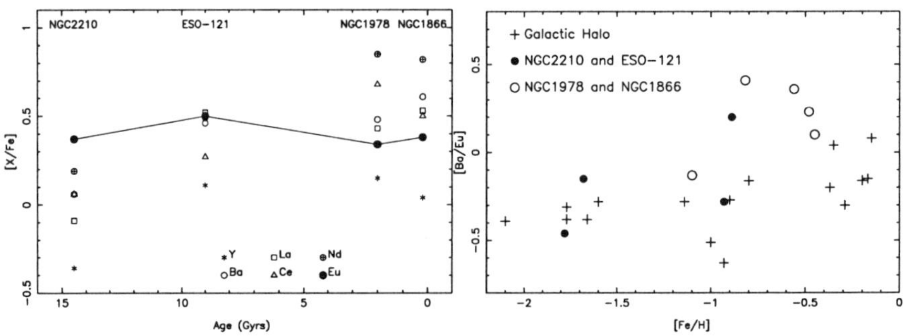

Figure 1. a) Mean [X/Fe] ratios of each cluster for heavy neutron-capture elements as a function of the cluster age. b) [Ba/Eu] ratio as a function of [Fe/H] for the stars in the two older clusters (filled circles), the stars in the two younger clusters (open circles), and Galactic giants from Gratton & Senden 1994 (plus)

in the LMC, the s-process dominates the neutron-capture process enrichment for ages younger than 9 Gyrs. Moreover, when compared to the Galactic Halo and Disk giants (Gratton & Sneden 1994), it is clear that the LMC [Ba/Eu] ratios are similar to those of the Galactic Halo at low metallicities, but increases faster with metallicity than in our Galaxy. The s- process therefore seems to be more efficient in the LMC than in our own Galaxy. In which case, the s- process would be responsible for the strong neutron-capture enhancement in the present-day matter of the LMC. This enhanced efficiency of the s- process in a low-metallicity environment (at similar ages, a LMC star is more metal-poor than a typical Galactic star, owing to the lower star formation rate in the LMC) can be understood if mixing is favoured in metal-poor stars (Leisy & Dennefeld 1996, Venn 1999), providing neutron-sources for the s-process in AGB stars.

References

1. Gratton R., & Senden C., 1994, 1994 A&A 287, 927
2. Hill V., François P., Spite M., Primas F., Spite F., 2000, A&A 364, L19
3. Hill V., 1997 A&A 324, 435
4. Leisy P., Dennefeld M., 1996, A&A 116, 95
5. Russell S.C., Bessell M.S., 1989, ApJS 70, 865
6. Spite M., Spite F., François P., 1989, A&A 210, 25
7. Venn K., 1999 ApJ 518, 405

GALACTIC WINDS IN IRREGULAR STARBURST GALAXIES

F. MATTEUCCI
Department of Astronomy, Trieste University, Via G.B. Tiepolo 11, 34100 Trieste, Italy
E-mail: matteucci@ts.astro.it

S. RECCHI
Department of Astronomy, Trieste University, Via G.B. Tiepolo 11, 34100 Trieste, Italy
E-mail: recchi@ts.astro.it

In this paper we present some results concerning the study of the development of galactic winds in blue compact galaxies. In particular, we model a situation very similar to that of the galaxy IZw18, the most metal poor and unevolved galaxy known locally. To do that we compute the chemo-dynamical evolution of a galaxy in the case of one istantaneous isolated starburst as well as in the case of two successive instantaneous starbursts. We show that in both cases a metal enriched wind develops and that the metals produced by the type Ia SNe are lost more efficiently than those produced by type II SNe. We also find that one single burst is able to enrich chemically the surrounding region in few Myr. Both these results are the effect of the assumed efficiency of energy transfer from SNe to ISM and to the consideration of type Ia SNe in this kind of problem. The comparison with observed abundances of IZw18 suggests that this galaxy is likely to have suffered two bursts in its life, with the previous being less intense than the last one.

1 One and two bursts of star formation

We assume a single instantaneous burst occurring at the center of a gas-rich dwarf surrounded by a dark matter halo. The mass of stars formed is $M_* = 6 \cdot 10^5 M_\odot$, the gas mass is $M_{gas} = 1.7 \cdot 10^7 M_\odot$ and the mass of the dark halo is $M_{dark} = 6.5 \cdot 10^8 M_\odot$. These parameters are chosen to reproduce, for what possible, the characteristics of IZw18 (unevolved local galaxy). The galactic region extends for 700 pc in the vertical (z) direction and for 1kpc in the radial direction (R). In order to study the chemo-dynamical evolution consequent to the burst, we adopt a 2-D hydrocode coupled with detailed chemical yields from: type II SNe, type Ia SNe and low and intermediate mass stars (see Recchi et al. 2001 for more details about this model). We follow the evolution of the abundances of H, He, C, N, O, Mg, Si, Fe in the gas. The evolution is followed for 375 Myr since the burst. We also consider a case with a second burst occurring after 300 Myr from the first one The mass in stars formed during the first burst is $M_* = 10^5 M_\odot$, whereas the mass in stars formed in the second burst is $M_* = 5.8 \cdot 10^5 M_\odot$. The initial gas mass and dark matter halo are the same as in the one-burst case. The simulation lasts 450 Myr in total since the first burst. The SN efficiencies of energy transfer into the ISM from SNe is the same in both cases, in particular: for type II SNe we assume that the efficiency is $\eta = 0.03$ and for type Ia SNe $\eta = 1$. This choice is due to results from Bradamante et al. (1998) for type II SNe indicating that the first SNe to explode lose a large fraction of their initial energy by radiation due to the cold and dense ISM in their surroundings, whereas for type Ia SNe the efficiency is maximum due to the fact that they explode into an ISM already hot and rarefied (see Recchi et al. 2001, for details).

2 Dynamical Results

We find that the starburst triggers indeed a galactic wind and the metals leave the galaxy more easily than the unprocessed gas confirming previous results (e.g. McLow and Ferrara, 1999). We find that SNe Ia eject their metals more efficiently than SNe II since they inject all of their initial energy into the ISM. This is a new result relative to previous studies since it is the first time that type Ia SNe are taken into account. At variance with previous studies (see e.g. Tenorio-Tagle 1996) we find that most of the metals are already in the cold gas phase after 8-10 Myr from the beginning of the burst, due to the fact that the superbubble created by the SNe does not break immediately and thermal conduction can act efficiently.

3 Chemical results

We find that one single instantaneous burst, occurring in a primordial gas (no metallicity), at an age of ~ 31 Myr can reasonably reproduce the abundances measured in IZw18 (see figure 1). From this one would conclude that perhaps this galaxy is experiencing its first burst of SF, although one cannot exclude a previous burst which enriched the gas no more than $1/50\ Z_{\odot}$. However, as evident in figure 1, the correct N/O ratio would last only for a very short time since for $t > 31$ Myr the N/O ratio will start to increase outside the permitted observational range. In addition, Color Magnitude Diagram studies (e.g. Aloisi et al. 1999) indicate the presence of an old underlying stellar population in IZw18, thus suggesting that the two-burst case is more realistic. An interesting result is that the $[\alpha/\mathrm{Fe}]$ ratios in the gas outside the galaxy (i.e. in the galactic wind) are lower than inside, due to the larger ejection efficiency by type Ia SNe (more iron is lost than α-elements). This creates an interesting dichotomy which can have important consequences on the evolution of the ICM, especially if this effect will be found also in elliptical galaxies.

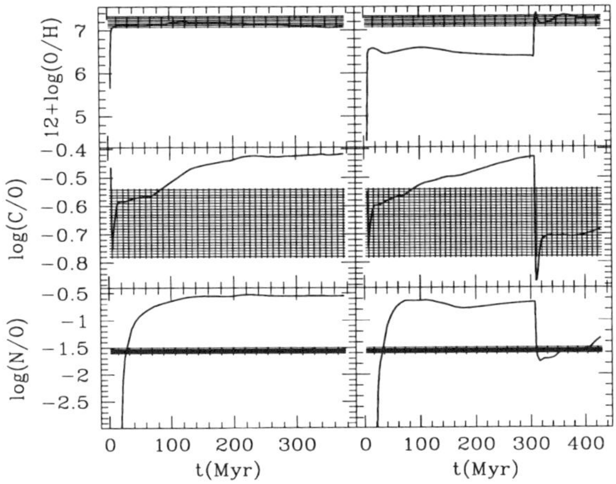

Figure 1: Evolution of O, C, N for the single-burst model (left panels) and the double-burst model (right panels). The superimposed grids represent the observative values found in literature for IZw18

In the two-burst case, at the onset of the second burst the metallicity in the burst region is $\sim Z_{\odot}/50$, and a very good agreement with the measured abundances in IZw18 is obtained in this case, especially for the C/O and N/O ratios and for a larger range of burst ages (see figure 1). Our results suggest that a first weak burst occurred more than 300 Myr ago and followed by a more intense one with an age between 20 and 70 Myr can reproduce most of the properties of this galaxy. The [α/Fe] ratios in the two burst case are always lower outside than inside the galaxy, but the effect is less evident than in the one-burst case.

Acknowledgments

We are gratefull to Annibale D'Ercole and Monica Tosi for their help in this work.

References

1. Aloisi A., Tosi M. and Greggio L., AJ **118**, 302 (1999).
2. Bradamante F., Matteucci F. and D'Ercole A., A&A **337**, 338 (1998).
3. Mac Low M.-M. and Ferrara A., ApJ **513**, 142 (1999).
4. Recchi S., Matteucci F. and D'Ercole A., MNRAS in press (2001).
5. Tenorio-Tagle G., AJ **111**, 1641 (1996)

NO DIFFUSE H_2 IN THE METAL DEFICIENT GALAXY I Zw 18

D. KUNTH, A. VIDAL-MADJAR, A. LECAVELIER DES ETANGS,
L. BENJAFFEL, R. FERLET, G. HÉBRARD AND M. LEMOINE
Institut d'Astrophysique de Paris, CNRS, 98 bis bld Arago, F-75014 Paris, France
E-mail: kunth@iap.fr

J. LEQUEUX
DEMIRM, Observatoire de Paris, 61 rue de l'observatoire, F-75014 Paris

J.C. HOWK, J.W. KRUK, H.W. MOOS AND K.C. ROTH
Department of Physics and Astronomy, Johns Hopkins University, Baltimore, MD 21218, USA

G. SONNEBORN
Laboratory for Astronomy and Solar Physics, NASA/GSFC, Code 681, Greenbelt, MD 20771, USA

D.G. YORK
Dept. of Astronomy & Astrophysics, University of Chicago, Chicago, IL 60637, USA

The metal deficient starburst galaxy I Zw 18 has been observed with the Far Ultraviolet Spectroscopic Explorer (*FUSE*), in a search for H_2 molecules. The spectrum obtained with an aperture covering the full galaxy shows no absorption lines of diffuse H_2 at the radial velocity of the galaxy. The upper limit for the diffuse H_2 column density is very low: $N(H_2) \leq 10^{15}$ cm^{-2} ($10\,\sigma$), unlike our Galaxy where H_2 is generally present for even low H I column densities. We cannot exclude that some H_2 could be in very dense, small and discrete clumps which cannot be detected with the present observation. The absence of diffuse H_2 in this metal-poor galaxy can be explained by the low abundance of dust grains, the high ultraviolet flux and the low density of the H I cloud surrounding the star–forming regions. Thus, it appears that the gas of the galaxy is dominated by H I, and that the high dynamical mass is not composed of cold and diffuse baryonic dark matter.

1 Introduction

I Zw 18 (Mkn 116) is a dwarf blue compact galaxy presently experiencing a strong burst of star formation which has the smallest known abundance of heavy elements. Its oxygen abundance is only $\sim 1/50$ of that of the Sun. The distribution and kinematics of neutral hydrogen indicate H I masses in the range 3—7 $\times\ 10^7 M_\odot$ and dynamical masses in the range 3—9 $\times\ 10^8 M_\odot$

(Lequeux & Viallefond 1980; van Zee et al. 1998). It has been suggested that objects with localized massive star formation surrounded by large H I envelopes might contain a significant reservoir of molecular hydrogen. Such material could represent a significant fraction of the dark matter (Lequeux & Viallefond 1980). Attempts to detect CO in H II galaxies have so far been unsuccessful (Gondhalekar et al. 1998 and references therein) but this lack of detection does not necessarily imply a lack of H_2: the CO excitation could be lower than in Galactic molecular clouds in or CO might be more photodissociated than H_2; but perhaps most importantly, C and O are highly underabundant in these metal deficient galaxies. Given the chemically unevolved nature of I Zw 18 and its lack of organized gas dynamics and/or spiral arms, we decided to observe it with *FUSE*, (Moos et al. 2000) with the aim of detecting cold molecular hydrogen lines in absorption against the stellar continuum of blue massive stellar clusters. In Sect. 2 we describe the observations and the data analysis, the results are discussed in Sect. 3. The full discussion of these results is given in Vidal-Madjar et al. (2000).

2 Data analysis

I Zw 18 has been observed on November 28, 1999 with *FUSE* through the large entrance aperture (30" x 30") fully covering the galaxy. The spectral range is ($\sim$ 980 to 1187 Å) and the resolution defined by both the instrument and the size of the galaxy (10") is about $\lambda/\Delta\lambda \sim 10000$ with a S/N ratio of $\sim$ 10 per resolution element. Many absorption lines are clearly detected. These can easily be identified respectively with a known high velocity cloud at -160 km s^{-1}, the clouds within the Galaxy expected at low radial velocity, and I Zw 18 itself with a redshift of 750 km s^{-1}. Only the component identified with galactic clouds show absorption lines from molecular hydrogen. On the other hand no line from H_2 has been observed at the radial velocity of I Zw 18 (Fig. 1).

We calculated the upper limits of the H_2 column densities toward I Zw 18 assuming an intrinsic width of the lines of $b = 18$ km s^{-1} (van Zee et al. 1998). These limits correspond to a total column density $N_{\rm tot}(H_2) \leq 10^{15}$ cm^{-2}. A different intrinsic width of the lines would not change the result significantly.

Using only the blue wing of the H I Lyman β absorption line we find N(H I)$\approx 2.1 \times 10^{21}$cm^{-2}. This value is consistent with that obtained with HST with a narrow slit (Kunth et al. 1994) and the peak column density obtained from the 21 cm emission line. Assuming a constant N(H_2)/N(H I) ratio across the whole galaxy, we scaled the upper limit on H_2 column density to a limit for the total mass of diffuse H_2, yielding $M_{H_2} \leq 30 M_\odot$.

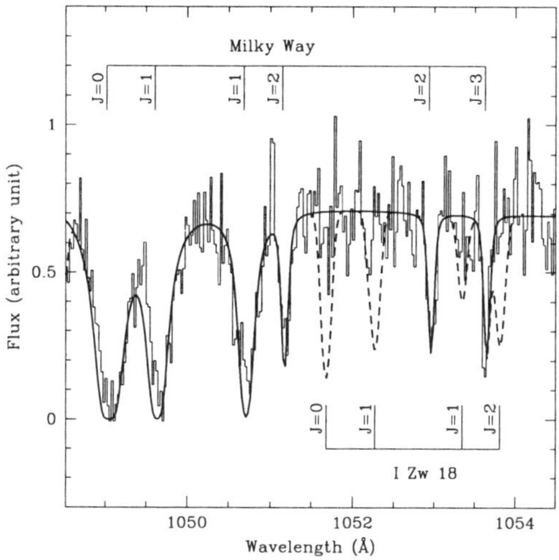

Figure 1. Plot of the 4-0 H_2 Lyman bands. Although the Galactic H I column density ($\sim 10^{20}$ cm^{-2}) is lower than the one from I Zw 18 ($\sim 10^{21}$ cm^{-2}), the Galactic H_2 is easily detected ($\sim 10^{20}$ cm^{-2}). No line of the H_2 bands is detected at the radial velocity of I Zw 18. The dashed lines shows the expected lines if the column density of H_2 had been 10^{15} cm^{-2} in the plotted J levels.

Other lines of atoms and ions like Ar I, N I, Fe II, and Si II were clearly detected in the I Zw 18 system at 750 km s^{-1}. Neither O VI, nor the electronic transitions of CO are detected at the I Zw 18 radial velocity.

3 Discussion

These observations are not sensitive to dense molecular clouds but to diffuse H_2 and its absence is very unusual. Note first that *FUSE* gives access to the average absorption over the full body of I Zw 18, providing $\geq 10^3$ lines of sight to stars emitting in the far–UV and gathered in a central region approximately 10" wide. Indeed in our Galaxy, H_2 is strongly detected for H I column densities larger than a few 10^{20} cm^{-2}, and is often detected for lower N(H I) (Dixon et al. 1998). Such an extreme situation has never been observed before. Even the Magellanic Clouds, with sub-solar metallicities and high far–UV radiation fields, show detectable H_2 along sightlines with lower

H I column densities (e.g., Friedman et al. 2000, Shull et al. 2000). This result also raises an interesting similarity with what is observed in the damped Lyα systems in QSO lines of sight at higher redshift.

The most plausible mechanism for the formation of H_2 is the combination of two H atoms on a dust grain (Hollenbach & Salpeter 1971).

We find a temperature around $T \simeq 15.5(a/0.1\,\mu\text{m})^{-1/6}$ K (where a is the radius of the grain) close to that in the diffuse Galactic interstellar medium thus allowing the formation of H_2 (Vidal-Madjar et al. 2000). However T increases as $R^{-1/3}$ closer to I Zw 18 and H_2 cannot form on grains in the inner parts of the H I cloud. On the other hand, the molecular hydrogen is also destroyed through absorption in the Lyman bands, and the resulting H_2 column density is as low as $N(H_2) \leq 1 \times 10^{-9}\, N(\text{HI}) \approx 2 \times 10^{12}$ mol. cm^{-2}, less than the observed upper limit by more than two orders of magnitude. This is because the UV flux is high and the dust-to-gas ratio is less than 1/50 of the Galactic value (Kunth et al. 1994). The lack of H_2 in the diffuse ISM of I Zw 18 is therefore a consequence of the low abundance of grains, of the high ultraviolet flux and of the low atomic density in the H I cloud surrounding I Zw 18. However it cannot be excluded that this medium contains very dense, discrete molecular clumps such as those proposed by Pfenniger et al. (1994) to account for the dark matter in our Galaxy. These clumps would escape detection since the associated absorption would be observed only in front of stars which, although very numerous, have a very small total surface coverage. However the suggestion of Lequeux & Viallefond (1980) that the dark matter seen dynamically in I Zw 18 is made of widespread diffuse molecular hydrogen is no longer tenable after the present observations.

References

1. Dixon, W.V.D., Hurwitz, M., & Bowyer, S. 1998, ApJ, 492, 569
2. Friedman, S. D., et al. 2000, ApJ , 538, L39
3. Gondhalekar, P. M., et al. 1998, A& A, 335, 152
4. Hollenbach, D. J. & Salpeter, E. E. 1971, ApJ, 163, 155
5. Kunth, D., et al. 1994, A& A 282, 282
6. Lequeux, J. & Viallefond, F. 1980, A& A, 91, 269
7. Moos, H.W., et al. 2000, ApJ this issue
8. Pfenniger, D., Combes, F., & Martinet, L. 1994, A&A, 285, 79
9. Shull,J. M., et al. 2000, ApJ, 538, L73
10. Vidal-Madjar et al. 2000, ApJ, 538, L77
11. van Zee, L., Westpfahl, D., & Haynes, M.P. 1998, AJ, 115, 1000

Part 5. Cosmic Evolution

The High Redshift Universe in Absorption

D. G. York
The University of Chicago, Department of Astronomy and Astrophysics,
5640 S. Ellis Ave., Chicago, IL 60637, USA
E-mail: don@oddjob.uchicago.edu

Absorption lines in the spectra of QSOs with $z_{abs} < z_{em}$ can, in principle, be used to study the build-up of elements over cosmic time. Images of the absorbing objects and their environments may allow direct observation of evolution in the morphology of galaxies over cosmic time. Preliminary work in these areas is promising. Fulfillment of the promise of these two techniques requires much larger sample sizes and a careful separation of the intergalactic absorbers from those absorbers that are actually in the circum-QSO regions, but moving at high ejection velocities. A new catalogue of thousands of QSO absorption lines, together with images in narrow band filters, tuned to the redshift of the absorbers, may allow us to achieve an empirical picture of galaxy formation as a function of cosmic time.

1 Introduction

QSO absorption line systems (QSOALS) have been known for some 35 years to occur in QSO spectra and have, since their discovery, been attributed to matter associated with galaxies (Bahcall & Spitzer[1]). A random sample of foreground galaxies is, in principle, probed by the very distant QSOs. Since abundances of the elements can readily be determined even from moderate resolution absorption line spectra (Meyer et al[2]), evolution of the elements should be directly observable over cosmic time, providing, of course, that the apparent redshift of the absorbers represents time.

A variety of studies are consistent with abundances [C/H]$\lesssim$10-2.5 at $z > 3$, with rising abundances of Fe and Si from $z = 3$ to $z = 1$, and with abundances [X/H]$\lesssim$1.0 at $z < 1$. In this paper, I use the standard notation [X/Y] to indicate the abundance of species X compared to species Y, dex, referenced to standard solar abundances (Grevesse and Noels[3]). The statistics of the absorbers are consistent with dM/dz, the number of heavy element absorption line systems per unit z, rising with redshift to $z \sim 2.5$, flattening near $z \sim 3$, then dropping to zero at higher z (York et al[4], Steidel[5], Van den Berk et al[6]).

A morphology of forming galaxies, which explains these statistics, is discussed by Yanny and York[7]. Consistent with current discussions of hierarchical galaxy formation (Yanny et al[8], Ivezic et al[9]), they imagine that, only dwarf galaxies exist at the highest redshifts. These dwarf galaxies are in groups that are destined to merge into modern large galaxies with a small number of associated dwarf galaxies, such as are found in the Local Group. Over time, the dwarfs merge, and subsequent star formation in the resulting large galaxies produces further element formation that yields "solar" abundances (but only for relatively small regions of space.) Mallouris et al[10] postulate that much of the cross-section in gas absorption at $z < 1$ is in tidal streams within groups of galaxies.

Thus, at high redshifts, where few stars have formed, the number of detectable systems is small because of very low abundances and resulting equivalent widths that are below current detection limits. As more stars are formed, more absorption lines will be above the detection limits at intermediate redshifts, the exact number depending on the geometry of the dwarf galaxies. At $z < 1$, the largest cross-section to absorption by background objects will be created by the dwarf galaxies themselves or by gas tidally stripped from them, that is, created in places with relatively low abundances of heavy elements (not many stars are formed in these locations.) QSOALS with near-solar abundance of heavy elements will only be formed when large disks are intercepted by the line of sight to background QSOs; the total cross-sectional area in these regions is small.

I outline here my work, with many collaborators, to confirm the picture described above. We hope to isolate a pure intergalactic sample of QSOALS (§2); to image the galaxies around the absorbers using narrow band imaging in Lyman α, H α or [O II] (§3); and to greatly expand the sample of QSOALS with known abundances.

2 A Pure Sample of QSOALS

It is well known that absorption lines occur in association with AGNs. Three categories can be readily distinguished: a) systems within 5000 kms^{-1} of the AGN (in the sense of an ejection velocity, positive moving away from the QSO) (Weymann et al[11], Hamann et al[12], Richards[13]); b) systems associated with Broad Absorption Line QSOs (BALs) (Becker et al[14]; Menou et al[15]); and c) systems at ejection velocities up to or exceeding 50,000 kms^{-1} (Jannuzi et al[16]). It is easy to remove lines of type a and b from the sample: we ignore lines within 5000 kms^{-1} of the QSO itself and we ignore objects that have BAL properties. The main concern is to purge lines of type c from the QSOALS sample.

Vanden Berk et al[6] showed that after defining a sub-sample of the heterogeneous catalogue of York et al[4], such that there is a uniform detection limit independent of QSO magnitude, the number of detected C IV absorbers at any z depends on the magnitude of the background QSO. Richards et al (1999)[17] found a larger number of C IV absorbers in optical spectra of QSOs with flat radio spectra than in the steep-spectrum sample of these objects. The effect is even stronger than the optical magnitude effect noted. Richards[13] and Richards et al[18] confirmed the radio result by obtaining uniform data sets in both the radio and the optical spectral regions. These results indicate that some of the absorbers must arise in conjunction with the QSO phenomenon, i.e., from near the QSO. In that case, the apparent redshift is not purely cosmological, but has a special relativistic component. Richards et al[7] show that up to 37% of the QSOALS with apparent $\beta = v/c > 0.016$ may belong to the circum-QSO environment.

QSOALS systems are often defined by the presence of C IV or Lyman α. For higher β, any C IV lines in a QSOALS are lost in the Lyman α forest (Vanden Berk et al[6]). Lyman α lines with $\beta > 0.16$ are merged with higher absorption lines of the Lyman series. Previous studies concluded that dM/dz does not show a gradient over these regions that might indicate a circum-QSO effect. The remaining possibility is that the high-velocity ejecta have a uniform distribution from 5000 kms^{-1} to 75,000 kms^{-1}, so that dM/dβ is flat.

How can the intergalactic and circum-QSO samples be separated? A number of suggestions are made by Jannuzi et al[19] that may be relevant. From my own view, there are at least three ways to limit the sample to pure, intergalactic lines.

a) Use systems that are at the highest apparent β possible: Mg II or C IV systems with $z_{em} - z_{abs} >$ A.) We take A=0.6, but real value needs to be derived from experiment. One can use high-resolution observations of the Lyman α forest in QSOs to separate out the lowest z C IV systems (Kulkarni et al[20], Khare et al[21]). If the contamination suggested above is correct, there may be a high velocity cutoff in the ejecta that can be discerned.)

b) Image the region along the line of sight to a QSO with a known QSOALS with narrow band filters, tuned appropriately to detect gas emission from galaxies in the same group as the absorber (Lyman α, H α and [O II].) If a given system is an ejected system at high velocity, the associated images will not be seen. (Obviously, this method must be checked.) The actual methodology is discussed below.

c) Deep images may reveal clusters on the sightlines to the QSOs with QSOALS. Multi-object spectroscopy can be used to see if the cluster is at the same redshift as the absorber. If so, the system is confirmed as intergalactic (Savage et al[22].)

3 Images at the Absorber Redshift

It is possible to detect galaxies associated with particular absorption line systems by using narrow band imaging. I assume the presence of a QSOALS directs us to a part of space where a near-perfect coincidence (< 1 arcsec) occurs between a QSO sightline and a gas rich dwarf galaxy. Other, small galaxies, and, possibly, a few large ones, depending on redshift, should be seen in the vicinity, if an observing technique is found to go faint enough to find candidates and if confirming redshifts can be obtained. I presume also that the detection of the associated galaxies is not affected by whether one uses C IV, Mg II, damped Lyman α or any other absorption line in a system to select the sight line to image. The particular species observed is a detail of where a QSO sightline happens to intersect a particular galaxy and all known lines in QSOALS occur naturally in low z galaxies.

If such galaxies can be seen, they may prove that the targeted absorber belongs to the pure sample we want for statistical purposes. Further, obtaining a large number of images of such groups of galaxies, spread over redshift, could give a picture of how groups evolve. If the picture of galaxy formation described above is true, we expect to find: fewer massive galaxies, relative to dwarf galaxies, at high redshift; more large galaxies and fewer dwarf galaxies in the groups at low redshift; and more high abundance QSOALS when large galaxies are seen. We expect, moreover, to find that QSOs behind massive galaxies will only be infrequently reddened, because of the small area of star forming clouds, in projection, in most galaxies. (Richards et al[23] find that only 1% of 2500 QSOs found in the Sloan Digital Sky Survey (*SDSS*) show even modest reddening. In at least half of those cases, the reddening seems to be intrinsic to the QSO, not intergalactic in orig! in.)

Searches for galaxies at the same sky position and redshift of the absorbers have been made. A particularly well defined case is found for the $z = 0.52$ Mg II absorber towards the BL Lac object AO0235+167. Smith et al[24] found a AGN at $z = 0.52$, two arcsec south of the BL Lac. Cohen et al[25] showed the BL Lac has $z = 0.92$. Following a suggestion by York et al[26], Yanny et al[27] observed an irregular, extended object in [O II] emission at $z = 0.52$, offset from the two AGNs by few arcseconds. Other, smaller [O II] emitters occur in a 30 arcsec field around the main galaxy. Yanny & York[7] summarize observations of 10 absorbers in [O II] emission, from $z = 0.5$ to 1.1, yielding dozens of small emitters in the fields, to limits of 1-2$\times 10^{-17}$ ergscm^{-2}s^{-1}. For a normal stellar mass function, this value corresponds to a star formation rate, summed over all emitters in a given field, of 1-2 solar masses per year.

Several successful broadband searches have been made for luminous galaxies near absorber sightlines. (Kulkarni et al[28] and references therein.) There are very few cases of a direct alignment between the luminous parts of a galaxy and the background QSO. The absorption is often then attributed to a "galaxy halo". Alternatively, since galaxies cluster, the absorbers could all arise in dwarf galaxies, themselves unseen because of the QSO image blur.

There are not yet many examples of purposeful detections of galaxies at z_{abs} based on Lyman α. A notable example is the $z = 2.308$ absorber in the spectrum of the QSO PHL957. Lowenthal et al[29] found a Lyman α emitter 48 arcsec from the QSO, which has a damped Lyman α line at the same redshift. Frances et al[30] targeted a part of a large clustering of absorbers at $z = 2.38$ and found three Lyman α emitting galaxies, which they postulate to be high redshift, QSO analogs of Seyfert II galaxies with high dust attenuation.

On the other hand, high z emission in Lyman α may be quite common. Pascarelle et al[31] found 35 Lyman α emitters, using *HST*, half near a $z = 2.4$ radio galaxy, the rest in three random fields. They conclude that the objects are small, dwarf galaxies with star formation rates (SFR)$\sim$1-10 solar masses per year, based on their optical luminosity ($V \sim 26$), their sizes (2-3 kpc) and their Lyman α flux (2×10^{-17} ergscm^{-2}s^{-1}.) Cowie & Hu[32] find 13,000 Lyman α emitters/unit redshift per sq. degree, based on 10 pointings using an 0.013 degree field at $z = 3.43$, above the same flux limit. Kudritzki et al[33], searching for [O III] emission from planetary nebulae in the Virgo cluster of galaxies, found, instead, 9 Lyman α emitters at $z = 3.13$.

The surface density of emitters agrees with that found by Cowie & Hu[32]. This surface density implies one Lyman α emitter per local group of galaxies for a targeted search, at a limit of a 2×10^{-17} ergscm^{-2}s^{-1}. There could be more or less depending on the star formation rate at the relevant redshifts and the effects of internal reddening on the Lyman α flux.

The 3.5-meter telescope at Apache Point Observatory is being used for a large survey of emission line objects near QSOALS sightlines. A newly modified Fabry Perot imager, built at the Goddard Space Flight Center (GSFC) is used as the tunable filter. With it, we can image in 200-400 kms^{-1} bands from 4000 Å to 10,000 Å. Using [O II], H α and Lyman α, we cover all redshifts to a limit of 2×10^{-17} ergscm^{-2}s^{-1}. The short term goals of the survey are as follows:

a) to confirm that the number of detections of galaxies associated with absorbers is independent of the species used to select the targeted absorbers;

b) to confirm that circum-QSO absorbers do not have associated objects in the near field (so that detection of a set of associated "galaxies" clearly implies the targeted absorber is intergalactic and part of the pure sample we seek);

c) to see if the morphology of gas emission in groups is related to redshift, as suggested above. Star formation rates can be derived, as a function of redshift.

The long term goal is to gain a picture of galaxy formation as a function of time and to correlate the patterns found with the heavy element abundances. To get a large sample, one must draw on a larger sample of QSOALS with $z_{em} - z_{abs} > 0.6$ in C IV, Lyman α and Mg II.

4 Obtaining a Large Sample of QSOALS

The Sloan Digital Sky Survey is an imaging and spectroscopic survey of the North Polar Cap. (Margon[34], Fukugita[35], Gunn & Weinberg[36], York et al[37], Burles[38].) It covers 10,000 square contiguous degrees and reaches 23.2 magnitudes in *r*' band (depending slightly on seeing). The four other filters are *u*', *g*', *i*' and *z*' (Fukugita et al[39]) Images in < 1.5 arcsec in 5 bands are being obtained in 53 second integration times (simultaneous to within 5 mintues.) Using the images, it is possible to select a nearly complete sample of QSOs for spectroscopic confirmation with about 70% efficiency.

Spectra from 3900 Å to 10,000 Å at a resolution of 140 kms^{-1} are being obtained for all QSO candidates to *g*'=20, expected to total 100,000 objects by the end of the survey. The spectra of all objects in the FIRST survey, co-extensive with SDSS, will be obtained as well. The signal to noise ratio in the QSO continuum at magnitude 18 is 20/1, so we expect to detect well over 10,000 QSOALS covering a z range in excess of 18,000 and to increase the number of known, high quality systems by at least a factor of 20. York et al[40] demonstrated the detection of species such as C IV, Si IV, Mg II, Mn II, Mg I, Fe II and Si II. The above properties of the *SDSS* mean that QSOALS can be detected adequately to allow studies of cosmic evolution of the elements. Thus, for the many Mg II (C IV) absorbers detected to $z = 2.6$ (5) with $z_{em} - z_{abs} > 0.6$, a sample of pure intergalactic QSOALS will be formed. The FIRST radio source survey data to! gether with the NVSS survey (Condon et al[41]) will allow confirmation that the above noted correlation of dM/dz with radio spectrum is not present for this sample. Abundance ratios [Fe/Zn] and [Fe/Cr] will be obtainable as values or good limits. Welty et al[42] show that Fe/Zn is a good stand-in for Fe/H. At redshifts above 1, [Si/Fe] will be obtainable using $\lambda\lambda$ 1808, 1526. Many of these systems will be observable with the new Cosmic Origins Spectrograph on the Hubble Space Telescope (Green et al[43]), allowing direct determinations of N(H I) and N(H_2) for the heavy element QSOALS.

The systems thus detected can then be imaged in [O II] and Lyman α to determine the morphology of the galaxies near the absorbers with known abundances. It can be expected that within 10 years, a sample of at least

2000 pure intergalactic absorbers will exist and that narrow band images of the environs of hundreds of these objects will be available. The combination of images of galaxies in formation and abundances for the gas in a sample of the pre-galactic fragments can be used to construct an empirical picture of galaxy formation. Such a large sample will also permit an analysis of the fraction of any local group of galaxies, at a given redshift, that has opaque dust clouds, thus determining how many high z QSOs we are missing because of reddening.

Acknowledgments

I thank Scott Burles, Arlin Crotts, Varsha Kulkarni, James Lauroesch, Martin Lemoine, Chris Mallouris, Jean Quashnock, James Rhodes, Gordon Richards, Povilas Polunas, James Truran, Dan Vanden Berk, Brian Yanny, Dan Welty and Bruce Woodgate for active discussions during the preparation of this contribution. Chris Mallouris helped with manuscript preparation.

References

1. Bahcall, J. N. and Spitzer, L., *ApJ* **156**, L63 (1969).
2. Meyer, D. *et al*, *ApJ* **343**, L37 (1989).
3. Grevesse, N. and Noels, A., in *Origin of the Elements*, ed. Prantzol *et al* (Cambridge, 1993).
4. York, D. G. *et al*, *MNRAS* **250**, 24 (1991).
5. Steidel C., *ApJS* **72**, 1 (1990).
6. Vanden Berk, D. *et al*, *ApJ* **469**, 78 (1996).
7. Yanny, B. and York, D. G., *ApJ* **391**, 569 (1992).
8. Yanny, B. *et al*, *ApJ* **540**, 825 (2000).
9. Ivezic, Z. *et al*, *AJ* **120**, 963 (2000).
10. Mallouris, C.*et al*, in *The Third Stromlo Symposium: The Galactic Halo*, eds. Gibson, B. K., Axelrod, T. S. & Putman, M. E., ASP Conference Series Vol. 165, p.459 (1999).
11. Weymann, R. J.*et al*, *ApJ* **234**, 22 (1979).
12. Hamann, F. W. `Astroph 0011030` (2000).
13. Richards, G. T., *ApJS*, in press (2001).
14. Becker, R. H., *ApJ* **479**, L93 (1997).
15. Menou, K. *et al*, *AJ*, in press (2001).
16. Jannuzi, B. T., *ApJ* **470**, L11 (1996).
17. Richards, G. T.*et al*, *ApJ* **513**, 576 (1999).
18. Richards, G. T. *et al*, *ApJ*, in press (2001).
19. Jannuzi, B. *et al*, *ApJ* **470**, L11 (1996).
20. Kulkarni, V. *et al*, *MNRAS* **279**, 197 (1996).

21. Khare, P. *et al*, *MNRAS* **285**, 167 (1996).
22. Savage, B. D. *et al*, *ApJ*, in press (2001).
23. Richards, G. T. *et al*, *AJ*, in press (2001).
24. Smith, G. *et al*, *ApJ* **218**, 611 (1977).
25. Cohen, L. *et al*, *ApJ* **318**, 577 (1987).
26. York, D. G. *et al*, *ApJ* **311**, 610 (1986).
27. Yanny, B. *et al*, *ApJ* **338**, 735 (1989).
28. Kulkarni, V. *et al*, *ApJ* **536**, 36 (2000).
29. Lowenthal J. D. *et al*, *ApJ* **377**, L73 (1991).
30. Francis P. J. *et al*, *ApJ* **482**, L25 (1997).
31. Pascarelle S. M. *et al*, *AJ* **116**, 2659 (1998).
32. Cowie, L. and Hu, E. M., *AJ* **115**, 1319 (1998).
33. Kudritzki, R. P. *et al*, *ApJ* **536**, 19 (2000).
34. Margon, B., in *PhilosophicalTransactions of the Royal Society of London A* **357**, 93 (1999).
35. Fukugita, M., in *Highlights of Astronomy 11A*, ed. J. Andersen **440** (1998).
36. Gunn, J. E. and Weinberg, D., in *Wide Field Spectroscopy and the Distant Universe*, eds. S. J. Maddox and A. Aragon-Salamanca (World Scientific: Singapore, **3**, 1995).
37. York, D. G. *et al*, BAAS **21**, xxx (1999).
38. Burles, S., in these *proceedings Cosmic Evolution*, ed. M. Lemoine and R. Ferlet (2001).
39. Fukugita, M. *et al*, *AJ* **111**, 1748 (1996).
40. York, D. G. *et al*, *AJ* **120**, 1579 (2000).
41. Condon, J. J. *et al*, *AJ* **115**, 1693 (1998).
42. Welty, D. *et al*, *ApJ* **589**, 672 (1997).
43. Green, J. *et al*, BAAS **32**, 1421 (2000).

CHEMICAL EVOLUTION IN THE EARLY UNIVERSE

J.X. PROCHASKA
Carnegie Observatories, 813 Santa Barbara St., Pasadena, CA, USA
E-mail: xavier@ociw.edu

I briefly summarize new metallicity measurements of damped Lyman *alpha* systems at $z > 3$ acquired with the Echellette Spectrograph and Imager (ESI) on the Keck II telescope. These observations lend further support to the following conclusions: (1) the cosmic metallicity in neutral gas does not evolve significantly from $z \sim 2$ to 4; (2) the unweighted metallicity exhibits a statistically significant decrease with increasing redshift; and (3) not a single system has a metallicity below [Fe/H] = −3.

1 Introduction

In a recent paper (Prochaska & Wolfe [5]), we investigated the chemical evolution history of the universe at $z > 1.7$ through an examination of the metallicities of the damped Lyman *alpha* systems – neutral hydrogen gas layers with HI column densities in excess of $2 \times 10^{20}\text{cm}^{-2}$. Our analysis revealed that the mean metallicity of the damped Lyman *alpha* systems exhibits no significant evolution from $z = 2$ to 4. This result is in contradiction with the majority of chemical evolution models, all of which significantly underpredict the observed enrichment at $z > 3$ (e.g. Pei *et al.* [4]; Mathlin *et al.* [3]; but see Cen & Ostriker [1]). For this epoch, however, the observational result was based on [Fe/H] measurements of only $\sim$ 10 protogalaxies and therefore suffered from small number statistics. To address the uncertainty arising from the small sample size at $z > 3$, we have initiated a program using the new Echellette Spectrograph and Imager (ESI) on the Keck II telescope.

2 Observations and Analysis

ESI is mounted at the cassegrain focus of the W.M. Keck II 10m telescope. The instrument has two spectrographic modes: (1) a prism-dispersed low resolution mode with resolution R decreasing linearly from 4800 to 1000; and (2) a multi-order echellette mode with $R \approx 10000$ over the entire spectrum. Both modes have continuous wavelength coverage from $\lambda = 0.4 - 1.1\mu$.

We observed several quasars from the Palomar Sky Survey (Djorgovski *et al.* [2]) which at the time of the observations were unknown to exhibit damped Lyman *alpha* systems. We acquired 400s exposures of each quasar in the low dispersion mode with a $1''$ slit and reduced the data in real time with the

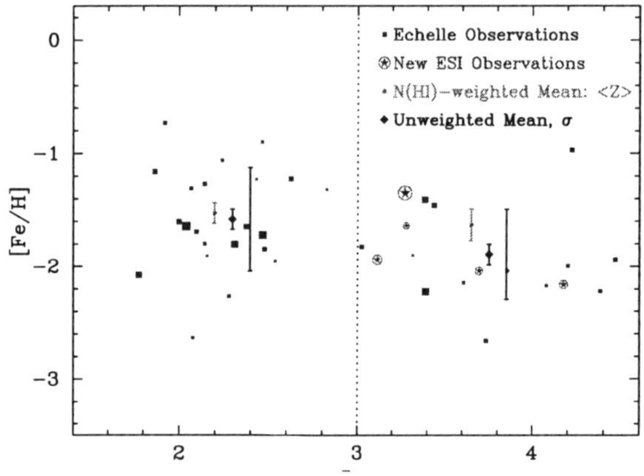

Figure 1. [Fe/H] vs. z plot of the 40 damped Lyman *alpha* systems which comprise our entire $z > 1.7$ sample. The squares indicate metallicity measurements acquired from high resolution echelle observations and the stars indicate the 5 new measurments obtained with the Echellette Spectrograph and Imager. For two samples ($z_{low} = [-1.8, 3.0)$ and $z_{high} = [3.0, 4.7]$) we present three statistics: (1) the HI-weighted mean metallicity $< Z >$ represented by the solid circle with bootstrap error; (2) the unweighted mean logarithmic metallicity $<$ [Fe/H] $>$ with bootstrap error depicted by the solid triangle; and (3) the scatter in the individual systems σ([Fe/H]) represented by the offset vertical error bar.

IRAF software package. We then obtained two or three 1200s exposures in the echellette mode of those quasars with promising damped Lyman *alpha* candidates; i.e., a Lyman *alpha* line with rest equivalent width $W_\lambda > 10$ Å. In this fashion we were able to identify a significant number of new $z > 3$ damped Lyman *alpha* systems and measure their metallicity in the same night of observing.

We searched each of the normalized echellette spectra for potential damped Lyman *alpha* systems and fitted the Lyman *alpha* profiles with Voigt profiles to estimate their HI column densities, N(HI). We then measured column densities for all of the metal-line transitions of the damped Lyman *alpha* systems that we identified.

3 Results

Here is a summary of the results presented in Figure 1:

- The new ESI instrument provides an ideal means for surveying the HI content and metal abundances of damped Lyman *alpha* systems at $z > 3$.
- We find no evidence for evolution in the N(HI)-weighted mean metallicity of the universe, $< Z > \equiv \log \frac{\Omega_{metals}}{\Omega_{gas}} - \log Z_{\odot}$, from $z = 2$ to 4.5. This result contradicts the majority of chemical evolution models and suggests a feedback mechanism which supresses star formation.
- There is a significant decline in the unweighted mean metallicity, $< [\mathrm{Fe/H}] > \equiv \frac{1}{N} \sum$ [Fe/H], indicating that the typical galaxy has lower metallicity at $z > 3$.
- Not a single damped system exhibits a metallicity [Fe/H] < -3. This limit is well above the detection limit and cannot be attributed to any observational selection effect. This lower metallicity threshold may be explained by Pop III enrichment in the early universe (e.g. Wasserburg & Qian [6]) or the rapid enrichment of these protogalaxies once they have a surface density large enough to yield a damped Lyman *alpha* system.

Acknowledgments

I wish to thank my collaborators E. Gawiser and A.M. Wolfe and acknowledge support from Carnegie Observatories.

References

1. Cen, R. & Ostriker, J.P. 1999, ApJ, 519, 109L
2. Djorgovski, S.G. et al. 1998, in "Wide Field Surveys in Cosmology", eds. S. Colombi & Y. Mellier, (astro-ph/9809187)
3. Mathlin, G.P., Baker, A.C., Churches, D.K., & Edmunds, M.G. 2000 (astro-ph/0009226)
4. Pei, Y.C., Fall, S.M., & Hauser, M.G. 1999, ApJ, 522, 604
5. Prochaska, J. X. & Wolfe 2000, ApJ, 533, L5
6. Wasserburg, G.J. & Qian, Y.-Z. 2000, ApJ, 538, L99

COSMOLOGICAL EVOLUTION OF THE UNIVERSE NEUTRAL GAS MASS MEASURED BY QUASAR ABSORPTION SYSTEMS

CELINE PEROUX, RICHARD G. MCMAHON, MIKE IRWIN
Institute of Astronomy, Madingley Road, Cambridge CB3 0HA, UK
e-mail: celine@ast.cam.ac.uk

AND LISA J. STORRIE-LOMBARDI
SIRTF Science Center, California Institute of Technology, MS 100-22, Pasadena CA 91125, USA

The cosmological evolution of neutral hydrogen is an efficient way of tracing structure formation with redshift. It indicates the rate of evolution of gas into stars and hence the gas consumption and rate star formation history of the Universe. In measuring HI, quasar absorbers have proven to be an ideal tool and we use observations from a recent survey for high-redshift quasar absorption systems together with data gathered from the literature to measure the cosmological comoving mass density of neutral gas. This paper assumes Ω_M=0.3, Ω_Λ=0.7 and h=0.65.

1 Quasar Absorption Systems

Quasar absorbers are systems along the line-of-sight between an observer and a quasar, and thus are observed in absorption in the spectra of the background source. They are classified according to their HI column density: Damped Lyman-α systems (hereafter DLA) have $N(HI) > 10^{20.3}$ atom cm^{-2}, Lyman Limit Systems (hereafter LLS) have $N(HI) > 10^{17.2}$ atom cm^{-2} and the lyman-α forest is composed of systems with smaller column densities. The work presented here is based on observations from a recent high-z survey (Peroux et al. 2001) together with a compilation of absorption systems from the literature (Storrie-Lombardi and Wolfe 2000). This included 86 quasars with $z > 4$ and resulted in a sufficiently large sample of quasar absorption systems to enable the study of their properties with redshift (z). Fig 1 shows the cumulative number of quasar absorbers per unit $dX(z)$, for two z ranges, absorbers with $z < 3.5$ and $z > 3.5$, where dX is the distance interval: $X(z) = \int_0^z (1+z)^2 \left[(1+z)^2(1+z\Omega_M) - z(2+z)\Omega_\Lambda\right]^{-1/2} dz$. The column density of LLS cannot be derived directly from their equivalent widths as they correspond to the non-linear part of the curve of growth. Nevertheless, the expected number of LLS can be calculated by fitting $N_o(1+z)^\gamma$ to the observed number of systems per unit z and integrating over the z-path surveyed: $LLS_{expected} = \sum_{i=1}^{n} \int_{z_{min}}^{z_{max}} N_o(1+z)^\gamma dz$. The observations presented

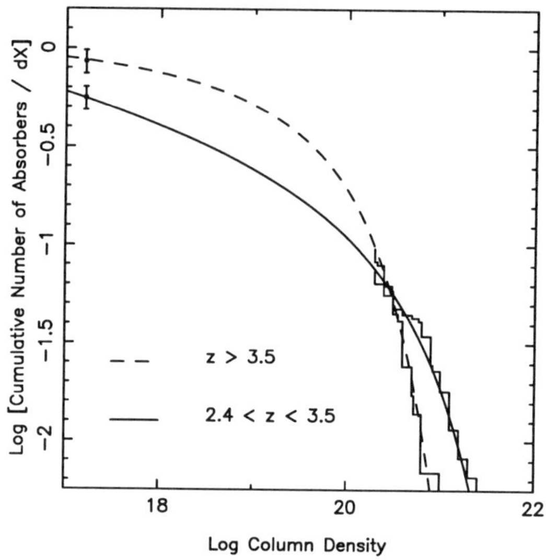

Figure 1. Cumulative number of absorbers per unit distance interval, dX, for two different z ranges. The "steps" correspond to the observed number of DLAs and the data points at $\log N(HI) = 17.2$ are the expected number of LLS derived from the observed number of LLS per unit z. Γ-functions are used for the fits to the data.

in Fig 1 show that there are fewer systems at the highest column density for $z > 3.5$ compared to the lower redshift range, which indicates that we are probing the epoch of assembly of high column density systems from smaller column density units.

2 Cosmological Evolution of Neutral Hydrogen

The neutral gas mass is calculated by integrating the quasar absorbers column density distribution, $f(N,z)$: $\Omega_{HI}(z) = \frac{H_o \mu m_H}{c \rho_{crit}} \int_{N_{min}}^{\infty} N f(N,z) dN$. We used a Γ-function to fit the data (Fig 1) in order to take into account the HI included below the formal DLA definition. Indeed, DLAs were previously thought to contain the majority of the total HI but our new study shows that 14% at $z < 3.5$ and up to 40% at $z > 3.5$ of the neutral gas is actually contained in smaller column density systems ($10^{17.2} < N(HI) < 10^{20.3}$). The total HI decreases with decreasing z thus matching the pattern expected from gas depletion due to star formation. Details on the method used and more discussion on the results presented here will be available in a subsequent paper.

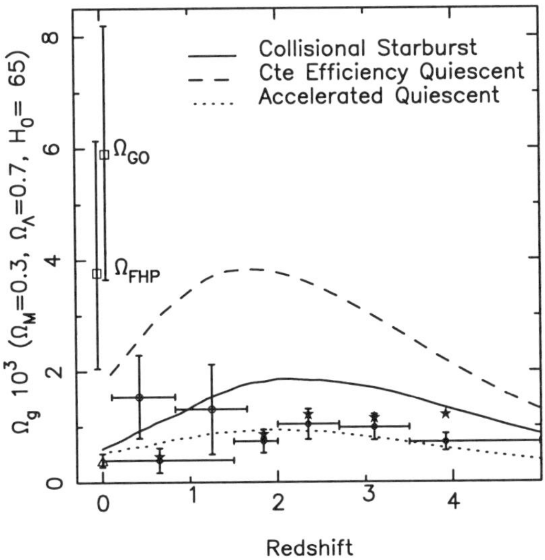

Figure 2. Cosmological evolution of HI. The circles show the neutral gas contained in DLAs (filled circles are this work, open circles are from Rao and Turnshek 2000 who used HST spectra of quasars with know MgII systems to look for low-z DLAs. Their results are surprisingly high but might be biased by small number statistics or the effect of gravitational lensing.) Vertical error bars correspond to 1σ uncertainties and the horizontal error bars indicate bin sizes. The stars are the total HI including a correction for the neutral gas contained in systems with column densities below the DLA formal definition. The triangle at $z = 0$ is the local HI mass measured by Rao & Briggs 1993. The squares, Ω_{FHP} and Ω_{GO} (Fukugita, Hogan & Peebles 1998 and Gnedin & Ostriker 1992 respectively) are $\Omega_{baryons}$ in local galaxies. Semi-analytical models from Somerville, Primack & Faber 2000 are overplotted.

References

1. Fukugita, M., Hogan, C. and Peebles, P. 1998, ApJ, 503, 518.
2. Gnedin, N. and Ostriker, J., 1992, ApJ, 400, 1.
3. Peroux, C., Storrie-Lombardi, L., McMahon, R., Irwin, M. and Hook, I., 2001, AJ, in press, astro-ph/0101179.
4. Rao, S. and Briggs, F. 1993, ApJ, 419, 515.
5. Rao, S. and Turnshek, D. 2000, ApJS, 130, 1.
6. Somerville, R., Primack, J. and Faber, S. 2000, MNRAS, in press, astro-ph/0006364.
7. Storrie-Lombardi, L. and Wolfe, A. 2000, ApJ, 543, 552.

VLT/UVES OBSERVATIONS OF THE DLA AT Z=3.39 TOWARDS QSO 0000-2621

P. Molaro, P. Bonifacio, M. Centurión, G. Vladilo
Osservatorio Astronomico di Trieste, Via G.B. Tiepolo 11, I 34131, Trieste Italy

S. D'Odorico
European Southern Observatory, Karl-Schwarzschild-Str. 2 D-85740 Garching Germany

S. A. Levshakov,
Department of Theoretical Astrophysics, Ioffe Physico-Technical Institute 194021 St Petersburg, Russia

UVES observations at ESO/VLT of the DLA at z=3.39 towards QSO 0000-2621 provided refined abundances for Fe, Si, N, Ni, and new abundances for O, Ar, P, Zn, Cr, H_2. These determinations confirm that the system has remarkable properties with the lowest metallicity, a very low H_2 molecular fraction and it has no detectable presence of dust. The $[\alpha/Fe - peak]$ ratio is ≈ 0.1, challenging current models of chemical evolution. The O/Ar and O/S ratios are the same as those observed in BCGs, suggesting a universal IMF.

1 Chemical abundances

UVES observations of QSO 0000-2621 (V=17.5 mag, z_{em}=4.108), which shows a DLA system at z_{abs}=3.3901, have been acquired at ESO/VLT Kueyen 8.2m during commissioning [1].

The Fe abundance is found [Fe/H]=-2.04, Cr is [Cr/H]=-2.01 while Ni is [Ni/H]=-2.27, somewhat below the other Fe-elements. The undepleted Zn is measured at [Zn/H]=-2.07 and so far it is the lowest in all DLAs showing that this material has indeed been processed very little. The Zn measure is also the one achieved at the highest redshift and suggests a mild chemical evolution when compared with extant data for Zn in the DLAs. However, no evolution is found when the column-density-weighted mean is considered instead [2]. Depleted elements such as Fe and Cr are at the same level of the undepleted Zn, showing that there is no detectable presence of dust. Si is found at [Si/H]=-1.91 or [Si/Fe]=0.13 $\pm$ 0.04, at variance with the mean value [Si/Fe]=+0.43$\pm$ 0.18 found in DLAs.

A special effort has been made to measure abundances by using transitions which fall within the Lyα forest where HI contamination and uncertainties in the continuum location make it a challenging task. The lines within the Lyα

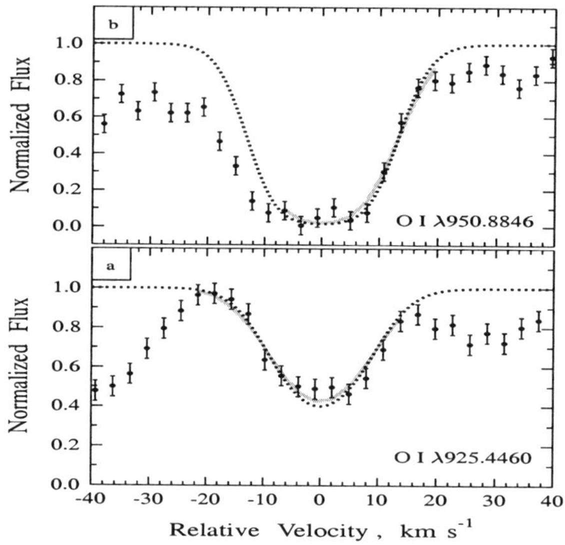

Figure 1: Oxygen lines

forest are fitted simultaneously with those outside the forest leaving the local continuum free to vary [3]. With this procedure we have been able to measure the abundances of O, Ar, P and H_2.

In the DLAs the OI λ 1302 Å is always found saturated, while the OI λ 1356Å is never detected. In our spectrum we succeed in detecting 7 OI lines mixed in the forest. The best cases for the elemental measure are the OI λ 950 and λ 925 Å shown in the figure. The O abundance is found [O/H]=-2.01. Note that this value has been slightly revised with comparison to the earliest measure [1] and gives better matching between the two transitions involved.

For the first time we detected the ArI λ 1048 and λ 1066 Å lines [3] which yield an abundance of [Ar/H]=-1.91. Argon is both an α element and a very volatile one. We detected the PII 963 Å line [3] obtaining [P/H]=-2.31. ^{31}P is an odd-element likely produced by n-capture in C and Ne shell burning and non-depleted in the ISM. This measure is unique at low metallicity and the [P/Fe]=-0.2 ratio implies metallicity-dependent yields.

Molecular hydrogen has been detected [4] from the L(4-0)R(1) and W(2-0)R(0) lines yielding a molecular fraction of $f(H2) = 6.8 \pm 2.0 \times 10^{-8}$ which is $\approx$ 100 times lower than the molecular fraction in the primordial gas at z $\leq$ 50. A correlation has been found between $f(H_2)$ and [Cr/Zn] in the DLAs, suggesting a negligible dust condensation in the z=3.39 absorber [4].

1.1 The α-elements/Fe-peak ratio and the IMF

Comparison of α elements with zinc, as a proxy for the iron peak elements, gives [O/Zn] = 0.06± 0.07 and [Ar/Zn]=0.16± 0.06. S has also been measured [5] providing [S/Zn]=0.16± 0.06. The average of all four α elements at our disposal with the Fe-peak elements available, without considering Ni which is found somewhat below, is $[\mathrm{O, Si, S, Ar/Zn, Fe, Cr}]$ = 0.10 ±0.06 and confirms early observations [6]. Although similar analysis are not presently available for other DLAs, there are indications that this result may be rather general. [S/Zn] $\approx$ 0 in the small sample where it has been measured [7]. [Si/Zn] $\approx$ 0 when Si has been corrected for the presence of dust [8], and low α abundances are found in low-dust systems [9].

Low α abundances at [Fe/H]$\approx$ -2 are not easy to explain. Burst(s) of star formation followed by quiescent periods where Type I SNe evolve have been considered [12]. However, for SN type I lifetimes of $\approx$ 1 Gyr to produce effects at z=3.4 we need stars formed at a very early time, namely z>7.

For the first time in a DLA system we have at our disposal information for O, Ar and S abundances. The S/O & Ar/O ratios measured in a variety of celestial objects are found to be independent from metallicity, which has been taken as evidence for a universal IMF [10]. In our DLA we find log(Ar/O)=-2.25 ± 0.06 and log(S/O)=-1.57± 0.06. These elemental ratios are almost identical to the -2.25 ±0.09 and -1.55 ±0.09 values measured respectively in Blue Compact Galaxies [11]. Thus the S/O & Ar/O constancy is probed for the first time in a possibly different type of objects and for a much lower O abundance than in the BCGs.

References

1. Molaro P., Bonifacio P., et al *ApJ* **541**, 54 (2000)
2. Vladilo G., Bonifacio P., Centurión M., Molaro P., *ApJ* **543**, 24 (2000)
3. Molaro P., Levshakov S.A., et al *ApJ*, (2001) in press astro-ph/0010434
4. Levshakov S.A., Molaro P., et al *astro-ph/0011470* , (2001)
5. Lu L., Sargent W. L. W., et al *ApJS* **107**, 475 (1996)
6. Molaro P., D'Odorico S., et al *A&A* **308**, 1 (1996)
7. Centurión M., Bonifacio P., Molaro P., Vladilo G., *ApJ* **536**, 540 (1999)
8. Vladilo G., *ApJ* **493**, 583 (1998)
9. Pettini M., Ellison S., L., et al *ApJ* **510**, 576 (1999)
10. Henry R. B. C., Worhtey G., *PASP* **111**, 919 (1999)
11. Izotov Y., & Thuan T., X., *ApJ* **511**, 639 (1999)
12. Matteucci F., Molaro P., Vladilo G., *A&A* **321**, 45 (1997)

DUST DEPLETION IN DAMPED LYMAN α SYSTEMS

J. L. HOU
Shanghai Astronomical Observatory, 80 Nandan Road, Shanghai, 200030, China
E-mail: hjlyx@center.shao.ac.cn

S. BOISSIER
Institute of Astronomy, Madingley Road, Cambridge CB3 OHA, England,
E-mail: boissier@ast.cam.ac.uk

N. PRANTZOS
Institute d'Astrophysique de Paris, 98bis, Bd. Arago, Paris, 75014, France,
E-mail: prantzos@iap.fr

We point out a previously unnoticed anti-correlation between the observed abundance ratio [X/Zn] (Zn is assumed undepleted and X stands for Fe, Cr and Ni) and metal column density ([Zn/H]+log(N_{HI}) = $\mathcal{F}$) in Damped Lyman α systems (DLAs). We suggest that this trend is an unambiguous sign of depletion.

1 A measure of dust depletion in DLAs

On the left panels of Fig. 1 we present the DLA data of [X/Zn] as a function of log(N_{HI})(data sources see Hou et al. 2001). No anti-correlation is found, contrary to the situation in gas clouds in our Galaxy, which is shown by solid lines in the panels of Mn and Fe (Wakker and Mathis 2000).

If we plot [X/Zn] as a function of $\mathcal{F}$ = [Zn/H]+log(N_{HI}) (right panels in Fig. 1), one can see that, there is no dependence of Si/Zn and S/Zn on $\mathcal{F}$, as expected for those two elements which are also found not to be depleted in local gas clouds, and there is a clear anti-correlation between [X/Zn] and $\mathcal{F}$ for Cr, Fe and Ni (dash lines) while the scarce data for Mn do not allow for a conclusion.

We suggest that a proper evaluation of depletion in DLAs is [X/Zn] instead of [X/H] as a function of column density. The reason is that DLAs have different metallicities either due to large depletion or little chemical evolution or both.

2 Depletion pattern in DLAs

Assuming that DLAs are (proto-)galactic disks and using detailed chemical evolution models with metallicity dependent yields, we explored the general

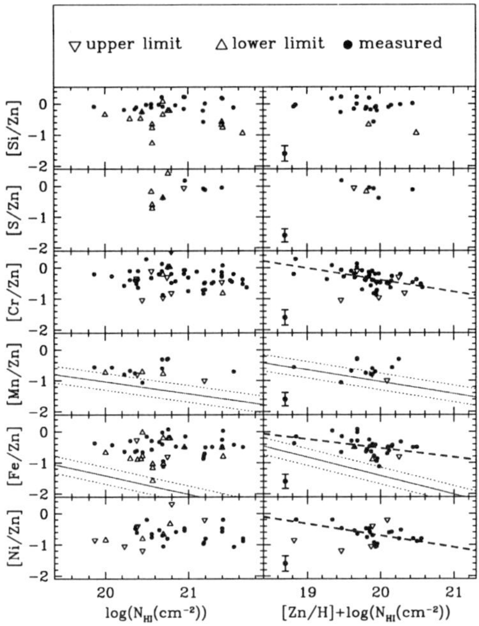

Figure 1: Gas phase abundance ratios [X/Zn]in DLAs as a function of log(N_{HI})(left panels) and of $\mathcal{F}$ (right panels). Dashed lines are obtained by a least squares fit to the data. In the panels of Mn and Fe solid and dotted lines show the corresponding trends observed in the Milky Way clouds and their associated uncertainties(Wakker and Mathis 2000).

dust depletion pattern in DLAs (Hou et al. 2001). For low $\mathcal{F}$ (no depletion), models compare fairly well to the data, while a solar pattern(that is the intrinsic DLA abundance pattern is simply assumed to be solar) has difficulties with Mn . At high $\mathcal{F}$ (small depletion), the solar pattern describes the data quite well, while our models have difficulties with S. We suggest that further measurements of those key elements, i.e. Zn, S and Mn, will help to gain more insight into the nature of DLAs.

References

1. J.L. Hou, S. Boissier & N. Prantzos, *A & Ap* , submitted (2001).
2. B.P. Wakker & J.S. Mathis, *ApJL* , astro-ph/0010045 (2000).

TYPE IA SUPERNOVAE AND THEIR IMPLICATIONS FOR COSMOLOGY

MARIO LIVIO

Space Telescope Science Institute, 3700 San Martin Drive, Baltimore, MD 21218

Models for Type Ia Supernovae (SNe Ia) are reviewed. It is shown that there are strong reasons to believe that most SNe Ia represent thermonuclear disruptions of C–O white dwarfs, when these white dwarfs reach the Chandrasekhar limit and ignite carbon at their centers. Different progenitor scenarios are reviewed critically and the strengths and weaknesses of each scenario are presented in detail. It is argued that theoretical considerations currently favor single-degenerate models, in which the white dwarf accretes from a subgiant or giant companion. However, it is still possible that more than one progenitor class contributes to the observed sample. The relation of the different models to the use of SNe Ia for the determination of cosmological parameters is discussed. It is shown that while the observed diversity of SNe Ia may argue for the existence of different progenitor classes, this does not affect the interpretation of an accelerating expansion of the universe. Crucial observational tests of the conclusions are suggested.

1 Introduction

During the past three years two groups [1,2] have presented strong evidence that the expansion of the universe is accelerating rather than decelerating [3,4,5] (see Livio [6] for a perspective; note however that an alternative interpretation, in terms of a quasi-steady state cosmology, has also been suggested by Banerjee et al. [7]. This surprising result comes from distance measurements to more than fifty supernovae Type Ia in the redshift range $z = 0.1$ to $z = 1$. The results are consistent with the cosmological constant (or vacuum energy) contributing to the total energy density about 60–70% of the critical density, which in turn, is consistent with recent measurements of the anisotropy of the cosmic microwave background [8,9,10].

This unexpected finding, as well as the use of supernovae Type Ia to measure the Hubble constant, [11,12] have focused the attention again on the frustrating fact that in spite of decades of research, the exact nature of the progenitors of supernovae Type Ia remains unknown. Until this problem is solved, one cannot be fully confident that supernovae at higher redshifts are not somehow different from their low redshift counterparts. In the present review I therefore examine critically models for supernovae Type Ia and their progenitors. Other recent reviews include Branch, [13] Livio, [14,15] Renzini, [16] Iben, [17] and Höflich & Dominguez [18].

2 SNe Ia Characteristics and the Basic Model

The *defining* characteristics of supernovae Type Ia (SNe Ia) are both spectral: (i) the *lack* of lines of hydrogen, and (ii) the *presence* of a strong red Si II absorption feature (λ6355 shifted to $\sim$ 6100 Å).

Once defined as SNe Ia, the following are several of the important *observational characteristics* of the class which may help in the search for progenitors:

(1) *Homogeneity*: Until very recently, it has generally been claimed that more than 80% of all SNe Ia form a homogeneous class (see however (2) below) in terms of their *spectra* [19], *light curves*, and *peak absolute magnitudes*. The latter are given by

$$M_{\rm B} \simeq M_{\rm V} \simeq -19.30(\pm 0.03) + 5\log(H_0/60 \text{ km s}^{-1} \text{ Mpc}^{-1}) \qquad (1)$$

with a dispersion of $\sigma(M_{\rm B}) \sim \sigma(M_{\rm V}) \sim 0.2$–$0.3$; [20,21] (and see Branch [22] for a review).

(2) *Inhomogeneity*: Some differences in the spectra and light curves have been known to exist for a while [23]. In terms of explosion strength, SNe Ia have traditionally been roughly ordered as follows: SNe Ia like SN 1991bg and SN 1992K represent the weakest events, followed by weak events like 1986G, followed by about 80% of all SNe Ia which are called "normals" (or sometimes "Branch normals"), to the stronger than normal events like SN 1991T. In a very recent work, however, Li et al. [24] find indications for a considerably higher *peculiarity* rate, a total of $(39 \pm 10)\%$; of which $(19 \pm 7)\%$ and $(21 \pm 7)\%$ are SN 1991bg-like and SN 1991T-like objects respectively.

(3) The *luminosity function* of SNe Ia was found in earlier studies to decline very steeply on the bright side [25]. Since selection effects cannot prevent the discovery of SNe which are brighter than the "normals" (unless they occur preferentially in high extinction regions), this is usually taken to imply that *the normals are essentially the brightest*. The recent study of Li et al. [24] seems to show, however, that the luminosity function is relatively flat at both the overluminous and underluminous ends.

(4) Near maximum light, the spectra are characterized by *high velocity* (8000–30,000 km s^{-1}) *intermediate mass elements* (O–Ca). In the late, nebular phase, the spectra are dominated by forbidden lines of iron [26,27,28,29,30].

(5) Fairly young populations appear to be very efficient at producing SNe Ia (e.g. they tend to be associated with spiral arms in spirals, [31,32] but relatively old populations ($\tau \gtrsim 4 \times 10^9$ yr) can also produce them. In particular, *SNe Ia do occur in ellipticals* [33]. In fact, the rates of SNe Ia in ellipticals appear to be similar to those in spirals, $\sim 0.18 SNu$ (where $SNu = 1 SN (100 \text{ yr})^{-1} \left(10^{10} L_{\odot}^{B}\right)^{-1}$. [34]. This immediately implies that *SNe Ia are not caused by the core collapse of stars more massive than 8 $M_{\odot}$*.

(6) There exist a number of correlations between different pairs of observables (see e.g. Branch [22] for a review). Of these, the most frequently used in the context of determinations of cosmological parameters is the correlation between the *absolute magnitude and the shape of the light curve*. Basically, brighter SNe Ia decline more slowly. A parameter commonly used to quantify the light curve shape is Δm_{15} [35] the decline in magnitudes in the B band during the first 15 days after maximum light. Hamuy et al. [20] find slopes $dM_{\rm B}/d\Delta m_{15} = 0.78 \pm 0.17$, $dM_{\rm V}/d\Delta m_{15} = 0.71 \pm 0.14$, and $dM_{\rm I}/d\Delta m_{15} = 0.58 \pm 0.13$. Using a stretch-factor s [1], one can write $M_{\rm B} = M_{\rm B}(s = 1) - \alpha * (s - 1)$, with $M_{\rm B}(s = 1) = -19.46$, [11] and $\alpha = 1.74$ [5]. Sophisticated techniques for using the different correlations in distance determinations have been developed [36,37].

The above characteristics can be augmented by the following suggestive facts:

(1) The *energy* per unit mass, $1/2(\sim 10^4 \text{ km s}^{-1})^2$, is of the order of the one obtained from the conversion of carbon and oxygen to iron.

(2) The fact that the event is explosive suggests that *degeneracy* may play a role.

(3) The spectrum appears to contain no hydrogen.

(4) The explosions can occur with long delays, after the cessation of star formation.

All the properties above have led to one agreed upon model: *SNe Ia represent thermonuclear disruptions of mass accreting white dwarfs*.

It is interesting that there exists a unanimous consensus on this model in spite of the fact that the essence of flame physics, burning front propagation, and the details of the (presumed) transition from deflagration to detonation (in particular the density at which the transition occurs), which are at the heart of the model, remain as major unsolved problems [38,39,40,18,41]. In fact, given these uncertainties, it is almost difficult to understand how the entire family of SNe Ia light curves can be fitted essentially with one parameter, (see Perlmutter et al, [1] although it is possible that all SNe Ia explode at the same WD mass (see §4), and that the entire observed diversity stems from different ^{56}Ni masses.

3 Why Is Identifying the Progenitors Important?

The fact that we do not know yet what are the progenitor systems of some of the most dramatic explosions in the universe has become a major embarrassment and one of the key unsolved problems in stellar and binary star evolution. There are several important reasons why identifying the progenitors has become more crucial than ever:

(i) The use of SNe Ia as one of the main ways to determine key cosmological parameters like H_0, and the contributions to the energy density (by matter and by the cosmological constant) Ω_M, Ω_Λ requires an understanding of the evolution of the luminosity, and the SN rate with cosmic epoch. Both of these depend directly on the nature of the progenitors.

(ii) Galaxy evolution depends on the radiative, kinetic energy, and nucleosynthetic output of SNe Ia [42].

(iii) Due to the uncertainties that still exist in the explosion mechanism itself, a knowledge of the initial conditions and of the distribution of matter in the environment of the exploding star are essential for the understanding of the explosion.

(iv) An unambiguous identification of the progenitors, coupled with observationally determined SNe Ia rates can help to place meaningful constraints on the theory of binary star evolution [43,44,45,46].

In particular, a semi-empirical determination of the elusive common-envelope-ejection efficiency parameter, α_{CE}, may be possible [47].

4 The Favored Model

On the basis of the discussion presented by Livio [15], the basic model can be further refined, and I tentatively conclude that: *Most SNe Ia represent thermonuclear disruptions of mass accreting C–O white dwarfs, when these white dwarfs reach the Chandrasekhar limit and ignite carbon at their centers*!

5 The Two Possible Scenarios

The next step, in which we search for the progenitor systems of SNe Ia is even more controversial. Two possible scenarios have been proposed: (i) The *double-degenerate* scenario, in which two CO WDs in a binary system are brought together by the emission of gravitational radiation and coalesce [48,49]. (ii) The *single-degenerate* scenario, in which a CO WD accretes hydrogen-rich or helium-rich material from a non-degenerate companion [50,51].

In the first scenario the progenitor systems are necessarily *binary WD systems* in which the total mass exceeds the Chandrasekhar mass, and which have binary periods shorter than about thirteen hours (to allow merger within a Hubble time).

In the second scenario the progenitors could be systems like: (i) *Recurrent novae* (both of the type in which the WD accretes hydrogen from a giant like T CrB, RS Oph, and of the type in which the WD accretes helium rich material from a subgiant like U Sco, V394 CrA, and Nova LMC 1990#2), (ii) *Symbiotic Systems* (in which the WD accretes hydrogen-rich material from a low mass red giant or a Mira variable), or (iii) persistent *Supersoft X-ray Sources* (in which the WD accretes at a high rate $\gtrsim 10^{-7}\ M_\odot/yr$ from a subgiant companion).

5.1 Overall Assessment of the Double-Degenerate Scenario

It has now been observationally demonstrated that many double-degenerate systems exist. The general agreement between the distribution of the observed properties (e.g. orbital periods, masses) and those predicted by population synthesis calculations, [52] suggests that the fact that no clear candidate (short period) system with a total mass exceeding the Chandrasekhar mass has been found yet, may merely reflect the insufficient size of the observational sample. Thus, there is very little doubt in my mind that statistics is not a serious problem. The most disturbing uncertainty is related to the outcome of the merger process itself. The discussion in Livio [15] suggests that *collapse to a neutron star is more likely than a SN Ia* [53].

5.2 Overall Assessment of the Single-Degenerate Scenario

The discussion in Livio [15] suggests that probably not all the scenarios for increasing the "phase space" of the single-degenerate channels work (if they did, we might have had the opposite problem of too high a frequency of SNe Ia!). However, these attempts serve to demonstrate that *the input physics to population synthesis codes still involves many uncertainties*. My feeling is therefore that given the many potential channels leading to SNe Ia, statistics should not be regarded as a serious problem.

Single-degenerate scenarios therefore appear quite promising, since unlike the situation a decade ago, a class of objects in which the WDs accrete hydrogen steadily (the Supersoft X-Ray Sources) has actually been identified. The main problem with single-degenerate scenarios remains the non-detection of hydrogen so far. While a difficult observational problem (see §6), *the establishment of the presence or absence of hydrogen in SNe Ia should become a first priority for SNe observers.*

6 Cosmological Implications: Could We Be Fooled?

One of the key questions that result from the uncertainties in the theoretical models and the fact that we do not know with certainty which systems are the progenitors of SNe Ia is clearly: *is it possible that SNe Ia at higher redshifts are systematically dimmer than their low-redshift counterparts*? In this respect it is important to remember that a systematic decrease in the brightness by ~ 0.25 magnitudes is sufficient to explain away the need for a cosmological constant. This question became particularly relevant when an analysis of the rise times of SNe Ia (which was based on preliminary estimates for the high-z sample) [54,55] seemed to show that high-redshift SNe have shorter rise times by 2.5 days than the low-z SNe [56]. A more recent analysis (which used more realistic error estimates than those used by Goldhaber), [54] however, found a better agreement (within 2σ) between the rise times of the low- and high-redshift SNe Ia [57].

In a recent work, Yungelson & Livio [58] calculated the expected ratio of the rate of SNe Ia to the rate of SNe from massive stars (Types II, Ia, Ic) as a function of redshift for *several* progenitor models. *The possibility of having different classes of progenitors contributing to the total SNe Ia rate should definitely be considered*, especially in view of the tentative finding by Li et al. [24] that there is a relatively high rate ($\sim 40\%$) of peculiar SNe Ia among the local sample. If confirmed, these findings suggest that homogeneity should no longer be considered a very strong constraint on progenitor models. I should note though that diversity among SNe Ia does not *necessarily* imply different progenitors, since even in the context of one progenitor model diversity may arise for example from changes in the carbon mass fraction of the WD, which in turn may depend on the environment [59]. Yungelson and Livio [58] showed (within the uncertainties of population synthesis models) that *if* different progenitor systems can contribute to the total SNe Ia rate (e.g. double-degenerates and single degenerates), then it is possible, in principle, that one class of progenitors (e.g. double degenerates) will dominate the rates of the local (low-z) sample, while a different progenitor class (e.g. single degenerate) will start to dominate at $z \lesssim 1$. This is a consequence of the fact that the SNe Ia rate from double degenerates is expected to decline quite steeply from $z = 0$ to $z \sim 1$, while the rate from single degenerates is expected to stay relatively flat in this redshift interval. However, at least within the assumptions of their model calculations, it appears that such a transition is not very likely, because of the following reason: If the contribution from physically different channels (like double degenerates and single degenerates, both at the Chandrasekhar mass) was indeed significant, with a transition from dominance by double degenerates to single degenerates occurring at $z \lesssim 1$, one would have expected to observe this division more clearly in the local and distant samples. For example, the local sample should be dominated by the double degenerate progenitors (with a ratio of double to single degenerates which should be consistent with the results of Li et al.) [24]. At the same time, however, the high-z ($z \lesssim 1$) sample should be dominated by single degenerates, but with the contributions from the single and double degenerate channels not being vastly different (in particular, the contribution should be equal at the transition point). This is *not* consistent with the observations of the high-z sample, the latter appearing to be (within the observational uncertainties) *very homogeneous* [24]. Consequently, I do not think it likely that the observed universal acceleration is an artifact of the observed SNe Ia sample being dominated by different progenitor classes at high- and low-z (this view is supported by the measurements of the anisotropy of the microwave background).

I should note that the most surprising aspect in the results of Li et al. [24] is the fact that although very bright SNe Ia (SN 1991T-like) constitute $(21\pm7)\%$ of the local sample, these bright objects appear to be totally absent from the high-z sample. One way in which one could (in principle) explain this fact is the following. Suppose that the SN 1991T-like events are caused by mergers (since a super-Chandrasekhar mass is possible in this case), while the "normals" are caused by single degenerates. In this case the local sample would be dominated by single degenerates ($\sim 60\%$ of the events to agree with Li et al.) [24], while double degenerates would contribute $\sim 20\%$ of the events (I ignore here the weak events since they cannot be seen at high-z). Now, since the rate of events from single degenerates stays quite flat till $z \sim 1$, while the rate of events from double degenerates declines quite steeply towards $z \sim 1$, the bright objects will be missing from the high-z sample. Note, however, that this potential explanation for the behavior of the diversity has no obvious implications for the finding of accelerating expansion, since the same class of progenitors dominates both the high-z and low-z samples. Nevertheless, a better understanding of the apparent diversity at low-z and apparent lack thereof at high-z is definitely needed.

Other evolutionary effects are still possible, in principle, [60,61] however, as far as I am aware, only one that is physically meaningful, likely, and mimics accelerated expansion, has been identified so far.

This one potential evolutionary effect that certainly deserves more work is *the effect of metallicity on the density at the point of carbon ignition*. Generally, it is expected that a lower metallicity will result in a lower central density [62]. This is because a lower metallicity results in a lower abundance of the Urca-active element ^{21}Ne, which in turn reduces the neutrino cooling and leads to an earlier ignition. A lower central density could (in principle at least) result in a more rapid light curve development (due to the lower WD binding energy), and a lower inferred maximum brightness.

It is important to note that in a recent work, Riess et al. [63] have shown that it is highly unlikely that the dimming of distant SNe Ia is caused by dust opacity (Galactic-type dust was rejected at the 3.4σ confidence level, and "gray" dust with grain size > 0.1 μm was rejected at the 2.3 to 2.6σ confidence level).

7 Tentative Conclusions and Observational Tests

On the basis of the analysis and discussion in the present work, the following tentative conclusions can be drawn:

(1) SNe Ia are almost certainly thermonuclear disruptions of mass accreting *C–O white dwarfs*.

(2) It is very likely that the explosion occurs *at the Chandrasekhar mass*, as *carbon is ignited at (or very near) the WD center*. The flame propagates either as a deflagration, or, starting as a deflagration which transitions into a detonation. Off-center ignition of helium at sub-Chandrasekhar masses may still be responsible for a subset of the SNe Ia which are subluminous, but this is less clear.

(3) The immediate progenitor systems are still not known with certainty. From the discussion in §5 however, I conclude that presently *single degenerate scenarios look more promising*, with hydrogen or helium rich material being transferred from a subgiant or giant companion (systems like Supersoft X-Ray Sources and Symbiotics). It is still possible, however, in view of the apparent diversity in the local sample, that more than

one progenitor class contributes to the total SNe Ia rate. In particular, a scenario in which single degenerates contribute $\gtrsim 60\%$ of the events and double degenerates $\sim 20\%$, appears to be consistent with the diversity of the $z \sim 0$ sample and the lack thereof in the distant sample.

(4) Definitive answers concerning the nature of the progenitors can be obtained from observations taken as early as possible in: *x-rays, radio, and high resolution optical spectroscopy. The establishment of the presence or absence of hydrogen in SNe Ia should be regarded as an extremely high priority goal for supernovae observers.* If hydrogen will not be detected at interesting limits (corresponding to $\dot{M}/v_{10} \sim 10^{-8}$ $M_\odot$ yr^{-1}), this will point clearly towards the double-degenerate scenario.

(5) Observations of SNe Ia at high redshifts can help to test particular ingredients of the models which are directly related to the nature of the progenitors. For example, most of the models aiming at improving the statistics of the single-degenerate scenarios rely on a strong wind from the accreting WD. These models thus predict an "inhibition" of SNe Ia in low-metallicity environments, and in particular a significant decrease in the rate of SNe Ia in spirals at $z \sim 2$ [64,59]. Furthermore, if the inferred cosmic star formation rate is used [65], then the SN Ia rate is expected to drop significantly at $z \sim 1.6$. At present, the detection of a very likely SN Ia at redshift $z = 1.32$ (SN 1997ff [66] in the Hubble Deep Field, and two more at redshifts 1.20 and 1.23 (Perlmutter et al., private communication and Tonry et al., private communication) appear to be at least mildly inconsistent with this prediction, but more observations will be required to give a more definitive answer.

(6) The potential "inhibition" of SNe Ia due to low metallicity should also manifest itself in an absence (or at least a significant decline in the rate) of SNe Ia in dwarf galaxies and in the outer regions of spirals. The statistics necessary to test this prediction are starting to accumulate.

(7) It is possible, in principle, that the local and high-z SNe Ia samples are dominated by different progenitor classes, thus mimicking accelerated expansion, however, this is neither very likely nor consistent with the observations of diversity. *With more detections of SNe Ia at redshifts $z \gtrsim 1$ it will probably become possible to directly confirm the transition in the expansion of the universe from deceleration to acceleration.* Such a transition would be difficult to mimic by systematic or evolutionary effects, and it would therefore confirm the accelerated expansion.

8 Acknowledgements

This research has been supported in part by NASA Grant NAG5–6857.
I am grateful to Adam Riess for helpful discussions.

References

1. S. Perlmutter, et al., *ApJ* **483**, 565 (1997).
2. B. Schmidt, et al., *ApJ* **507**, 46 (1998).
3. A. G. Riess, et al., *AJ* **116**, 1009 (1998).
4. S. Perlmutter, et al., *Nature* **391**, 51 (1998).
5. S. Perlmutter, et al., *ApJ* **517**, 565 (1999).
6. M. Livio, *Science* **286**, 1689 (1999).
7. S. K. Banerjee, J. V. Narlikar, N. C. Wickramasinghe, F. Hoyle, & G. Burbidge, *ApJ* **119**, 2583 (2000).
8. A. D. Miller, et al., *ApJ* **524**, L1 (1999).
9. G. W. Wilson, et al., astro-ph/9902047, (1999).
10. P. D. Mauskopf, et al., astro-ph/9911444, (1999).
11. A. Sandage, A. Saha, G. A. Tammann, L. Labhardt, N. Panagia, & F. D. Macchetto, *ApJ* **460**, L15 (1996).
12. A. Saha, A. Sandage, L. Labhardt, G. A. Tammann, F. D. Macchetto, & N. Panagia, *ApJ* **486**, 1 (1997).
13. D. Branch, M. Livio, L. R. Yungelson, F. R. Boffi, & E. Baron, *PASP* **107**, 717 (1995).
14. M. Livio, in *Supersoft X-Ray Sources*, ed. G. Greiner (Springer, Berlin, 1996a).
15. M. Livio, in *Type Ia Supernovae: Theory and Cosmology*, eds. J. C. Niemeyer & J. W. Truran, (Cambridge Univ. Press, Cambridge, 2000).
16. A. Renzini, in *Supernovae and Supernova Remnants*, eds. R. McCray & Z. Wang (Cambridge Univ. Press, Cambridge, 1996).
17. I. Iben, Jr., in *Thermonuclear Supernovae*, eds. P. Ruiz-Lapuente, R. Canal & J. Isern (Kluwer, Dordrecht, 1997).
18. P. Höflich & I. Dominguez, in *The Greatest Explosions since the Big Bang: Supernovae and Gamma-Ray Bursts*, eds. M. Livio, N. Panagia, & K. Sahu (Cambridge Univ. Press, Cambridge, 2001).
19. D. Branch, A. Fisher, & P. Nugent, *AJ* **106**, 2383 (1993).
20. M. Hamuy, M. M. Phillips, N. B. Suntzeff, R. A. Schommer, J. Maza, & R. Avilés, *AJ* **112**, 2398 (1996a).
21. G. A. Tamman & A. Sandage, *ApJ* **452**, 16 (1995).
22. D. Branch, *ARAA* **36**, 17 (1998).

23. M. Hamuy, M. M. Phillips, N. B. Suntzeff, R. A. Schommer, J. Maza, et al., *AJ* **112**, 2438 (1996b).
24. W. Li, A. V. Filippenko, A. G. Riess, R. R. Treffers, J. Hu, & Y. Qiu, *preprint* (2000).
25. T. E. Vaughan, D. Branch, D. L. Miller, & S. Perlmutter, *ApJ* **439**, 558 (1995).
26. R. P. Kirshner, et al., *ApJ* **415**, 589 (1993).
27. J. C. Wheeler, R. P. Harkness, A. V. Khokhlov, & P. Hoeflich, *Phys. Rep.* **256**, 211 (1995).
28. P. Ruiz-Lapuente, R. P. Kirshner, M. M. Phillips, P. M. Challis, B. P. Schmidt, A. V. Filippenko, & J. C. Wheeler, *ApJ* **439**, 60 (1995).
29. G. Gómez, R. López, & F. Sánchez, *AJ* **112**, 2094 (1996).
30. A. V. Filippenko, *ARAA* **35**, 309 (1997).
31. M. Della Valle & M. Livio, *ApJ* **423**, L31 (1994).
32. O. S. Bartunov, D. Yu Tsvetkov, & I. V. Filimonova, *PASP* **106**, 1276 (1994).
33. M. Turatto, E. Cappellaro, & S. Benetti, *AJ* **108**, 202 (1994).
34. M. Turatto, E. Cappellaro, & A. R. Petrosian, in *Active Galactic Nuclei & Related Phenomena*, IAU Symp. 194, eds. Y. Terzian, D. Weedman, & E. Khachikian (ASP, San Francisco, 1999).
35. Phillips, M. M., *ApJ* **105**, L413 (1993).
36. A. G. Riess, W. H. Press, & R. P. Kirshner, *ApJ* **473**, 88 (1996).
37. A. G. Riess, P. Nugent, A. V. Filippenko, R. P. Kirshner, & S. Perlmutter, *ApJ* **504**, 935 (1998).
38. A. M. Khokhlov, E. S. Oran, & I. C. Wheeler, *ApJ* **478**, 678 (1997).
39. S. E. Woosley, in *Thermonuclear Supernovae*, eds. P. Ruiz-Lapuente, R. Canal & J. Isern (Kluwer, Dordrecht, 1997).
40. M. Reinecke, W. Hillebrandt, & J. C. Niemeyer, *A&A* **347**, 739 (1999).
41. A. Khoklov in *The Greatest Explosions since the Big Bang: Supernovae and Gamma-Ray Bursts*, eds. M. Livio, N. Panagia, & K. Sahu (Cambridge Univ. Press, Cambridge, 2001).
42. G. Kauffmann, S. D. M. White, & B. Guiderdoni, *MNRAS* **264**, 201 (1993).
43. M. Livio, in *Evolutionary Processes in Binary Stars*, eds. R. A. M. J. Wijers, M. B. Davies & C. A. Tout, (Kluwer, Dordrecht, 1996).
44. X.-D. Li & E. P. J. van den Heuvel, *A&A* **322**, L9 (1997).
45. L. Yungelson & M. Livio, *ApJ* **497**, 168 (1998).
46. I. Hachisu, M. Kato, & K. Nomoto, *ApJ* **522**, 487 (1999).
47. I. Iben, Jr. & M. Livio, *PASP* **105**, 1373 (1993).
48. R. F. Webbink, *ApJ* **227**, 355 (1984).

49. I. Iben, Jr. & A. V. Tutukov, *ApJS* **54**, 355 (1984).
50. J. Whelan & I. Iben, Jr. *ApJ* **186**, 1007 (1973).
51. K. Nomoto, *ApJ* **253**, 798 (1982).
52. R. A. Saffer, M. Livio, & L. R. Yungelson, *ApJ* **502**, 394 (1998).
53. R. Mochkovitch, J. Guerrero, & L. Segretain, in *Thermonuclear Supernovae*, eds. P. Ruiz-Lapuente, R. Canal & J. Isern (Kluwer, Dordrecht, 1997).
54. G. Goldhaber, in *Gravity: From the Hubble Length to the Planck Length* (SLAC, Stanford, 1998).
55. D. E. Groom, *BAAS* **193**, 111.02 (1998).
56. A. G. Riess, A. V. Filippenko, W. Li, & B. P. Schmidt, *AJ* **118**, 2668 (1999b).
57. G. Aldering, R. Knop, & P. Nugent, astro-ph/0001049 (2000).
58. L. R. Yungelson & M. Livio, *ApJ* **528**, 108 (2000).
59. K. Nomoto, H. Umeda, I. Hachisu, M. Kato, C. Kobayashi, & T. Tsajimoto, in *Type Ia Supernovae: Theory and Cosmology*, eds. J. Truran & J. Niemeyer (Cambridge Univ. Press, Cambridge, 2000).
60. P. S. Drell, T. J. Loredo, & I. Wasserman, astro-ph/9905027 (1999).
61. W. Hillebrandt, in *Type Ia Supernovae: Theory and Cosmology*, eds. J. C. Niemeyer & J. W. Truran (Cambridge Univ. Press, Cambridge, 2000).
62. K. Nomoto, et al. in *Thermonuclear Supernovae*, eds. P. Ruiz-Lapuente et al. (Kluwer, Dordrecht, 1997).
63. A. G. Riess, et al., astro-ph/0001384 (2000).
64. C. Kobayashi, T. Tsujimato, K. Nomoto, I. Hachisu, & M. Kato, *ApJ* **503**, L155 (1998).
65. M. Pettini, et al., *ApJ* **508**, 539 (1998).
66. R. I. Gilliland, P. E. Nugent, & M. M. Phillips, *ApJ* **521**, 30 (1999).

TYPE IA SUPERNOVA RATE AT A REDSHIFT OF 0.1

G. BLANC, THE EROS* COLLABORATION
SPP, DAPNIA, CEA Saclay, F-91191 Gif-sur-Yvette, France, and
LBNL, 1 Cyclotron Road, MS 50-232, Berkeley, CA 94720, USA
E-mail: gblanc@cea.fr

We measured the type Ia supernova (SNIa) rate at a redshift of $\sim$ 0.1 from a systematic SN search performed by EROS in early 1999.

1 A worldwide SN search

In the beginning of 1999 (February and March), the *Supernova Cosmology Project* (SCP) coordinated a worldwide nearby (z $\sim$ 0.01 - 0.2) SN search, and provided spectroscopic and photometric follow-up. Nine groups were involved, including EROS[a]. 41 SNe were found, with 19 SNIa near maximum.

To perform this SN search, EROS monitored about 420 deg^2 of sky and discovered 16 SNe (IAUC 7092, 7117, 7136), 12 of which are of type Ia.

2 Search strategy

A set of reference images were taken around the January 1999 new moon. The search took place one and two months later. The detection of objects is done online on images obtained by subtracting the reference frames from the ones taken during the search. This operation is preceded by a geometric and photometric alignment and a seeing matching. An automatic algorithm vetoes known backgrounds (cosmic rays, variable stars, asteroids). A visual scan removes the last spurious candidates. Confirmation images of selected candidates are taken the following night. But only spectroscopy of the candidates can really confirm them by giving their redshift and type.

3 Supernova rate

Under the assumption that the SN rate per galaxy is proportional to the galactic luminosity, dN/dt $= R_{\rm SN} \cdot L^{\rm gal}$, the rate per unit galactic luminosity

*EROS is a collaboration primarily consisting of DAPNIA-CEA (Saclay), LAL-IN2P3 (Orsay), and IAP-INSU (Paris).

[a]This experiment uses a fully automated 1 meter telescope based at La Silla (ESO) observatory in Chile fitted with two 1 deg^2 field CCD cameras with wide non-standard band filters. Each camera has 8 CCDs of 2k $\times$ 2k pixels of size 0.62".

(in SNu[b]) is:

$$R_{\rm SN} = \frac{N_{\rm SNfound}}{\sum_{\rm images}\sum_{\rm galaxies}\int_0^{+\infty} dz \cdot p(z|m_{\rm gal}) \cdot L_{\rm B}^{\rm gal}(z, m_{\rm gal}) \int_{-\infty}^{+\infty} dt \cdot \epsilon(t,z)} \quad (1)$$

where p is the probability distribution of redshift z as a function of galaxy apparent magnitude $m_{\rm gal}$, L the galaxy luminosity, and ϵ the detection efficiency as a function of z and time after maximum light t.

We use the SExtractor[1] software, tuned in order to match our galaxy population with the published galaxy counts. Next we determine our SN detection efficiency ϵ as a function of SN magnitude: simulated SNe (bright stars picked up from the same image, rescaled in flux and moved by an integer number of pixels to the required position) are superimposed on each galaxy of each image, before the detection software runs on it. The simulated positions of the SNe are distributed according to the galaxy surface brightness.

The integral calculation (eq. (1) denominator) is computed by a Monte-Carlo method. SN phase t is randomly chosen; then we use a set of well sampled published SNIa light curves to get a magnitude and deduce the efficiency taking into account the calibration of the images.

4 Preliminary result

The SN redshift distribution as obtained by the simulation is very close to the distribution of the 10 SNIa kept after the cuts (a Kolmogorov-Smirnov test gives a 50% probability). The mean redshift of our SN sample is 0.11.

The rate[c] we obtain is $0.34 \pm 0.11\ \mathrm{h}^2 \cdot \mathrm{SNu}$ (The error bar is only statistical[d].), which is in good agreement with the previous EROS such measurement[2]: $0.44^{+0.35}_{-0.21}{}^{+0.13}_{-0.07}\ \mathrm{h}^2 \cdot \mathrm{SNu}$. Multiplying this by the galaxy luminosity density[3]: $1.4 \pm 0.1 \cdot 10^8\ \mathrm{h} \cdot \mathrm{L}_{\rm R}^{\odot} \cdot \mathrm{Mpc}^{-3}$, we get: $4.42 \pm 1.43 \cdot 10^{-5}\ \mathrm{h}^3 \cdot \mathrm{Mpc}^{-1} \cdot \mathrm{yr}^{-1}$.

References

1. Bertin and Arnouts, 1996, A&ASS, 117, 393
2. Hardin et al., 2000, A&A, 362, 419
3. Lin et al., 1996, ApJ, 464, 60

[b] 1 SNu = 1 SN / 10^{10} $\mathrm{L}_{\rm B}^{\odot}$ / century

[c] The Hubble constant is written: $\mathrm{H}_0 = 100\mathrm{h}\ \mathrm{km} \cdot \mathrm{s}^{-1} \cdot \mathrm{Mpc}^{-1}$.

[d] The systematic error should be of the same order of magnitude than the statistical one.

COSMIC SUPERNOVA RATE HISTORY AND TYPE IA SUPERNOVA PROGENITORS

C. KOBAYASHI[1], K. NOMOTO[1], and T. TSUJIMOTO[2]

[1] *Department of Astronomy, School of Science, University of Tokyo, Bunkyo-ku, Tokyo 113-0033, Japan*

[2] *National Astronomical Observatory, Mitaka, Tokyo 181-8588, Japan*

Adopting a single degenerate scenario for Type Ia supernova progenitors with the metallicity effect, we make a prediction of the cosmic supernova rate history as a composite of the supernova rates in spiral and elliptical galaxies, and compare with the recent observational data up to $z \sim 0.55$.

1 Metallicity Effects on Type Ia Supernova Progenitors and Chemical Evolution of Galaxies

We adopt a single-degenerate (SD) scenario, which assumes that a C+O white dwarf accretes H-rich meterials from the binary companion star and grows its mass to the Chandrasekhar mass to explode as a Type Ia supernova (SN Ia). There are two progenitor systems: One is a red-giant (RG) companion with the initial mass of $M_{\rm RG,0} \sim 0.9 - 1.5 M_\odot$ and an orbital period of tens to hundreds days [1 2 3]. The other is a near main-sequence (MS) companion with an initial mass of $M_{\rm MS,0} \sim 1.8 - 2.6 M_\odot$ and a period of several tenths of a day to several days [4 3]. Optically thick winds from the mass accreting WD play an essential role in stabilizing the mass transfer and escaping from forming a common envelope. Since the optically thick winds are driven by a strong peak of iron lines, the occurrence of SNe Ia depends strongly on the iron abundance. If the iron abundance is as low as [Fe/H] $\lesssim -1.1$, then the wind is too weak for SNe Ia to occur [5].

Our model successfully reproduces [O/Fe]-[Fe/H] relation observed by oxygen forbidden lines, which has a constant [O/Fe] of ~ 0.45 at [Fe/H] $\lesssim -1$ and a decrease in [O/Fe] with increasing metallicity. If we do not include the metallicity effect, the largest companion star with $M \sim 2.6 M_\odot$ produce SNe Ia at the age of ~ 0.5 Gyr and decrease [O/Fe] too early to be compatible with the observations. For the metallicity dependent SD scenario, SNe Ia occur at [Fe/H] $\gtrsim -1$, which naturally reproduce the observed break in [O/Fe] at [Fe/H] ~ -1 [5].

We construct the galaxy models to meet the latest observational constraints of chemical and photometrical properties, and can reproduce the present supernova rates in the galaxy. For spirals, owing to the presence of the *two* kinds of the SN Ia progenitor systems (MS+WD and RG+WD) with shorter (0.5 – 1.5 Gyr) and longer lifetimes (2 – 20 Gyr), respectively, we can explain the difference in the relative ratio of the SN Ia to SN II rate $\mathcal{R}_{\rm Ia}/\mathcal{R}_{\rm II}$ between early and late types of spirals. For ellipticals, owing to the presence of the RG+WD systems with over 10 Gyr lifetime, SNe Ia can be seen even at present in ellipticals where the star formation has already ceased more than 10 Gyr before because of a supernova-driven galactic wind [6].

2 Cosmic Supernovae Rate

Galaxies that are responsible for the cosmic SFR have different timescales for the heavy-element enrichment, and the occurrence of supernovae depends on the metallicity therein. Therefore we calculate the cosmic supernova rate by summing up the supernova rates in spirals and ellipticals with the ratio of the relative mass contribution [6]. Here we adopt $H_0 = 65$ km s^{-1} Mpc^{-1}, $\Omega_0 = 0.3$, $\lambda_0 = 0.7$, and the formation epoch of $z_{\rm f} = 5$.

Figure 1 shows the cosmic supernova rates in cluster galaxies. The SN Ia rate in spirals drops at $z \sim 1.9$ because of the low-metallicity inhibition of SNe Ia. We can precisely test the metallicity effect by finding this drop of the SN Ia in spirals, if high-redshift SNe Ia at $z \gtrsim 1.5$ and their host galaxies are observed with the Next Generation Space Telescope. In ellipticals, the chemical enrichment takes place so early that the metallicity is large enough to produce SNe Ia at $z \gtrsim 2$. The two peaks of SN Ia rates at $z \sim 2.6$ and $z \sim 1.6$ come from the MS+WD and the RG+WD systems, respectively. The SN Ia rate in ellipticals decreases at $z \sim 2.6$, which is determined from the shortest lifetime of SNe Ia of ~ 0.5 Gyr. Thus, the total SN Ia rate decrease at the same redshift as ellipticals, i.e., $z \sim 2.6$.

We also predict the cosmic supernova rates assuming that the formation of ellipticals in field took place for over the wide range of redshifts, which is imprinted in the observed spectra of ellipticals in the Hubble Deep Field. The adopted SFRs are the same as the case of cluster galaxies, but for the formation epochs $z_{\rm f}$ of ellipticals distribute in the range of $0 \leq z \leq 5$. Figure 2 shows the cosmic supernova rates in field galaxies. As in Figure 1, the SN Ia rate in spirals drops at $z \sim 1.9$. The averaged SN Ia rate in ellipticals decreases at $z \sim 2.2$ as a result of ~ 0.5 Gyr delay of the decrease in the SFR at $z \gtrsim 3$. Then, the total SN Ia rate decreases gradually from $z \sim 2$ to $z \sim 3$.

Although the error bars are large, there is a hint that the observed SN Ia rate decreases from $z \sim 0.4$ to the present. If this is confirmed, it could imply that the rate of SNe Ia from long-lived $0.9 M_\odot$ companions is lower than

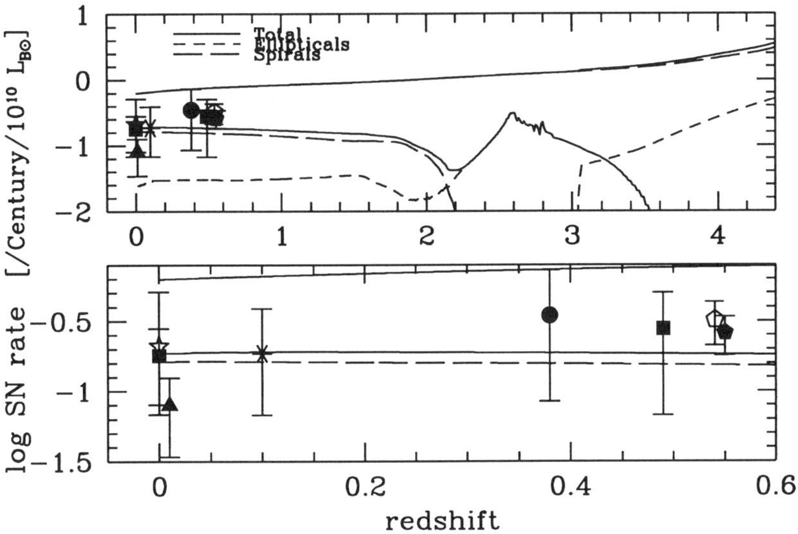

Figure 1: The cosmic supernova rates (solid line) as the composite of ellipticals (short-dashed line) and spirals (long-dashed line). The upper three lines show SN II rates, the lower three lines show SN Ia rates. Observational data sources: triangle [7], star [8], asterisk [9], circle [10], open pentagon [11], closed pentagon [12], and square [13]

that assumed in our model. From $z \sim 0.4$, the observed SN Ia rate seems to slightly decrease toward higher redshifts. To discuss their implications in terms of the progenitors' evolution, we should exclude the luminosity uncertainties in the unit of supernova rates and introduce more detail galaxy models including internal structure of a galaxy and galaxy number evolution.

The rate of SNe II in ellipticals evolves following the SFR without time delay. Then, it is possible to observe SNe II in ellipticals around $z \sim 1$. The difference in the SN II and Ia rates between cluster and field ellipticals reflects the difference in the galaxy formation histories in the different environments.

3 Discussion

We make the following predictions that can test this metallicity effect. 1) SNe Ia are not found in the low iron abundance environments such as dwarf galaxies and the outskirts of spirals. 2) The SN Ia rate in spirals drops at $z \sim 2$ due to the low-iron abundance, while SNe Ia can be found at $z \gtrsim 2.5$ in cluster ellipticals, where the timescale of metal enrichment is sufficiently short. 3) If the formation of field ellipticals is protracted to lower redshifts, the SNe Ia

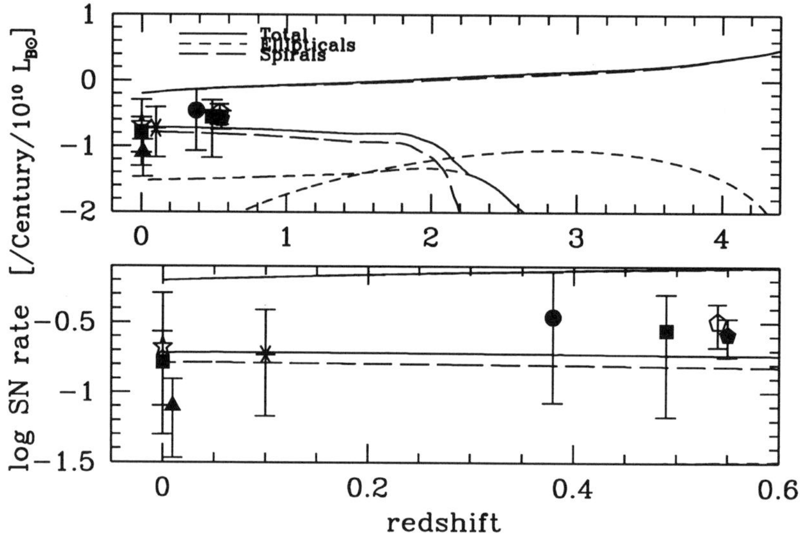

Figure 2: The same as Figure 1, but for the formation epochs of ellipticals span at $1 \lesssim z \lesssim 4$, which corresponds to field ellipticals.

rate decreases from $z \sim 2$ in the field ellipticals.

Our predicted cosmic SN Ia rate is consistent with the recent observation within error bars up to $z \sim 0.55$. If stronger dependences on redshifts are found, it would provide important constraints on the mass-range (i.e., lifetime) of the WD's companions and metallicity effects.

References

1. Hachisu, I., Kato, M., & Nomoto, K. 1996, *Astrophys. J.*, 470, L97
2. Hachisu, I., Kato, M., & Nomoto, K. 1999b, *Astrophys. J.*, 522, 487
3. Hachisu, I., Kato, M., Nomoto, K., & Umeda, H. 1999a, *Astrophys. J.*, 519, 314
4. Li, X. -D., & van den Heuvel, E. P. J. 1997, *Astron. Astrophys.*, 322, L9
5. Kobayashi, C., Tsujimoto, T., Nomoto, K., Hachisu, I, & Kato, M. 1998, *Astrophys. J.*, 503, L155
6. Kobayashi, C., Tsujimoto, T., & Nomoto, K. 1999, *Astrophys. J.*, 539, 26
7. Cappellaro, E., Evans, R., & Turatto, M. 1997, *Astron. Astrophys.*, 351, 459
8. Hamuy, M. & Pinto, P. A. 1999, *Astron. J.*, 117, 1185
9. Hardin, D. et al. 2000, *Astron. Astrophys.*, 362, 419
10. Pain, R., et al. 1996, *Astrophys. J.*, 473, 356
11. Pain, R. 1999, talk at the Type Ia Supernova workshop (Aspen Center for Physics)
12. Pain, R., et al. 2000, in preparation (see Ruiz-Lapuente, P. & Canal, R. astro-ph/0009312)
13. Reiss, D. J. 2000, PhD thesis, Univ. of Washington

QUINTESSENCE COSMOLOGY

S. A. BLUDMAN

DESY, Hamburg & University of Pennsylvania, Philadelphia

We illustrate quintessence cosmology by the evolution of a scalar field with linear equation of state $P/\rho \equiv w_Q$ = constant in a background of radiation+cold dark matter. For a flat universe presently dominated by smooth energy, either cosmological constant (LCDM) or quintessence (QCDM), we find that, as function of hypothetical present ratio of smooth energy to matter energy, the asymptotic collapsed mass fraction peaks at the observed ratio $\mathcal{R}_0 \sim 2$. Identifying the normalized collapsed fraction as a conditional probability for habitable galaxies, we infer that the prior distribution is flat in $\mathcal{R}_0$ or Ω_{m0}. Interpreting this prior as a distribution over *theories*, rather than as a distribution over unobservable *sub-universes*, leads us to a heuristic prediction about the class of future fundamental theories for quintessence or cosmological constant.

1 Rolling Scalar Field as Quintessence

We have recently learned that the density of clustered matter is low $\Omega_{m0} \sim 1/3$, the Universe is flat, and the residual unclustered or *smooth energy density* $\Omega_{Q0} \sim 2/3$. The flat cosmology is therefore determined by one parameter $\mathcal{R}_0 \equiv \Omega_{Q0}/\Omega_{m0}$, which is remarkably ~ 2 and by the nature of the smooth energy, which may be static (cosmological constant) or slowly varying (quintessence). Quintessence has been invoked in order to explain dynamically why the cosmological "constant" is now so small.

By quintessence we mean a rolling scalar field ϕ which, for a very large class of initial conditions, converges towards a *tracker field* mimicing the background radiation or matter. For scalar potential we choose

$$V(\phi)/V_0 = [\sinh(\alpha\phi_0)/\sinh(\alpha\phi)]^\beta,$$

where $n_Q \equiv 3(1+w_Q)$ is fixed, but the background adiabatic index $n_B \equiv 3(1+w_B)$, $\alpha \equiv 3(w_B - w_Q)/(2\sqrt{n}_Q)$, $\beta \equiv 2n_Q/3(w_B - w_Q)$ all change with background pressure/density ratio, from $w_B = 1/3$ in the radiation era to $w_B = 0$ in the cold matter era. This potential is paramatrized by a constant w_Q and normalized so that the present value $V(\phi_0) = V_0$. Because $V''V/V'^2 = 1 + 1/\beta\cosh^2(\alpha\phi) > 1$ this potential is always tracking. But

$$V'/V = -\sqrt(n_Q \coth(\alpha\phi), \quad V''/V = (n_Q/\beta)[(\beta+1)\coth^2(\alpha\phi) - 1],$$

shows that it is slow-rolling only for $n_Q < 1$, $w_Q < -2/3$ and only late, when the Q-field is dominating.

Before tracking, $P(\phi)/\rho_Q(\phi)$ oscillates between ≈ 1 and -1 and enters a long epoch in which the energy density $\rho_Q(\phi) = \dot{\phi}^2/2 + V(\phi)$ is frozen at a small value. Ultimately, this field converges onto a tracker solution in which the scalar field pressure is given by the linear equation of state $P(\phi)/\rho_Q(\phi) = w_Q$: $n_Q = 3(1+w_Q)$, $\rho_Q = \rho_{Q0}a^{-n_Q}, \rho_B = \rho_{B0}a^{-n_B}$ the ratio $\mathcal{R} \equiv \Omega_Q/\Omega_B = \mathcal{R}_0 a^{3(w_B-w_Q)}$.

The evolution of the tracking field with cosmological scale $a = 1/(1+z)$ is fixed by the scaling $\mathcal{R}(a) = \sinh^2(\alpha\phi)$ which makes

$$\Omega_Q = \tanh^2(\alpha\phi), \quad \Omega_B = 1/(\cosh^2(\alpha\phi)), \quad V/V_0 = a^{3\beta(w_Q-w_B)/2} = \rho_Q/\rho_{Q0}.$$

For the present ratio $\mathcal{R}_0 = \sinh^2(\alpha\phi_0) \approx 2$, $\alpha\phi_0 = 1.146$. (We express ϕ in Planck units $8\pi G \equiv M_P^{-2} = 1$.)

For $w_Q = -1$, $\beta = 0$ the field is static (cosmological constant). The data on the presently accelerating universe allow only the upper bounds

$$w_Q \leq -1/2, \ \beta \leq 2, \ \alpha \geq 0.612, \ \phi_0 \leq 1.87.$$

Tracking takes place only in the matter-dominated era and is never quite slow-rolling. Nevertheless, $\rho_Q(a)$ remains subdominant to the radiation density all the way back into the era of Big Bang nucleosynthesis. In the next section, we will distinguish the two allowed limits for the smooth energy $w_Q = -1$ (LCDM) and $w_Q = -1/2$ (QCDM).

If w_Q were not constant, we could still reconstruct the Q-potential from $w_Q(z)$. Indeed $dw_Q(z)/dz > 0$ for quintessence and < 0 for k-essence, an alternative in which, instead of a potential, the scalar field has a non-linear kinetic energy. In principle, $w_Q(z)$ is observable in high red-shift supernovae, cluster evolution and gravitational microlensing. Over the small red-shift range that is practically observable, however, the effects of varying $w_Q(z)$ are generally smaller than the uncertainties in cosmological model. We can interpret the constant $w_Q = -1/2$ we choose for QCDM as some average of $w(z)$ back to the last scattering surface. The present value w_{Q0} would then be somewhat less than -1/2 for quintessence and somewhat greater for k-essence.

2 Evolution of Large Scale Structure

Structure evolves only in a matter-dominated universe, so that it begins about radiation-matter equality ($z \sim 12,000$) and practically ends when, at $z \sim 0.6$ or 0.3, quintessence or cosmological constant overtakes the matter density . Using an improved Press-Schecter formalism, we[1] have calculated that the fraction of mass that would collapse would be a maximum for $\mathcal{R}_0 \sim 2$ in QCDM, $\mathcal{R}_0 \sim 3.5$ in LCDM! This means that, among all a priori possible flat universes, ours maximizes (for QCDM) or nearly maximizes (for LCDM) the probability of evolved galaxies and other structures and, presumably, the probability of intelligent observers.

3 The Anthropic as a Heuristic Principle

Quintessence explains why the cosmological "constant" Ω_Q is so small, but not why just now it is comparable to Ω_B and the Universe is accelerating. Hopefully, the one remaining cosmological parameter $\mathcal{R}_0 \sim 2$ will someday be derived from fundamental theory. Meanwhile, in the face of limited data, what is the best inference to be now made from the observed datum $w_Q \sim 2$, at or near the peak of the logarithmic asymptotic mass distribution? Following Vilenkin's *Principle of Mediocrity*, assume that, out of a conceivable ensemble of flat universes or cosmologies, our universe or cosmology is typical of those permitting intelligent life. Then the observed datum suggests that the entire prior distribution in $\mathcal{R}_0$ must be flat, at least for $\mathcal{R}_0 = \mathcal{O}(1)$.

It is important to remember that the prior is a functional of (subjective) *hypotheses*, not of the data sample. Proponents of the Anthropic Principle[2] have assumed a prior distribution of real universes, with different values of $\mathcal{R}_0$, almost all of which are inhospitable to life. Since these many other universes are practically unobservable, this application has no practicable predictive value, and is rightfully criticized as unscientific.

Instead of such a prior distribution of *universes*, we[1] recognize a flat prior probability distribution for (future) fundamental *theories*. Applied this way, the Anthropic Principle suggests for the quintessence potential a generic form like $V(\phi) = V_1 f(\lambda\phi)$, where $V_1 \sim \mathcal{O}(M_P^4)$ is a large energy density, $f(x)$ is a dimensionless function involving no very large or very small parameters, and λ is a very small dimensional parameter[3]. Because this prediction of the quintessence potential is theoretically and observationally falsifiable, our application of the Anthropic Principle to theories is scientific, while the application to universes is not, unless these other universes can be discovered. This application of the Anthropic Principle is only heuristic. So was the Equivalence Principle which guided us towards General Relativity and the Correspondence Principle which guided us towards quantum mechanics.

References

1. S. Bludman and M. Roos, Astroph. J. **547**, (2001)
2. G. Efstathiou, MNRAS **274**L73(1995); M. Tegmark and M. J. Rees, Astroph. J. **499**526(1998); J. Garriga *et al.*astro-ph/9906210
3. S. Weinberg, PRD **61**, 103505 (2000)

COSMOLOGICAL HISTORY OF STARS AND METALS

R. SADAT
Observatoire de Midi-Pyrénées, 14 av. Edouard Belin- 31400 Toulouse, France
E-mail: rsadat@ast.obs-mip.fr

B. GUIDERDONI
Institut d'Astrophysique de Paris, 98bis Boulevard Arago, F-75014 Paris, France

J. SILK
Department of Physics, Astrophysics, 1 Keble Road, Oxford OX1 3NP, England

We investigate various classes of cosmic star formation rates histories consistently with current estimations by means of self-consistent chemo–spectrophotometric models that take into account the metallicity and dust obscuration effects to study the global evolution of the stellar, chemical and radiation content of the galaxies averaged over large comoving volume independently of their individual details and properties.

1 Introduction

The history of star formation rate in the universe has been inferred with an unprecedented way by combining the SFR in the local universe, the results of ground-based spectroscopic redshift surveys out to $z \sim 1$ and the results of deep photometric surveys such as the *HDF*. The result of such a cosmic star formation history (CSFR) shows a rapid increase up to $z \sim 1$ and then a decrease at higher z with a peak at $z \sim$1.5-2 (Madau et al 1996). However, such interpretation is based on UV observations and is heavily biased because of dust effects so that the exact shape of the CSFR is still a matter of debate. Figure 1a depicts the various CSFR histories we investigate in this work using self-consistent chemo-spectrophotometric models with detailed modelling of dust absorption and re-radiation (see Sadat et al. 2001 for details). In all what follows the adopted IMF is a Salpeter ($x = 1.35$) IMF with $m_{up} = 100 M_\odot$. $H_0 = 50$ km s^{-1} Mpc^{-1}, a density parameter $\Omega_0 = 1$ and $\Omega_\Lambda = 0$.

1.1 Constraining the CSFR with the CIB

The background radiation results from integrating over the cosmic time the emission coming from various extragalactic objects on the line of sight. Therefore the history of CSFR can be observed in its imprint on the Background Radiation.Although the overall shape seems to be well reproduced by all the

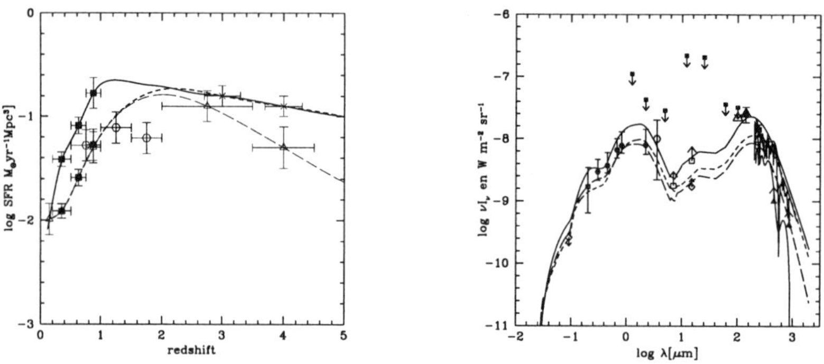

Figure 1. *Left panel*: SFR histories: *solid line* and *small and long dashes* corresponds respectively to H and L_1 and L_2 models. The data are from Lilly *et al.* (1996 ; *filled squares*) (lower points) and Flores *et al.* (1998 ; upper points); Madau *et al.* (1998 ; *open triangles*) and *crosses* by Steidel *et al.* (1998). *Right panel*: the predicted diffuse backgrounds in the optical and FIR/submm as compared to current limits and detections.

models regardless of their very different CSFR histories, only mid–infrared derived CSFR at low–z (H model) is able to correctly match the cosmic infrared background (CIB) at 140 μm, while UV-derived low–z CSFR (L_1 and L_2 models), falls considerably short of the COBE detections at $\lambda = 140$ μm (Fig. 2b). Moreover, the CIB is less sensible to the CSFR shape at high–z. This result suggests that the bulk of the energy in the FIR/submm is due to moderately obscured normal galaxies lying at low and moderate redshifts ($0 \leq z \leq 1$) and not by a population of heavily extinguished star forming regions located at very high redshifts as previously claimed (Blain *et al.* 1998).

2 Cosmological history of metals

We find that in the closed-box assumption our best-fit to the CIB at 140 μm model clearly fails to reproduce the data, it overpredicts the mean metallicity by a factor of ~ 3 at $2 \leq z \leq 3$. Relaxing the closed-box approximation by allowing for some outflowing metal–enriched gas out off the galaxies, leads to a better agreement with the data (Fig. 2a). Assuming that the ejected metals are responsible for IGM pollution, we find that the derived mean comoving density of metals in the IGM si in good agreement within the error bar with the observations at high-z of CIV and Si IV absorption lines in the Lyα forest reported by Songaila (1997).
Fig. 2b shows the inferred evolution of the stellar content as a function of the

cosmic epoch for each of the models of CSFR. We can see, that the best–fit model to the 140 μm data point predicts a present-day stellar mass density larger by at least a factor of two than the current value. If the contribution of the AGNs is negligible then this would be an indication in favour of an IMF which is biased toward massive stars.

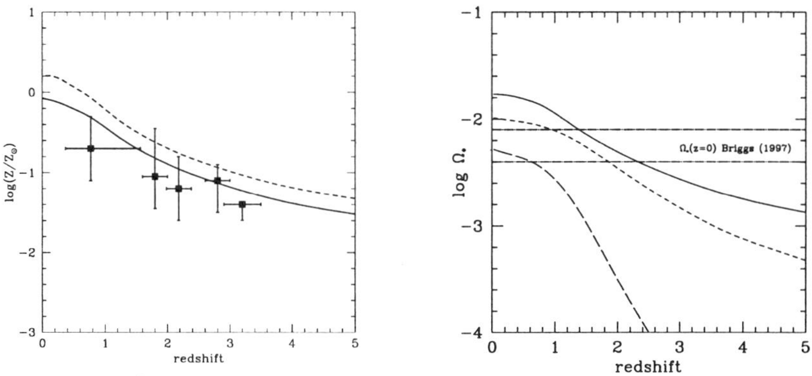

Figure 2. Evolution with redshift of the mean comoving ISM metallicity as predicted by the best-fit model (*solid line*). The *dotted line* curve corresponds to the case of closed box evolution . On the *right panel* is shown the evolution of the comoving stellar mass density. Coding of the models are as in Fig.1.

3 Conclusion

We show that the point at 140 μm is crucial for constraining the CSFR history and seems to favour a higher comoving star formation rate at low redshifts than UV/optical derived one while the shape of the CSFR at high redshifts is not well constrained by the CIB measurements suggesting that the main contribution to the CIB is due to normal galaxies, lying at $0 \leq z \leq 1$.Our best–fit model predicts ISM/IGM metal-enrichment in fair agreement with the observations. However, it overpoduces the present–day stellar mass density $\Omega_*(0)$.

References

1. A.W. Blain *et al*, MNRAS **302**, 632 (1998).
2. P. Madau, *et al*, MNRAS **283**, 1388 (1996).
3. R. Sadat, B. Guiderdoni and J. Silk, to appear A&A , (2001).

THE FORMATION OF GALAXIES

JOSEPH SILK
Department of Astrophysics, 1 Keble Road,
Oxford OX1 3RH, England
and Institut d'Astrophysique de Paris
98 bis bd Arago, 75014 Paris, France
E-mail: silk@astro.ox.ac.uk

The ingredients of a successful theory of galaxy formation must include prescriptions for cosmological initial conditions, dark matter, and star formation.These represent the essential core of any theory. I will describe how these prescriptions are implemented in the various types of theory that have been developed: numerical simulations, semi-analytical approaches, and analytical discussions.

1 Introduction

The *ab initio* approach to large-scale structure has met with great success. Bottom-up growth by gravitational instability of inflation-boosted quantum fluctuations occurs in cold dark matter. A natural scale is imprinted on the density fluctuations with a peak in the power spectrum at $\sim$ 15 Mpc, that corresponds top the horizon scale at the epoch of equal densities of matter and radiation. This results in galaxy cluster formation, once COBE normalization is adopted, at the present epoch, with galaxies forming at $z \sim 10$. There is general agreement on the choice of cosmological model parameters ($\Omega_m = 0.3 \pm 0.2; \Omega_b = 0.05 \pm 0.03; \Omega_\Lambda = 0.7 \pm 0.2$ and $h = 0.7 \pm 0.1$) that result in a successful theory of large-scale structure formation.

The success of the theory is demonstrated by accounting for several properties of the observed universe: the approximately scale-invariant cosmic microwave background fluctuations (spectral index $n = 1 \pm 0.2$), the two-point correlation function of galaxies ($\xi(r) = (r/r_0)^{1.8\pm0.1}, r_0 = 8\pm1\text{Mpc}, 0.1\text{Mpc} \lesssim r \lesssim 20\text{Mpc}$), and the mass function of galaxy clusters ($dN/dM \propto M^{-\alpha}$ at $M \lesssim 10\text{M}_\odot$, with $\alpha \approx 1$). Inclusion of gas in the simulations has led to an explanation of the distribution of intergalactic gas, spanning damped Lyman alpha clouds ($N_{HI} \sim 10^{20} - 10^{23}\text{cm}^{-2}$) to the Lyman alpha forest ($N_{HI} \sim 10^{12} - 10^{15}\text{cm}^{-2}$).

However galaxy formation has proven to be much less tractable. Cold dark matter clumps on all substructure scales, and high numerical resolution is needed to follow the interactions and evolution of the clumps. Star formation requires implementation of yet another level of complexity, since gas evolution requires a multiphase interstellar medium in order to properly include the effects of feedback.

2 The problem with disk sizes

An outstanding problem in understanding the formation of galaxy disks is the origin of the characteristic disk scale length. In the absence of a theory that determines this parameter, one can place little credence in simulations of numerical or semi-analytical galaxy formation. Yet until recently, the elements of a robust theory for disk scales appeared to be in place.

Tidal torques between neighbouring density fluctuations generate an initial dimensionless angular momentum $\lambda_i \equiv J|E|G^{-1}M^{-5/2} \approx 0.05$, with a broad dispersion $\Delta\lambda_i \sim \lambda_i$. Baryons cool and contract once dark halos virialize, conserving specific angular momentum to form a centrifugally supported disk when $\lambda \sim 1$. The final disk radius is $r_d \sim \lambda_i r_h$, where r_h is the halo radius. This yields a disk scale length of several kpc for an L_* galaxy, as observed (Fall and Efstathiou 1980).

However with the advent of numerical hydrodynamical simulations with adequate resolution, it has become apparent that angular momentum is effectively transferred outwards as dense baryon clumps sink into the central regions of the forming galaxy. Clumpiness in the dark halo also tidally perturbs the disk and aggravates the angular momentum transfer problem. The result of the simulations is that the disk sizes are invariably too small (Steinmetz and Navarro 1999), by a factor ~ 5.

Photoionization of the intergalactic medium destroys small-scale gaseous substructure (Navarro and Steinmetz 1997), but there is little effect on the formation of massive galaxies. Feedback from star formation via supernovae can delay gas cooling to $z \sim 1$ and result in formation of disks of the appropriate size and mass (Weil, Eke and Efstathiou 1998). However late disk formation implies strong evolution in disk size, luminosity and surface brightness, but there is little evidence for any evolution, other than passive stellar population ageing (Lilly et al. 1998; Vogt 2000).

3 Semi-analytic galaxy formation

Numerical simulations have succeeded in resolving the first stellar mass clumps (Abel, Bryan and Norman 2000). The physics is reasonably straightforward, with only the complication of molecular hydrogen cooling to be included. Fragmentation occurs to masses of about $100\mathrm{M}_\odot$. However the simulations are unable to resolve more than the formation of the first star-like objects in a single molecular cloud. Nor can one be confident for that matter that continued subfragmentation does not occur.

To proceed further with star formation during the process of galaxy formation, a semi-analytic approach to galaxy formation has been introduced (Lacey & Silk 1991; White & Frenk 1991; Kauffmann et al. 1999; Somerville & Primack 1999; Cole et al. 2000). N-body simulations of a volume containing up to 10^9Mpc^3 allow one to locate the sites of forming galaxies, and high resolution rezoning provides a detailed view of galaxy halos. Monte-Carlo realisations of the merging histories are constructed of dark matter halos which have collapsed under their own self-gravity. Baryons are incorporated, as disk and spheroid components, and form stars according to simple rules based on local cooling and dynamical time-scales. The size of a galaxy is determined by ensuring that angular momentum is conserved in the collapsing cloud. The cloud forms with $\lambda_i \approx 0.05$, and attains virial equilibrium at a fraction λ_i of the radius at maximum extent. A simple Schmidt law is used for the star formation rate in the resulting disk, and cold gas infall continues to feed the disk. Spiral galaxies are relatively isolated galaxies that have not undergone a merger (that is, of comparable mass) within the past 10 Gyr or so. Elliptical galaxies are assumed to form in major mergers on a dynamical time scale: stars are assumed to form in a starburst. Semi-analytical galaxy formation is able to make realistic maps of the galaxy redshift distribution, and results in striking images of a cosmological volume containing hundreds of galaxies of varied morphological types and evolutionary states. Feedback is incorporated in order to prevent star formation from being too efficient. Once the feedback parameter is carefully adjusted, the galaxy luminosity function and cosmic star formation history can be derived. One of the most interesting results of semi-analytic galaxy formation theory has been to demonstrate that the Lyman break galaxies are highly clustered.

However, it is far from clear whether the model is robust or even physically meaningful, since exploration of the large parameter space for the gas and star formation physics has not hitherto proved to be feasible. The theory has not fared especially well at reproducing the low luminosity end of the galaxy luminosity function, when more realistic physics is incorporated. Not only do high resolution studies find that disk sizes are far too small because of angular momentum loss, but the rotation curves of low surface brightness dwarf galaxies, which are everywhere dark matter-dominated, cannot readily be explained. Cold dark matter simulations are generically cuspy and fail to explain the generally soft cores of low surface brightness dwarfs.

4 Analytic galaxy formation

A general discussion of analytic galaxy formation begins with present day star formation on galactic scales, and attempts to run star formation backwards in time. Several ingredients are needed: the initial stellar mass function (IMF), the star formation efficiency (SFE), and the star formation rate (SFR).

Consider the IMF. There is a state of confusion in the literature. First, there is little in the way of theory. Recent reviews claim a flattening below $\sim 0.3\mathrm{M}_\odot$, (Kroupa 2001, Gilmore 2001) from optical surveys and cluster data, yet the recent data from field NIR surveys find evidence for free-floating brown dwarfs characterized by a rise in $dN/dlogM$ roughly as M^{-1} (Luhmann et al. 2000).

Consider next the SFE. Theory has progressed considerably in recent years. However the adaptive mesh simulations can only resolve a single "star" and can say nothing about the efficiency of star formation. Feedback is considered to be an essential ingredient of galaxy formation simulations, in order to avoid such catastrophes as overcooling and excessive angular momentum loss. However no physically plausible paradigm has yet been quantified for physical feedback in terms of realistic interstellar physics that actually resolves these problems. On the phenomenological side, the star formation efficiency is considered to be high for spheroid formation and low for disk formation, an argument that goes back a long way (Gott and Thuan 1976). Major mergers are responsible for forming massive spheroids, and minor mergers and accretion fuel disk formation. Small spheroids plausibly form via secular instability of disks. At the same time spheroid formation precedes disk formation, as is evident from the increasing frequency of major mergers towards higher redshift, and substantial amounts of gas must remain available both to form disks and to give the frosting of recent star formation that many spheroids require, especially at high redshift.

Related to both the IMF and SFE is the star formation rate (SFR). It is usually assumed that the star formation rate in a self-gravitating gas-rich disk is regulated by gravitational instability according to whether the Toomre parameter

$$Q_g \approx \frac{\delta \kappa \sigma_g}{\pi G \Sigma_g}$$

is smaller or larger than unity, for instability or stability respectively. Here, κ is the epicyclic frequency, σ_g is the gas-velocity dispersion, Σ_g is the gas surface density, and δ is a parameter, of order unity for a gas-rich disk, that allows for the stellar contribution to the self-gravity of the gas component of the disk. Hα, HI and CO mapping of disks shows that $Q_g \sim 1$ in the star-forming regions of disk galaxies, suggesting that star formation is self-regulated by gravitational instability (Kennicutt 1989). This is plausible, since disk instabilities heat the disk and raise Q_g via tidal acceleration of clouds, whereas gas cloud collisional dissipation decreases the gas velocity dispersion, decreasing Q_g. Star formation occurs in the clouds that form by gravitational instability, grow in mass by coalescence, and become unstable to fragmentation. Q_g self-regulation predicts that there is a threshold to star formation. Star formation ceases when the gas surface density drops below a critical value

$$\Sigma_{crit} \approx \delta\kappa\sigma_g/\pi G.$$

Now $\kappa \approx 2^{\frac{1}{2}}\Omega$ for a flat rotation curve, $\Omega \approx 10^{-15}$ s^{-1} and $\sigma_g \sim 10$ kms^{-1} for the Milky Way, as a typical example, so that $\Sigma_{crit} \approx 10 M_{\odot}\mathrm{pc}^{-2}$. Clouds of giant HII regions, both in our galaxy and in nearby spirals, support this star formation threshold (Kennicutt 1998).

However Hα mapping (Ferguson et al. 1998) has shown that substantial star formation occurs at $\Sigma < \Sigma_{crit}$. The gravitational instabilty theory is too simplistic. Presumably, the star formation rate depends on gas volume density, rather than surface density. In this way, disk flaring would inhibit, and density waves would stimulate, star formation. Another complication must come from supernovae, which control the energy and momentum input that heats and stirs the interstellar medium. The rate of supernovae is controlled by the star formation rate, and therefore provides a natural means of self-regulating star formation on small scales. To properly implement self-regulation of star formation on large and small scales, one needs a two-phase model of the interstellar medium, consisting of the hot gas directly heated by supernova remnants and the colder gas that aggregates via gravitational instabilities and is compressed by supernova blast waves into dense shells and clumps where the bulk of the star formation occurs.

Cloud-cloud collisions provide an effective turbulent viscosity which is responsible for the formation and contraction of the disk. Viscous galactic disk models have a long history, commencing with the original proposal by Silk and Norman (1981), the study by Lin and Pringle (1987a) that derived exponential stellar surface density profiles, and the chemical evolution investigations of Clarke (1989,1991). In the cosmological context of cold dark matter halos, turbulent viscosity has been incorporated into disk models by Firmani et al. (1996). Non-axisymmetric gravitational instabiliies of a cold disk provide an effective turbulent viscosity (Lin and Pringle 1987b) that results in a turbulent velocity $\sigma_g \approx \Omega l_t$, where l_t is the characteristic gravitational instability scale and Ω is the disk rotational velocity.

The viscous drag time-scale is of the same order as the star formation time-scale and the dynamical friction time-scale that controls the loss of angular momentum from baryons to dark matter (Silk 2001). Since all time-scales are initially comparable, it is likely that self-regulation maintains this situation, since the gas supply controls star formation, viscous drag and the ability of the disk to form massive gas clouds by gravitational instability. Slyz et al. (2001) argue for implementing star formation on a viscous timescale in hydrodynamical simulations of disk galaxy formation and evolution. Modelling two-dimensional isolated disk galaxies, they verify that if the characteristic time scale for star formation is equal to the viscous time scale of disks, the resulting stellar profile is exponential on several scalelengths whatever the initial gas and dark matter profile. This casts new light on both numerical and semi-analytical disk formation simulations which either commence star formation in an already exponential gaseous disk, begin a disk simulation with conditions known to lead to an exponential, i.e. the collapse of a spherically symmetric nearly uniform sphere of gas in solid body rotation under the assumption of specific angular momentum conservation, or in simulations performed in a hierarchical context, try to tune feedback and star formation parameters to delay disk formation until the dark matter halos are slowly evolving and without much substructure so that the gas has the chance to collapse under conditions known to give exponential profiles. No matter what the initial conditions are in the protodisk, viscous evolution drives the final stellar profile towards an exponential on a timescale of order 1 Gyr. Star formation in these models generally follows a Schmidt law. With star formation prescribed on a viscous timescale, gas and star fractions after $\sim$ 12 Gyr are found to be consistent with observations without having to invoke an efficiency parameter for star formation. These results suggest that despite our gap in understanding the exact link between star formation and viscosity, the viscous timescale is the natural timescale for star formation.

Acknowledgments

It is a great pleasure to acknowledge my many years of fruitful collaborations with Jean Audouze. I look forward to many more.

References

1. T. Abel, G. Bryan and M. Norman, *Astrophys. J.* **540**, 39 (2000).
2. C. Clarke, *M.N.R.A.S.* **238**, 283 (1989).

3. C. Clarke, *M.N.R.A.S.* **249**, 704 (1991).
4. S. Cole,, C. G. Lacey, C. M. Baugh, & C. S. Frenk, *M.N.R.A.S.* **319**, 168 (2000).
5. R. Gott and T. Thuan, *Astrophys. J.* **204**, 649 (1976).
6. S. M. Fall and G. Efstathiou, *M.N.R.A.S.* **193**, 189 (1980).
7. A. Ferguson, R. Wyse, J. Gallagher and D. Hunter, *Astrophys. J.* **506**, 19 (1998).
8. G. Gilmore, to appear in *Starbursts: Near and Far*, Ringberg Castle, Germany, 10-15 September 2000 (eds. L.J. Tacconi and D. Lutz) astro-ph/0102189 (2001).
9. G. Kauffmann, J. M. Colberg, A. Diaferio,& S. D. M. White, *M.N.R.A.S.* **303**, 188 (1999).
10. R. Kennicutt, *Astrophys. J.* **344**, 685 (1989).
11. R. Kennicutt, *Astrophys. J.* **498**, 541 (1998).
12. P. Kroupa, to appear in *Modes of Star Formation*, E.Grebel, W.Brandner (eds.), ASP Conf. Series, astro-ph/0102155 (2001).
13. C. Lacey, & J. Silk, *Astrophys. J.* **381**, 14 (1991).
14. S. Lilly et al. *Astrophys. J.* **500**, 75 (1998).
15. D. Lin and J. Pringle, *Astrophys. J.* **320**, 87 (1987a).
16. D. Lin and J. Pringle, *M.N.R.A.S.* **225**, 607 (1987b).
17. K. Luhman et al., *Astrophys. J.* **540**, 1016 (2000).
18. J. Navarro and M. Steinmetz *Astrophys. J.* **478**, 513 (1997).
19. J. Silk, *MNRAS*, in press (2001).
20. A. Slyz et al., preprint (2001).
21. R. S. Somerville, & J. R. Primack, *M.N.R.A.S.* **310**, 1087 (1999).
22. M. Steinmetz and J. Navarro, *Astrophys. J.* **513**, 550 (1999).
23. N. Vogt, in *Dynamics of Galaxies: from the Early Universe to the Present,* Combes, G. A. Mamon, V. Charmandaris (eds.) ASP Conf. Series, Vol. 197, p. 435 (2000).
24. M. Weil, V. Eke and G. Efstathiou, *M.N.R.A.S.* **300**, 773 (1998).
25. S. D. M. White, & C. S. Frenk, *Astrophys. J.* **379**, 52 (1991).

THE GZK CUTOFF IN THE LIGHT OF THE GALAXY DISTRIBUTION

PASCAL CHARDONNET

LAPP LAPTH, BP 110 F-74941 Annecy-le-Vieux Cedex, FRANCE
E-mail: chardon@lapp.in2p3.fr

1 Introduction

The past decade has seen tremendous progress in the search for the high energy cosmic rays [1]. Conclusive detections from different experiments present evidences indicating that the events above the GZK bound are statistically relevant. It has long been hypothesized that the GZK cutoff could arise from a local overdensity in our neighborhood. Our ignorance of the cosmic rays sources distribution is the principal obstacle for this mystery. Therefore, if we assume that these sources follow the distribution of matter in the universe we can examine the question more quantitatively. In order to trace the distribution of matter we will analyze how the galaxies are distributed. Such analyze was made and the answer was that the absence of the GZK cutoff could be explained by the overdensity observed in the CfA survey [2]. Emphasis will be given here that there is no such local overdensity as claimed. I put forward an interpretation that the error was an incorrect computation of the number of galaxies and the effect of the abundant dwarf galaxies that make a bias effect.

2 The overdensity computed with the UZC survey

I present here a study of the local density based on the optical survey UZC [3]. This survey is magnitude limited up to 15.5. I have computed the overdensity as a function of redshift. This parameter is crucial for the cosmic ray propagation. The method used is the standard one with a Schecheter luminosity function $\Phi(L) = \Phi_\star\,(\frac{L}{L_\star})^{-\alpha}\,exp(-\frac{L}{L_\star})$. Here I have used the following parameters: $\alpha = -1.0$, $M_\star = -18.80$ and $\Phi_\star = 0.040 Mpc^{-3}$. Finally, the expected number of galaxies in a given volume is $dN(z) = n(z)\,dV\,\int_{L_{min(z)}}^{\infty}\Phi(L)\,dL$ where $n(z)$ is the spatial density of galaxies at the redshift z and $L_{min(z)}$ is a function of the distance $r(z) \simeq \frac{cz}{H_o}$, where H_o is the Hubble constant and c the celerity of light.

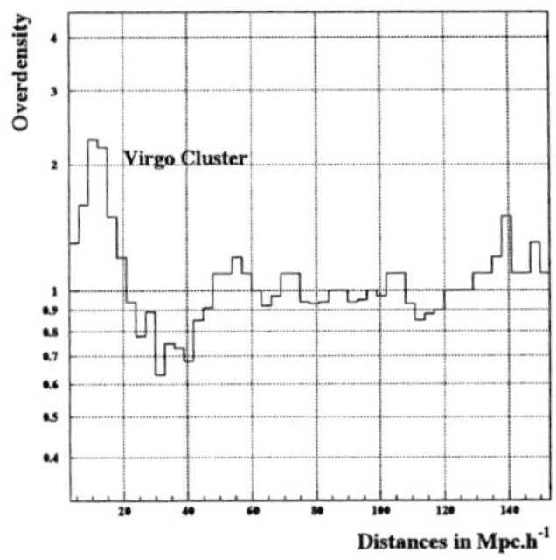

Figure 1. The overdensity computed with the UZC survey. Faint galaxies excluded.

3 Results and conclusions

Dwarf galaxies are the most numerous type of galaxies in the universe. Then, we naturally expect a local overdensity due to both facts that they are the most abundant galaxies type and that we miss the distant one. I have made a cut in absolute magnitude to reject the faint galaxies in the distribution. Figure 1 shows the overdensity computed with the UZC survey with the faint galaxies excluded. We conclude that there is no overdensity that can explain the presence of the GZK cutoff contrary to what is claimed previously. For a detail study of this problem see reference 4. As a final remark we may express a wish for the cosmic ray community to develop a network like it was done for GRBs to search the cosmic rays also in the other wavelengths and perhaps find precious information about the distance of cosmic rays events.

Acknowledgments

It's a pleasure to thank V. de Lapparent for useful discussions. I want to thank my collaborators V. Berezinsky and A. Doroshkevich. This work was supported by a délégation CNRS.

References

1. Nagano, M. and Watson, A.A. 2000, Rev. Mod. Phys., 72, 689
2. Medina-Tanco, G. (1999) ApJL , 510, 91
3. Falco, E. (1999) PASP, 111, 438
4. Berezinsky, V., Chardonnet, P. and Doroshkevich, A. work in preparation

OBSERVATIONAL SIGNATURES OF THE FIRST STARS

APARNA VENKATESAN

CASA, Department of Astrophysical and Planetary Sciences,
University of Colorado, 389 UCB, Boulder, CO 80309-0389, USA
E-mail: aparna@casa.colorado.edu

JAMES W. TRURAN

Department of Astronomy and Astrophysics, 5640 S. Ellis Ave.,
University of Chicago, Chicago, IL 60637, USA
E-mail: truran@nova.uchicago.edu

At present, there are several feasible observational probes of the first stars in the universe. Here, we examine the constraints on early stellar activity from the metallicity of the high-redshift Ly-α clouds, from the effects of stellar ionizing photons on reionization and the cosmic microwave background (CMB), and from the implications of gravitational microlensing results for the presence of stellar remnants in galactic halos. We also discuss whether the above signatures are consistent with each other, i.e., if they reflect the same population of stars.

The epoch and sites of formation of the first stars in the universe are some of cosmology's most intriguing questions today. Observations of the most distant quasars and galaxies to date imply that hydrogen reionization of the intergalactic medium (IGM) occurred before redshifts, $z \sim 6$. While a variety of astrophysical objects can reionize the IGM, the apparent decrease in the space density of quasars beyond $z \sim 3$ implies that early stars must play some role in reionization; furthermore, they may account for the persisting trace levels of metal enrichment seen in the high-z Ly-α clouds (see Venkatesan[1] for the relevant references). Here, we will focus on three specific probes of early stellar activity: stellar remnants, which may contribute to galactic halo dark matter (DM) in the form of Massive Compact Halo Objects (MACHOs); the influence of stellar ionizing photons on the CMB; and, the metal enrichment of the IGM. In particular, one may ask how each of these constrains $f_\star$, the fraction of baryons that participate in any early episodes of star formation.

• **Stellar Remnants**: Recent observations of gravitational microlensing towards the Large and Small Magellanic clouds from several collaborations, including MACHO and EROS, indicate that the lenses, if they are present in the Galactic halo, have a most likely mass $\sim$ 0.5–1 $M_\odot$, and a halo mass fraction of about 0.2–0.3 (see, e.g., Kerins[2], and references therein). This has renewed theoretical interest in the viability of baryonic DM candidates, which include white dwarfs (WDs; see, e.g., Fields, Mathews & Schramm[3]), and neutron stars/black holes (NSs/BHs; Venkatesan, Olinto & Truran[4]; see also Fields, Freese & Graff [5] for a review of various models for baryonic DM). Although these models (just barely) satisfy a number of constraints, almost all of the baryons must either participate in an early burst of star formation that creates the MACHOs, or assist in dilution of the stellar ejecta to avoid overpollution in ^{4}He or metals; in effect, $f_\star \geq 0.8$, when dilution requirements are included.

• **Stellar Radiation and the CMB**: The first stars generate ionizing photons that increase the free electron population in the IGM; these electrons then interact with the CMB via Thomson scattering, which results in an overall damping of the CMB's primary temperature anisotropies. Also created are new secondary temperature anisotropies and a linear polarization signal, which is a relatively clean probe of the reionization epoch. Analysis of CMB data can yield constraints on a chosen set of cosmological parameters (CPs) and the optical depth to reionization, τ_e. If we further adopt a stellar reionization model for τ_e, then well-known degeneracies in CMB parameter extraction can be broken; also, limits on CPs can be used to constrain astrophysical quantities such as $f_\star$ (see Venkatesan[1], and references therein). Strong constraints on $f_\star$ within individual star-forming halos are possible, in principle, by this method with polarization data from future CMB experiments such as $PLANCK$ (see, e.g., Figure 13 of Venkatesan[1]).

• **Metal Production**: The IGM at $z \sim 3$ has an average enrichment, detected in carbon, of $Z_{\rm IGM} \sim 10^{-2} Z_\odot$; this requires $f_\star \sim 1\%$, if we assume that most of the baryons at $z \sim 3$ are in the Ly-α clouds. It can be shown that this value of $Z_{\rm IGM}$ implies an average number of stellar ionizing photons per baryon in the universe, $N_{\rm Lyc}/b \sim 10$, which again points to a possibly significant part played by the first stars in reionization. Note, however, that if this calculation were repeated using only the carbon (and not total metal) yields from low-Z massive stars, then $N_{\rm Lyc}/b \sim 360$! This implies that either the reionizing stars made a negligible contribution to the IGM carbon detected at $z \sim 3$, or that in enriching the IGM, they generated a tremendous amount of ionizing radiation.
There are other drawbacks to relating stellar ionizing activity to $Z_{\rm IGM}$ through carbon: the photons relevant for reionization come from massive stars (≥ 10 M$_\odot$), while carbon is produced dominantly by 2–6 M$_\odot$ stars. Thus, different regions of the stellar IMF are being probed, and may not be mutually constraining if the IMF was different in the past. Secondly,

for burst-driven star formation (i.e., not continuous), most of the carbon is produced an order of magnitude in time *after* the epoch, set by the timescales associated with massive stars, of ionizing photons and Type II supernovae (which could drive galactic outflows). The question then arises of the mechanism by which the carbon is expelled from individual halos and distributed so ubiquitously in the IGM by $z \sim 3$.

To summarize: [1] In order to explain the observations of MACHOs as stellar remnants, most of the baryons must either participate in an early burst of star formation or assist in the dilution of enriched stellar ejecta to $\sim$ solar levels. For both WD and NS/BH MACHO models, the required $f_\star$ far exceeds that needed for stellar reionization. [2] Strong limits on $f_\star$ from reionization alone are possible with polarization data from future CMB experiments such as $PLANCK$, but this may be hampered by foregrounds. [3] Combining reionization and metal enrichment constraints, it is possible for the same population of stars to reionize the universe in hydrogen by $z \sim 6$, and produce $Z_{\rm IGM} \sim 10^{-2} Z_\odot$ in carbon at $z \sim 3$, if $f_\star \sim$ 1–2% and star formation is continuous. [4] If star formation occurred in bursts, it is unclear how to relate ionizing stellar activity (from short-lived massive stars) to the carbon abundance in the IGM (from intermediate-mass stars). The detection of elements which are the products of massive stars, e.g., oxygen or silicon, in absorption-line systems at $z \geq 3$, and a comparison of the values of $Z_{\rm IGM}$ implied by their abundances with that measured in carbon should help to consistently address whether the reionizing stellar population generated the metals observed in the high-z IGM. The details of the calculations on this connection can be found in our forthcoming work (Venkatesan & Truran[6]).

Acknowledgments

A.V. gratefully acknowledges support from NASA LTSA grant NAG 5-7262.

References

1. Venkatesan, A. 2000, ApJ, 537, 55
2. Kerins, E. 2000, in Cosmological Physics with Gravitational Lensing, ed. J.-P. Kneib et al., in press (preprint at astro-ph/0007137)
3. Fields, B. D., Mathews, G. J., & Schramm, D. N. 1997, ApJ, 483, 625
4. Venkatesan, A., Olinto, A. V., & Truran, J. W. 1999, ApJ, 516, 863
5. Fields, B. D., Freese, K., & Graff, D. S. 1998, New Astronomy, 3, 347
6. Venkatesan, A. & Truran, J. W. 2001, in preparation

EVOLUTION OF LOW-MASS POPULATION III STARS

ACHIM WEISS AND HELMUT SCHLATTL
Max-Planck-Insitut für Astrophysik, Karl-Schwarzschild-Str. 1, 85748 Garching, Germany
E-mail: weiss@mpa-garching.mpg.de; schlattl@mpa-garching.mpg.de

SANTI CASSISI
Osservatorio Astronomico di Collurania, Via M. Maggini, I-64100 Teramo, Italy
E-mail: cassisi@astrte.te.astro.it

MAURIZIO SALARIS
Astrophysics Research Institute, Liverpool John Moores University, Twelve Quays House, Egerton Wharf, Birkenhead CH41 1LD, UK
E-mail: ms@astro.livjm.ac.uk

We have followed the evolution of low-mass, initially metal-free stars from the zero-age main sequence to the white dwarf stage under the assumption of pollution with external, metal-rich material and in the presence of diffusion. We identify conditions under which helium-flash induced mixing between the envelope and carbon-rich layers occurs and compare the resulting surface abundances and post-flash stellar parameters with those of observed extremely metal-poor stars.

1 A scenario for the population of ultra metal-poor halo stars

We are following the idea that the ultra metal-poor stars of the halo, as observed in the HK-survey[1] and follow-up projects, are low-mass stars of the first generation. Assuming that such stars formed out of primordial material, the fact that they now have $[Fe/H] \geq -4$ we take as evidence for post-formation pollution with stellar material, for example from nearby type II supernovae, such that only the (convective) envelope gets metal-enriched. In our first paper[11] we followed the evolution of such models up to the red giant tip, which for zero central metallicity is at much lower luminosity than compared to standard Pop. II stars. In this contribution we report about the results of calculations, where we followed the evolution through the core helium flash terminating the RGB-phase, and about the resulting surface abundances. The intention is to compare post-flash models with observed carbon-enriched metal-poor stars. The observational evidence[1] is that 10% of stars with $[Fe/H] \leq -2.5$ show a carbon overabundance of $[C/Fe] \geq 1.0$, this fraction rising to 25% at $[Fe/H] \leq -3.0$, with the excess increasing to $[C/Fe] = 2$ at the lowest metallicities[7]. In addition, nitrogen is enhanced to

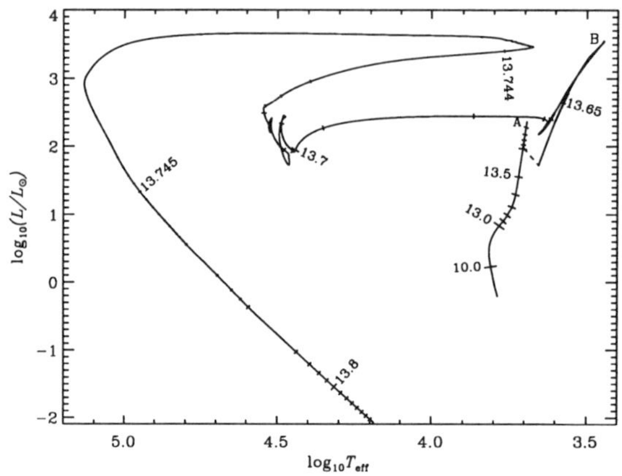

Figure 1. Evolutionary track of our most probable ultra metal-poor object candidate. At point A, the primary helium flash and the induced mixing occurs. Later phases depend crucially on the assumed mass loss rate (Reimers with $\eta = 0.3$ here) and should be taken only as a possibility. Ages in Gyr are given along the track.

similar levels, while oxygen remains normal with respect to iron, i.e. enhanced by ≈ 0.4 dex as is typical for Pop. II. No standard stellar source is known to produce such abundances[10] and neither do so massive Pop. III supernovae (A. Heger, private communication). Extended results will be published in a forthcoming paper[8].

2 Helium-flash mixing and comparison with observations

Hollowell et al.[4] found that the core helium flash in metal-free low-mass stars can lead to a mixing between the convective envelope and the helium-burning regions, resulting in surface abundances strongly enhanced in carbon and nitrogen, the latter element being produced in CN-cycle reactions of protons mixed from hydrogen-rich layers into the carbon-enhanced helium-layer. Fujimoto et al.[3] showed that lowest initial metallicities and low stellar mass favour the occurrence of this Helium-Flash Mixing (HEFM). In our calculations we investigated this event for our polluted stars and found that it depends on additional parameters: a higher initial helium content ($Y_{\rm p} = 0.24$ instead of 0.23, our standard value) and pollution tend to disfavour it, while diffusion enhances the probability for it. A giant star presumed to be of cosmological age ($\approx$ 13 Gyr) will have a mass of about $0.8\,M_\odot$. With a pollution of an additional $3 \cdot 10^{-4}$ to $0.01\,M_\odot$ ($Z = 0.02$) and diffusion, which is our *most probable* candidate for an observed ultra metal-poor object, we find that the HEFM *does* occur. A sample evolutionary track is shown in Fig. 1.

For a first preliminary comparison we chose the star CS22892-052. Tab. 1 shows that models immediately after the helium flash (second line) or at a

Table 1. Observational properties of the extremely metal-poor star CS22892-052 and a theoretical model after the HEFM. $\log L/L_\odot$ of CS22892-05 was derived from $T_{\rm eff}$ and $\log g = 1.5$ taken from Sneden et al.[10] by assuming the model mass of $0.82\,M_\odot$. Its age estimate is from Cowan et al.[2], and abundances are from Norris et al.[5,6].

	Age [Gyr]	$T_{\rm eff}$ [K]	$\log L/L_\odot$	[C/Fe]	[N/Fe]	[Fe/H]
CS22892-052	15.6 ± 4.6	4760	2.5	1.1	1.0	-3.0
$M/M_\odot = 0.82$	13.65	4500	1.7	3.9	4.1	-3.3
	13.68	4700	2.4	4.0	4.2	-3.3

luminosity as estimated for CS22892-052 (third line) are approximately at the right locations in the HRD. The pollution was assumed to be such that the observed [Fe/H] is obtained. Although [C/N] is in good agreement with the observations, there clearly is too much of both elements produced. Our calculations indicate that this amount is rather independent of the other parameters (as composition of pollution material), and thus our models fail quantitatively. However, qualitatively the results are in the desired direction. As there is no other source of stellar nucleosynthesis resulting in carbon and nitrogen abundances as those observed, we think it worthwhile to continue our line of models.

Acknowledgments

We are grateful to J. Audouze and J. Truran for their contributions to the subject of this conference. A. Weiss thanks the Institute for Advanced Study for its hospitality and the Fulbright Foundation for financial support.

References

1. T. C. Beers, Astrophys. and Space Science, Vol. 265, 547 (1999).
2. J.J. Cowan, S.G. Pfeiffer, K.-L. Kratz, *et al.*, ApJ 521, 194 (1997).
3. M.Y. Fujimoto, Y. Ikeda and I. Iben, Jr., ApJ 529, L25 (2000).
4. D. Hollowell, I. Iben, Jr. and M.Y. Fujimoto, ApJ 351, 245 (1990).
5. J.E. Norris, S.G. Ryan and T.C. Beers, ApJ 488, 350 (1997).
6. J.E. Norris, S.G. Ryanand T.C. Beers, ApJ 489, L169 (1997).
7. S. Rossi, T.C. Beers and C. Sneden, in ASP Conf. Ser. 165, 264 (1999).
8. H. Schlattl, S. Cassisi, M. Salaris and A. Weiss, ApJ, submitted (2001).
9. C. Sneden, A. McWilliam, G.W. Preston, *et al.*, , ApJ, 467, 819 (1996).
10. F.X. Timmes, S.E. Woosley, S. E. and T.A. Weaver, ApJS 98, 617 (1995).
11. A. Weiss, S. Cassisi, H. Schlattl and M. Salaris, ApJ 533, 413 (2000).

COSMIC EVOLUTION: A FEW CONCLUDING REMARKS

W. HILLEBRANDT
Max-Planck-Institut für Astrophysik, Karl-Schwarzschild-Str.1, D-85748 Garching, Germany
E-mail: wfh@mpa-garching.mpg.de

Cosmic evolution, as we see it, involves a variety of different length- and time-scales, ranging from the micro-cosmos we experience on Earth to scales of billions of years and light-years which constitute the macro-cosmos around us. This Conference was devoted to many questions relevant on all those scales: Nuclear reactions that are responsible for the generation of energy in stars, but can be studied in the laboratory; stellar evolution from birth to death, studied observationally and by computer simulations; formation and evolution of galaxies, large scale structures, and the entire Universe from the Big Bang until today. The common link between all these seemingly rather different topics is our curiosity, best reflected in the scientific lives of two eminent colleagues, Jean Audouze and Jim Truran, who's 60th birthdays were celebrated. An attempt is made to give a brief account of some of their most important contributions to all these fields which, due to the lack of space, has to be a very incomplete summary. Moreover, some of the recent developments, presented at the conference, will be mentioned briefly.

1 Introduction

A conference on cosmic evolution, as well as its summary, obviously has to address the following questions: *Where do we come from? Where do we stand? Where do we go to?* For the two jubilees, in contrast to most other known cosmic objects (but in accord with many of their terrestrial colleagues) these questions are easy to answer: One grows up in a group of seniors and peers (Fig.1) where one learns to think, and to behave well. Next a more mature state is reached where one spreads words of wisdom (Fig.2), meant to last for many generations to come. Finally, a couple of more years down the road, one is asked to give summary talks and/or after-dinner speeches, considers conferences as being social events, places to meet old friends over a glass of wine (Fig.3), and remembers the good old days..... But of course, progress in science is not too different from this general scheme, as will be demonstrated in the following sections.

Figure 1. The Institute of Theoretical Astronomy, Cambridge, Summer 1970. *Front row from left:* P.C.N. Vaugon, S.W. Hawking, M.J. Rees, F. Axisa, E.M. Burbidge, F. Hoyle, V.I. Trimble, G.R. Burbidge, G.P. Steigmann, R.F. O'Connell. *Middle row:* J. Bergeron, J. Audouze, J. Danziger, J.M. Sutton, S.J. Aarseth, W.B. McAdam, K. Fricke, J.V. Narlikar, N.C. Wickramasinghe, J.N. Islam, W.D. Arnett. *Back row:* D.C. Heggie, S.H. Plagemann, P.C.W. Davies, T. De Graaf, S. Bonometto, J.E. Felten, P.P. Eggleton, A.M. Wolfe, G. Setti, D.O. Gough. (From Douglas Heggie's homepage.)

2 Cosmology, the Big-Bang, and Light Element Nucleosynthesis

In the past several years cosmology has evolved into an empirical science. First high-resolution maps of small parts of the sky in micro-waves by the balloon experiments MAXIMA and BOOMERANG indicate that the Universe is Euclidian on large scales, with an uncertainty of a few percent only [1] (see also Melchiori, this volume). Observations of Type Ia supernovae, in addition, provide evidence that this flat Universe is made of about 30% matter, mostly in form of yet unknown weakly interacting "dark" particles, and

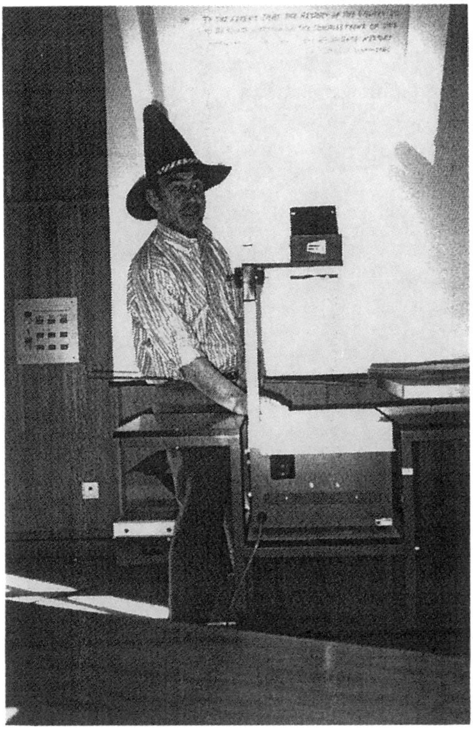

Figure 2. Jim Truran, presenting his deep thoughts at a Ringberg Castle workshop in 1998.

of 70% either in form of a cosmological constant or an (also unknown) energy with negative pressure [2,3,4]. Moreover, there can be no doubt that the contribution of baryons to the energy and mass content of the Universe is small, of the order of 1 to a few percent only (e.g., Olive, this volume). The latter conclusion is reached from the now generally accepted primordial ^{7}Li abundance ($\simeq (1-4) \times 10^{-10}$ (Bonifacio & Molaro [5]; see also Cayrel and Lambert, this volume) and the still somewhat controversial ^{2}H abundance ($\simeq (2-5) \times 10^{-5}$ (Tytler *et al* [6]) or $(1-2) \times 10^{-4}$ (Rugers & Hogan [7]), giving $\Omega_B \simeq 0.05$ and 0.01, respectively, although most observers present at the conference seemed to favor low D-values), together with standard Big-Bang nucleosynthesis. Weaker constraints are obtained from the primordial abundances of ^{4}He and ^{3}He, but recent ^{3}He determinations again seem to favor the low-D high-Ω_B solution [8]. Many talks at the meeting dealt with these

Figure 3. A group of senior scientists (from right: Friedel Thielemann, Dmitrij Nadyozhin, Jim Truran, and the author) thinking about life, the Universe, and everything.

numbers and related topics (e.g. Kolb, Salati, Melchiori, Madjumdar, Olive, Jedamzik, Vidal-Madjar, O'Meara, Livio, Silk, Wasserburg, and others), and their summaries can be found in these proceedings.

In fact, it is interesting to compare our present knowledge on primordial light element abundances with what was known about 30 years ago when Jean and Jim wrote their first papers on this subject. An approximate value of the deuterium abundance at the time of formation of the solar system was known, but only upper limits for its value in the interstellar medium, and nothing for QSOs [10]. The solar value of ^{3}He was known, but it was totally unclear whether it survived from the big bang was destroyed in matter ejected stars [11], and the situation was not much better for the other light elements under consideration. Nevertheless, in those early papers, it was stated with certainty that D, ^{3}He, ^{4}He, and, possibly, ^{7}Li originated in the Big Bang, and ^{6}Li, ^{9}Be, and ^{10,11}B stem from interactions between the galactic cosmic rays and interstellar matter [9,10], but with a bit more caution in Audouze & Tinsley [12], leaving some room for galactic chemical evolution.

While the general picture has survived until today, new problems emerged which cast first shadows on the standard model of cosmology. The models of inflation need a lot of fine-tuning in order to work. Searches for the dark matter have not been successful yet. The cosmological constant (or dark energy), if interpreted as the energy of the vacuum, is so embarrassingly small that one can hardly belief in it. Here one is reminded of the situation prior to the discovery of Special Relativity when one had to invent an "ether" in order to explain the Michelson experiments in the frame-work of Newtonian physics. As far as Big Bang Nucleosynthesis is concerned, the question remains with us, whether or no systematic errors affect the abundance determinations and whether or no we have all possible evolutionary effects under control (e.g., Rood, Tosi, and Galli; these proceedings).

3 Nucleosynthesis and Chemical Evolution

As compared to the early days of nucleosynthesis and chemical evolution, notably documented in Jean's and Jim's early papers, our knowledge of nucleosynthesis and the chemical evolution of galaxies and, in particular, of our Milky Way Galaxy, has increased considerably over the past several years. This was mainly due to new telescopes on ground and in space (Keck, HST, VLT, ...) which together with high-resolution spectroscopy offer a huge data base of abundances in nearly all astronomical objects, from individual stars to clusters of stars and beyond, and in the diffuse interstellar and intergalactic gas, in all wave-bands of the electromagnetic spectrum (which has the side effect that there is an increasing need for people taking data, having a quick look at them, and then archiving them, with no time and manpower left for a real and deep understanding). At the same time accelerator astrophysics and theoretical modeling has increased the nuclear data base considerably, the NACRE library of nuclear reaction rates being an outstanding example [13]. Finally, due to the ever increasing computer power and improved "input" physics, e.g. multi-D hydrodynamics and radiative transfer, also good stellar and galactic evolution models have become available.

Many talks at the conference demonstrated this progress, from new abundance determinations of light elements and their interpretation (see, e.g., the contributions of Gayrel, Gratton, D'Antona, Asplund, Lambert, Cassé, Cunha, Duncan, Primas, Suzuki, Nadyozhin), heavy element nucleosynthesis in massive stars, novae and supernovae (e.g. Paul, Turck-Chièze, Hernanz, Thielemann, Knödlseder), to the s- and r-process (e.g. Sneden, Matthews, Cowan, and Arnould), but also left many questions open. On the observational side, it is still unclear whether or no the huge scatter in the elemental abundances, seen in individual stars of seemingly similar age, location and mass, is real or reflects uncertainties in the abundance determinations (and this question has been with us since many years [14]). The new high-quality data do indicate that most of the scatter is real which, actually, does not come as a big surprise since chemical evolution is a local process, after all.

Therefore, more information seems to be provided by the scatter than by some "average" trends. For example, it could tell us about additional parameters that govern stellar nucleosynthesis and surface abundances, such as rotation.

As far as nuclear reactions rates are concerned, according to Thielemann, Arnould & Truran [15], most of them have are known to within a factor of two since already about 15 years ago. Therefore the question is: Do we need better rates? Here a recent comparison of stellar abundances predicted from two sets of commonly used reaction rate libraries has demonstrated that there are still significant deficiencies, even in seemingly simple cases such as (n,γ)-rates of heavy nuclei [16]. Therefore, despite of the recent progress, the quest for better reaction rates is still with us.

The same also holds for stellar evolution and explosion models. It has become apparent that spherically symmetric models are largely over-simplifying the problem. Time-dependent non-local convection is the best known example, mixing by shear instabilities is an other one, and Rayleigh-Taylor instabilities in novae and supernovae may serve as a third example. All these examples have in common that due to the big length-scales in stars (and even more so in the interstellar matter) the Reynolds number, which measures the relative importance of inertia and viscous dissipation, is huge, of the order of 10^{14} or larger. This, in turn, means that astrophysical flow-fields are nearly always highly turbulent and non-linear, and that the relevant length-scales typically span up to 10 orders of magnitude in linear dimension. Direct (3-dimensional!) numerical simulations, therefore, are impossible, even on the fastest computers one can imagine, and one has to *model* the unresolved scales which, however, is hardly ever done in astrophysical simulations.

As a consequence nuclear abundances predicted by stellar models should be taken with care, and putting them into evolutionary models of the Milky Way galaxy is even more risky. Several detailed models have been presented at this conference, trying to shed light an a variety of questions (e.g., by Pagel, Henry, Prantzos, Burkert, Matteucci, and others) but because of the inherent uncertainties discussed above, the "predictive power" of such models, unfortunately, is still not very high. In fact, most models do not aim at predictions but rather try to "fit the observations".

4 Cosmic Evolution and the High-Z Universe

The evolution of the high-redshift Universe, including the formation of galaxies and large-scale structures, and the formation and evolution of the first generation of stars, for many years has mainly been a field for theorists but, recently, begins to be driven by observations. Here, in particular, the Sloan

Figure 4. Jean and Jim over a full-size glass of beer at a monastery in Southern Bavaria. That is were and how many great ideas are born!

Digital Sky Survey needs to be mentioned, early results being presented by Burles and York, or new observations with the ESO-VLT and with Keck. Several of these questions were addressed during the Conference (e.g., in the talks of White, Burkert, Silk, Venkatesan, Weiss, and others), but they remain largely unanswered. Still we do not know when and where the first stars formed, how the super-massive black holes in the centers of galaxies came into being, what the γ-ray bursters are, if there was a first generation of super-massive stars, and one could continue with this list.

Hope is left that we will be able to resolve several of these problems in the coming years, mainly through new observational techniques. One example may exemplify the gain of information expected to become available in a few years. Cobe gave the fist all-sky map of the CMB temperature fluctuations with a resolution of about 7° at a level of 10^{-5}. Planck-Surveyor, to be launched in 2007, in contrast, will have a 50 times higher resolution and ten times more sensitivity. It will not only help to determine cosmological parameters with much higher accuracy then possible today (which, however, may not be a serious issue any more by that time), but it will, in addition, detect thousands of galaxy clusters and distant radio sources in between us and the redshift of recombination.

5 A Final Word

For me, like for most of us, Jean and Jim are (and have always been) colleagues whom we admire because of their outstanding ability to ask the right questions at the right time, and to find ways to answer them. And, even more important, they are also good friends who not only share with us their love of science, but also their willingness to enjoy life.

Happy Birthday, Jean and Jim!

References

1. A.H. Jaffe *et al*, astro-ph/0007333; *Phys. Rev. Lett.* , (2000), in press.
2. S. Perlmutter *et al*, APJ **517**, 565 (1999).
3. B.P. Schmidt *et al*, APJ , 507 (46)1998.
4. A.G. Riess *et al*, *Phys. Rev.* D **45**, 4088 (1992).
5. P. Bonifacio, P. Molaro, MNRAS **285**, 847 (1997).
6. D. Tytler *et al*, Physica Scripta **85**, 12 (2000).
7. M. Rugers, C. Hogan, APJ **459**, L1 (1996).
8. G. Gloeckler, J. Geiss, Space Sci. Rev. **98**, 275 (1996)
9. M. Menegguzzi, J. Audouze, H. Reeves, A&A **15**, 337 (1971)
10. H. Reeves *et al*, APJ **179**, 909 (1973)
11. J.W. Truran, A.G.W. Cameron, Astrophys. Space Sci. **14**, 179 (1971)
12. J. Audouze, B.M. Tinsley, APJ **192**, 487 (1974)
13. C. Angulo *et al*, Nuc. Phys. **A656**, 3 (1999)
14. J.C. Wheeler, C. Sneden, J.W. Truran, ARA&A **27**, 279 (1989)
15. F.-K. Thielemann, J.W. Truran, M. Arnould, in Advances in nuclear astrophysics; Proceedings 2nd IAP Workshop, Paris, France, Editions Frontieres, p. 525, 1986.
16. R.D. Hoffman *et al*, APJ **521**, 735 (1999)

LIST OF PARTICIPANTS

NAME	FIRST	INSTITUTE	COUNTRY	EMAIL
ALIBÉS	Andreu	Universidad de Barcelona	SPAIN	aalibes@am.ub.es
ANDERSEN	Johannes	Niels Bohr Institute	DENMARK	ja@astro.ku.dk
ARNOULD	Marcel	Université Libre de Bruxelles	BELGIUM	marnould@astro.ulb.ac.be
ASPLUND	Martin	Astronomical Observatory	SWEDEN	martin@astro.uu.se
AUDOUZE	Jean	Palais de la Découverte/I.A.P.	FRANCE	audouze@iap.fr
BALSER	Dana	N.R.A.O.	USA	dbalser@nrao.edu
BANIA	Thomas	Inst. Astrophys. Res.	USA	bania@bu.edu
BASDEVANT	Jean-Louis	Ecole Polytechnique	FRANCE	basdevan@poly.polytechnique.fr
BASTERO-GIL	Mar	Scuola Normale Sup. Pisa	ITALY	
BERTOUT	Claude	I.A.P.	FRANCE	bertout@iap.fr
BJORNSSON	Gunnlaugur	University of Iceland	ICELAND	gulli@raunvis.hi.is
BLANC	Guillaume	C.E.A./Saclay	FRANCE	gblanc@cea.fr
BLUDMAN	Sydney	DESY	GERMANY	bludman@mail.desy.de
BRAGAGLIA	Angela	Osserv. Astron. di Bologna	ITALY	angela@bo.astro.it
BROUSSE	Edouard	INSA	FRANCE	
BURKERT	Andreas	M.P.I. für Astronomie	GERMANY	burkert@mpia-hd.mpg.de
BURLES	Scott	FermiLab/Chicago	USA	sburles@fnal.gov
CALOI	Vittoria	Istituto Astrofisica Spaziale	ITALY	caloi@ias.rm.cnr.it
CANAL	Ramon	Universidad de Barcelona	SPAIN	ramon@am.ub.es
CARRETTA	Eugenio	Osserv. Astron. di Padova	ITALY	carretta@pd.astro.it
CASSÉ	Michel	SAp/C.E.A./Saclay and I.A.P.	FRANCE	casse@iap.fr
CAYREL	Roger	Observatoire de Paris	FRANCE	roger.cayrel@obspm.fr
CHANG	Ruixiang	Shanghaï Observatory	CHINA	crx@center.shao.ac.cn
CHARDONNET	Pascal	L.A.P.T.H.	FRANCE	chardon@lapp.in2p3.fr
CHIÈZE	Jean Pierre	Sap/CEA	FRANCE	chieze@cea.fr
CLAUDI	Riccardo	Osserv. Astron. di Padova	ITALY	claudi@pd.astro.it
COC	Alain	C.S.N.S.M.	FRANCE	coc@csnsm.in2p3.fr
CORDIER	Bertrand	S.Ap/C.E.A./Saclay	FRANCE	bcordier@cea.fr
COWAN	John	University of Oklahoma	USA	cowan@mail.nhn.ou.edu
CUNHA	Katia	Observatorio Nacional	BRAZIL	katia@gauss.on.br
D'ANTONA	Francesca	Osserv. Astron. di Roma	ITALY	dantona@coma.mporzio.astro.it
DELBOURGO-SALVADOR	Pascale	C.E.A./Bruyères Le Châtel	FRANCE	delbo@bruyeres.cea.fr
DELIYANNIS	Constantine	Indiana University	USA	con@athena.astro.indiana.edu

NAME	FIRST	INSTITUTE	COUNTRY	EMAIL
DUNCAN	Douglas	University of Chicago	USA	duncan@oddjob.uchicago.edu
FALL	Michael	Space Telescope Science Inst.	USA	fall@stsci.edu
FERLET	Roger	I.A.P.	FRANCE	ferlet@iap.fr
FERRINI				
FRIEDJUNG	Michael	I.A.P.	FRANCE	friedjung@iap.fr
GALLI	Daniele	Osserv. Astronomico di Arcetri	ITALY	galli@arcetri.astro.it
GARCIA-PEREZ	Ana Elia	Astronomical Observatory	SWEDEN	aegp@astro.uu.se
GLASNER	Shimon Ami	The Hebrew University	ISRAEL	ami@saba.fiz.huji.ac.il
GORIELY	Stéphane	Université Libre de Bruxelles	BELGIUM	sgoriely@astro.ulb.ac.be
GRATTON	Raffaele	Osserv. Astronomico di Padova	ITALY	gratton@pd.astro.it
H'EBRARD	Guillaume	I.A.P.	FRANCE	hebrard@iap.fr
HENRY	Richard	University of Oklahoma	USA	henry@mail.nhn.ou.edu
HERNANZ	Margarita	I.E.E.C./C.S.I.C.	SPAIN	hernanz@ieec.fcr.es
HILL	Vanessa	E.S.O.	GERMANY	vhill@eso.org
HILLEBRANDT	Wolfgang	M.P.I. für Astrophysik	GERMANY	wfh@mpa-garching.mpg.de
HOU	Jinliang	Shanghai Observatory	CHINA	jinliang@iap.fr
IYUDIN	Anatoli	M.P.E. Garching	GERMANY	ani@mpe.mpg.de
JEDAMZIK	Karsten	M.P.I. für Astrophysik	GERMANY	jedamzik@mpa-garching.mpg.de
JOSE	Jordi	Univ. Politec. Catalunya	SPAIN	jordi.jose@upc.es
KARLSSON	Torgny	Astronomical Observatory	SWEDEN	torgny@astro.uu.se
KNÖDLSEDER	Jürgen	C.E.S.R.	FRANCE	knodlseder@cesr.fr
KOBAYASHI	Chiaki	University of Tokyo	JAPAN	chiaki@astron.s.u-tokyo.ac.jp
KOLB	Edward	Fermilab/Chicago	USA	rocky@fnal.gov
KOVETZ	Attay	Tel Aviv University	ISRAEL	attay@etoile.tau.ac.il
KUNTH	Daniel	I.A.P.	FRANCE	kunth@iap.fr
KYLAFIS	Nick	University of Crete	GREECE	kylafis@physics.uoc.gr
LAMBERT	David	University of Texas	USA	dll@astro.as.utexas.edu
LAURENT	Philippe	S.Ap./C.E.A./Saclay	FRANCE	plaurent@cea.fr
LEMOINE	Martin	Observatoire de Paris-Meudon/I.A.P.	FRANCE	Martin.Lemoine@obspm.fr
LIVIO	Mario	Space Telescope Science Instit.	USA	mlivio@stsci.edu
LOCKMAN	Felix	N.R.A.O.	USA	jlockman@nrao.edu
LUBOWICH	Donald	Hofstra University	USA	dlubowic@aip.org
MACERONI	Carla	Osserv. Astronomico di Roma	ITALY	maceroni@coma.mporzio.astro.it

NAME	FIRST	INSTITUTE	COUNTRY	EMAIL
MAIGNAN	Georges	Eurocontrol	FRANCE	gmaignan@compuserve.com
MAJUMDAR	Subhabrata	Indian Institute of Science	INDIA	sum@rri.ernet.in
MALHOTRA	Sangeeta	Johns Hopkins University	USA	san@pha.jhu.edu
MATHEWS	Grant	University of Notre Dame	USA	gmathews@nd.edu
MATTEUCCI	Francesca	University of Trieste	ITALY	matteucci@ts.astro.it
MELCHIORRI	Francesco	Universita La Sapienza	ITALY	francesco.melchiorri@roma1.infn.it
MELLIER	Yannick	I.A.P.	FRANCE	mellier@iap.fr
MERSINI	Laura	Scuola Normale Sup./Pisa	ITALY	
MIRABEL	Félix	S.Ap/C.E.A./Saclay	FRANCE	mirabel@discovery.saclay.cea.fr
MOCHKOVITCH	Robert	I.A.P.	FRANCE	mochko@iap.fr
MOLARO	Paolo	Osserv. Astron. di Trieste	ITALY	molaro@ts.astro.it
MOLLÁ	Mercedes	Univ. Autonoma de Madrid	SPAIN	mercedes@pollux.ft.uam.es
MONTALBAN	Josefina	Osserv. Astronomico di Roma	ITALY	montalbn@coma.mporzio.astro.it
MOUCHET	Martine	Observatoire de Paris-Meudon	FRANCE	martine.mouchet@obspm.fr
NADYOZHIN	Dmitrij	I.T.E.P.	RUSSIA	nadezhin@vxitep.itep.ru
NAKAMURA	Takayoshi	University of Tokyo	JAPAN	nakamura@astron.s.u-tokyo.ac.jp
NORDSTROM	Birgitta	Niels Bohr Institute	DENMARK	birgitta@astro.ku.dk
O'MEARA	John	U.C. San Diego	USA	jomeara@ucsd.edu
OLINTO	Angela	University of Chicago	USA	olinto@oddjob.uchicago.edu
OLIVE	Keith	University of Minnesota	USA	fvs6308@vx.acs.umn.edu
PAGEL	Bernard	Sussex University	UK	bejp@star.cpes.susx.ac.uk
PALACIOS	Ana	LA-OMP	FRANCE	apalacio@ast.obs-mip.fr
PASACHOFF	Jay	Williams College	USA	pasachoff@williams.edu
PAUL	Jacques	S.Ap/C.E.A./Saclay	FRANCE	jpaul@cea.fr
PÉROUX	Céline	Institute of Astronomy	UK	celine@ast.cam.ac.uk
PITOIS-PUJADE	Marie-Laurence	Minist. Finances et Industrie	FRANCE	marie-laurence.pitois-pujade@dpma.finances.gouv.fr
PRANTZOS	Nicolas	I.A.P.	FRANCE	prantzos@iap.fr
PRIALNIK	Dina	Tel Aviv University	ISRAEL	dina@planet.tau.ac.il
PRIMAS	Francesca	ESO	GERMANY	fprimas@eso.org
PROCHASKA	Jason	Carnegie Observatories	USA	xavier@ociw.edu
REEVES	Hubert	C.N.R.S.	FRANCE	hreeves@club-internet.fr
ROLAND	Jacques	I.A.P.	FRANCE	roland@iap.fr
ROMANO	Donatella	SISSA/ISAS	ITALY	romano@sissa.it

NAME	FIRST	INSTITUTE	COUNTRY	EMAIL
ROOD	Robert	University of Virginia	USA	rtr@virginia.edu
RYDE	Nils	Astronomical Observatory	SWEDEN	ryde@astro.uu.se
SADAT	Rachida	Observatoire Midi-Pyréné es	FRANCE	rsadat@ast.obs-mip.fr
SALATI	Pierre	L.A.P.T.H.	FRANCE	salati@lapp.in2p3.fr
SCHAEFFER	Richard	SPhT/C.E.A	FRANCE	schaeffer@cea.fr
SCHANNE	Stéphane	C.E.A./Saclay	FRANCE	s.schanne@cea.fr
SILK	Joe	University of Oxford	UK	silk@astro.ox.ac.uk
SKILLMAN	Evan	University of Minnesota	USA	skillman@astro.umn.edu
SMITH	Verne	University of Texas El Paso	USA	verne@barium.physics.utep.edu
SNEDEN	Chris	University of Texas	USA	chris@verdi.as.utexas.edu
SPITE	François	Observatoire de Paris-Meudon	FRANCE	Francois.Spite@obspm.fr
SPITE	Monique	Observatoire de Paris-Meudon	FRANCE	Monique.Spite@obspm.fr
SUZUKI	Takeru	NAOJ/Univ. of Tokyo	JAPAN	stakeru@th.nao.ac.jp
THIELEMANN	Friedrich-Karl	Universität Basel	SWITZERLAND	fkt@quasar.physik.unibas.ch
TOSI	Monica	Osservatorio di Bologna	ITALY	tosi@bo.astro.it
TRURAN	James	University of Chicago	USA	truran@nova.uchicago.edu
TSUJIMOTO	Takuii	National Astronomical Obs.	JAPAN	taku.tsujimoto@nao.ac.jp
TUCHMAN	Yitzchak	Hebrew University	ISRAEL	tuchma@vms.huji.ac.il
TURK - CHIÈZE	Sylvaine	S.Ap/C.E.A./Saclay	FRANCE	sylvaine.turk-chieze@cybercable.fr
VANGIONI- FLAM	Elisabeth	I.A.P.	FRANCE	flam@iap.fr
VAUCLAIR	Sylvie	Observatoire de Toulouse	FRANCE	
VENKATESAN	Aparna	University of Colorado	USA	aparna@casa.colorado.edu
VENTURA	Paolo	Osserv. Astronomico di Roma	ITALY	paolo@coma.mporzio.astro.it
VIDAL - MADJAR	Alfred	I.A.P.	FRANCE	alfred@iap.fr
VIGROUX	Laurent	S.Ap/C.E.A./Saclay	FRANCE	lvigroux@cea.fr
WASSERBURG	Gerry	Caltech	USA	isotopes@gps.caltech.edu
WEISS	Achim	M.P.I. für Astrophysik	GERMANY	weiss@mpa-garching.mpg.de
WHITE	Simon	M.P.I. für Astrophysik	GERMANY	swhite@mpa-garching.mpg.de
YORK	Don	University of Chicago	USA	don@oddjob.uchicago.edu